SHIPBOARD ANTENNAS

The Artech House Antenna Library

Antennas by L.V. Blake
Shipboard Antennas by Preston E. Law, Jr.
Antenna Design Using Personal Computers by David M. Pozar
Microstrip Antennas by I.J. Bahl and P. Bhartia
Interference Suppression Techniques for Microwave Antennas and Transmitters by Ernest R. Freeman
Principles of Electromagnetic Compatibility by Bernhard E. Keiser

SHIPBOARD ANTENNAS

preston e. law

Artech

Cover Design by
Rosemarie Morelli-Crocker,
Crockergraphics
Needham, MA

ARTECH HOUSE, INC.
610 WASHINGTON ST.
DEDHAM, MA 02026

Printed and bound in the United States of America.

International Standard Book No.: 0-89006-211-0
Library of Congress Catalog Card No. 86-70448

10 9 8 7 6 5 4 3 2 1

CONTENTS

THE AUTHOR

Since 1974 Preston E. Law, Jr. has been Head of the Topside Electromagnetics Design Branch of the Naval Sea Systems Command. He began his specialized work in shipboard antenna systems design with the Navy in 1968. His experience includes employment as a Communications Systems Engineer with the Defense Communications Agency in Thailand; the Voice of America in Washington, D.C.; and the Federal Aviation Agency in Anchorage, Alaska. Mr. Law is an electrical engineering graduate of the University of Maryland and a registered Professional Engineer in the State of Maryland. He is a Senior Member of the Institute of Electrical and Electronics Engineers (IEEE), a member of the American Society of Naval Engineers (ASNE), Armed Forces Communications Electronics Association (AFCEA), Association of Scientists and Engineers (ASE) of NAVSEA, and Tau Beta Pi and Eta Kappa Nu Engineering Honor Societies.

Preface

The first time ever I recall seeing a shipboard antenna was on a cold, gray December day in 1952. I was high on the mast of a destroyer, out on the end of a yardarm, cleaning and repairing the coaxial cable connector of a UHF dipole antenna. As a seaman apprentice and yet untrained striker in the electronics shop of the destroyer tender *USS Grand Canyon,* I knew nothing about antennas.

Four years later, after intensive study and schooling in electronics; after years of first-rate experience on four different ships; after advancing in rank to Electronics Technician second class, I had completed my Navy enlistment and was honorably discharged and I still knew next to nothing about antennas. Why? Because the Navy placed very little importance on the subject of antennas; as a matter of fact, I do not recall that it was ever brought up in the service schools. Aboard ship the extent of our working with antennas was confined to making sure that the proper level of lubricant was maintained in the rotating joints of radar antennas, to wire brushing the soot and salt from feed-wire and tie-down terminations of long-wire antennas, and to repairing or replacing damaged coaxial connectors at the base of VHF/UHF antennas. In the main, however, our full attention was devoted to below decks equipment and systems —radio transmitters, receivers, and radar sets — those systems that really mattered to us. Antennas were but incidental, low priority accessories to the ship's sophisticated electronics equipment, we thought. We were wrong.

In 1968 when I returned to the Navy, this time as a civilian electronics engineer and specialist in antennas, I was very much aware how truly essential shipboard antennas are to the ship's electronic systems. Antennas are in fact a crucial element to the satisfactory performance of the manifold ship missions — a fact which cannot be overemphasized. Yet, though 15 years of solid improvements in antenna systems had transpired, I was surprised and disappointed to learn that no single reference book existed to present the philosophy and application of naval shipboard antenna engineering. To be sure, there were scattered bits of information here and there, much of it quite helpful: a set of manuals giving an overview of theory, installation details, technical data sheets, and testing and maintenance procedures for shipboard communications antennas; drawers of laboratory reports for individual ship antenna systems design; technical manuals for the various antennas; separate antenna system arrangement drawings; and a multiplicity of research and development studies. For the most part, however, the principles and practices of shipboard antenna engineering were collected and stored in the aggregate minds of my colleagues, some of whom had a wealth of experience and know-how gained over many years in this field of work. As time slipped by, with more information filed away, and my colleagues one by one retiring or transferring to other disciplines, the why and wherefores of shipboard antennas remained scattered.

The principles and practices of shipboard antenna systems are unique in electronic engineering. Far from being nebulous, ambiguous, or arbitrary, they are instead quite definite and established. Thus there should be a single compendium setting down those principles and practices for the use and reference of all brave souls whose task it is to work in this field. To be that comprehensive source, and thereby to facilitate shipboard antenna engineering itself, is the purpose of this book.

No attempt is made herein to present the rigorous electromagnetic theory of antennas. The literature abounds with superb works of classical antenna theory and certainly there is no reason to reiterate it in this book. It is assumed that the reader is a graduate engineer or a highly trained technician advanced to the stage of engineering application, and, therefore, already familiar with the basic theory and formal treatment of antennas as an electromagnetic radiating (or sensing) device. For those desiring to review or delve more deeply into the mathematical aspects, an extensive bibliography is included. What this book hopes to offer is a satisfaction of the need for a practical and useful reference source as to (1) *what* types of antennas are used on ships, particularly on U.S. Navy vessels, (2) *why* these antennas are used, and (3) *how* they are used, i.e., how they are integrated into the complex topside environment of a ship. Finally, then, the user as well as the designer will have an understanding and appreciation of shipboard antenna systems engineering.

Preface to Second Edition

Dedicated to the Memory of Marcelle Van Amsterdam of Artech House, Inc.

It is with sincere and deep appreciation that I note the welcome manner in which the first edition of Shipboard Antennas was received. I am especially overjoyed by the kind remarks of colleagues and readers. The many favorable comments certainly have made the effort worthwhile. I have taken care, too, to note all suggestions for improvements, and it is to that end that portions of the text have been revised, enhanced, and upgraded for this new edition.

First and foremost is the addition of an Index, crucial to proper use of the book as a reference source. Secondly, Chapter 8 has been significantly expanded to (a) more fully describe the antenna systems integration process through the coordinated workings of the Topside Design Team and the EMC Advisory Board, (b) include more detailed information on ship antenna systems scaled and computer-aided modeling, (c) discuss the impact of important new radiation hazard exposure level restrictions, and, (d) provide insight into possible future needs in shipboard antenna design and technology.

Thirdly, new antennas recently adapted for shipboard use are presented; e.g., the LWCA lightweight HF communications fan antenna and the LTN-211 OMEGA loop antenna. Also, updated information is given on the AS-3503/SKR ASWM telemetry antenna, and the AS-3399/WSC SHF SATCOM antenna. And, of course, several typographical errors appearing in the original text have been corrected.

In working on this edition I acknowledge again the need for others and their friendly support. Many have kindly assisted. Special thanks are due Mr. Dwight Heasty of Eldyne, Inc. for his peerless experience and contributions in the history of RF radiation hazards to personnel and the problems that recently adopted Navy standards may impose on the systems designer; to Mr. Paul K. Wheeler of Eldyne, Inc., principal designer of the Navy's new lightweight HF fan antenna; to Mr. L.M. Black of Ball Aerospace for his help in securing the latest information on the AS-3503 Telemetry antenna; to Messrs Morton F. Spears and Laurence G. Burke of Spears Associates, Inc. for their kindness in supplying new data on the LTN-211 OMEGA antenna, and to Thomas P. Morgan and Billy S. Douglas of NAVSEA for their expertise in computer-aided ship systems design.

Finally, my heartfelt gratitude to Mr. William M. Bazzy and his skillful staff at Artech for the superbly professional and cheerful manner in which they have supported and encouraged me at all times.

P.E. Law, Jr.
17 June 1986

Acknowledgements

One does not embark alone upon a venture such as this. It is reassuring, and certainly comforting, to have friends. In the case at hand, it is all the more encouraging when friendships have developed among one's colleagues who are highly regarded specialists in their respective disciplines. It is they, the individual specialists, who have so importantly contributed to this work. Without their cheerful cooperation the task would have proved futile.

First there are the teacher friends. Beginning in July 1968 and continuing right to the present time, I have had the profound good fortune of having the finest naval shipboard antenna engineers as my mentors. I therefore wish to salute Mr. Kennett D. Wilson and Mr. Richard W. Pride of the Naval Electronic Systems Command, Washington, D.C.; Mr. Martin L. Musselman and Mr. Richard K. Royce of the U.S. Naval Research Laboratory, Washington, D.C.; Mr. Charles A. Nelson,, Dr. John W. Rockway, and their expert associates of the Naval Ocean Systems Center, San Diego, California; Mr. Ivan Faigen, Mr. Valor C. Smith, and Mr. John Watson of Chu Associates, Inc., Littleton, Massachusetts and El Cajon, California; and Mr. Lynnwood L. Lay of Presearch Incorporated, Arlington, Virginia.

Second I express my deep appreciation to my coworkers, those specialists spread throughout the naval community who have so willingly shared information pertaining to their individual areas of responsibility and knowledge: Roy J. Biondi, Gordon A. Bates, and Anthony P. Merkouriadis of NAVSEA (Radar); Benjamin Lee, NAVELEX (IFF); Nicholas T. Yannarell and Albert S. Cheung, NAVSEA (Electronic Warfare); William T. Fousheee, NAVELEX (Navigation); Peter E. Larsen and Andrew K. Simmons, NAVAIR (Meteorology); Gary N. Skillicorn, NAVELEX (Telemetry); John Roman, NAVSEA (EM Compatibility); Frances M. Prout, NAVSEA (EM Assessment), and my immediate colleagues Roy T. Okamoto,

John W. Eadie, Joseph Juras, Lawrence S. Cohen, and Seiji Kuniyoshi (EM Systems Design and Integration). Special thanks are due Nicholas Yannarell for careful proofreading and providing excellent comments to improve the manuscript text.

Finally there are many resource personnel crucial to any effort of this magnitude. My sincere gratitude to: Deborrah J. Jones, Linda M. Diamond, Karen R. Deller, and Sherry D. Hilton for the expert typing and proficient clerical help in compiling the original draft manuscript; to Susan E. Weidner of NAVSEA's Public Affairs Office for aid and advice in having the manuscript cleared for public release; to the NAVSEA Technical Library staff, in particular Patricia A. Prentice, for invaluable assistance over these many months in ferreting out a hundred dusty manuals containing critical information; and to Herbert K. Old of Automation Industries' Vitro Laboratories Division, my special thanks for magically providing that last straw on several occasions when all other sources were bone dry of much needed documents. A friend in need is a friend indeed.

Introduction

What is a ship? The question would very likely evoke a wide range of answers depending on the perspective, and perception, of the person being asked. Nevertheless, in its simplest form a ship must certainly be thought of as a large sea-going vessel used to transport things over water. It might haul crude oil, a cargo of wheat, or vacationing passengers — or it might be bearing guns, missiles, and airplanes. Then if we narrow down our scope of view to a naval warship we realize that its primary purpose is to carry things that enable it to fight; to attack; to defend. If so, then a naval ship in reality becomes a platform — a floating, moving, weapons platform used to support guns, missiles, and aircraft while maneuvering freely about the open seas. It is crucially important to understand, however, that in the modern, grand scale war operations of today, that platform is uselessly blind and utterly vulnerable unless it is capable of sensing where the enemy is and who he is, then taking some type of action: search, detect, track, attack, or evade. Nowadays all of these things are done with the aid of a highly sophisticated, and nearly always computerized, mass of electronic systems. The great majority of these systems lead to and from *antennas* — the ship's emitters and sensors — the eyes and ears of a ship.

In order to fight, a warship must be able to navigate the seas with confidence, communicate with its own forces and headquarters command, direct accurately its fire power, and to launch, control, and land its helicopters and planes. An aircraft carrier, for example, is simultaneously a floating airport, a navigation station, a radio communications facility, and a weapons foundation. At any one moment during the heat of battle, a user of any of these functions —a pilot, a ship navigator, a communicator, a gunner — might believe that *his* need is paramount. Therefore, his system must be up and operating to his satisfaction. So what does this mean to us, the shipboard antenna systems design engineers? We know that not one of these ship systems would be working effectively without a properly performing antenna — a reliable, dependable, error-free, sensitive antenna — to receive and return information vital to each mission. The result of this critical need is that the ship, that single, self-contained, multipurpose platform, becomes an electromagnetic jungle bristling with sensors: antennas for radar, navigation, communications, electronic countermeasures, and air traffic control, all competing for a good working spot on the small piece of floating real estate. Giant carriers have as many as 130 antennas installed in the topside. The problem is acute.

The problem is that the topside of a ship is extremely hostile to anything of an electromagnetic nature. Yet a warship has an indispensable need for a very large number of electromagnetic sensors and simply cannot exist otherwise. It is the old case of "can't live with and can't live without." The first and most obvious constraint is the severely limited space. Antennas cannot be placed just anywhere in the topside. An aircraft carrier would not tolerate antennas sticking up in the way of air operations on the flight deck. Thus, its largest piece of topside real estate is immediately eliminated. As a consequence most antennas must be bunched on the carrier's island and the large monopoles required for HF communications must be tilted horizontally around the flight deck. Cruisers and destroyers would not stand for antennas in the way of their guns or missile launchers, so their antennas are clustered about the central superstructure and folded down off the fantail. Cargo ships will not abide antennas in the way of their loading booms, so available space must be found on top of the pilot house, on the mast, and out on yardarms, along with a crowd of other things sharing these locations.

Ships are constructed of, and liberally endowed with, metallic structures throughout the topside. These metal objects capture, reflect, scatter, and reradiate the ship's own intentionally transmitted energy causing badly distorted radiation patterns, high losses from impedance mismatches, and pick up of undesirable sporadic noise emissions. To make matters worse, the ship must transmit and receive simultaneously, so that high power levels of transmitted energy are subject to being received by its own close-by antennas

—overloading and sometimes burning out inadequately protected equipment. It is for these reasons that receiving antenna facilities on land are separated by several miles from transmitting stations. No such luxury exists on ships.

So, here we have a paradox — antennas are vital to the ship's operations, but the ship is intolerant of their presence. We see that individual antennas, which may work well when isolated, are likely to have their performance seriously impaired when made to work in concert with a host of competitors sharing the unfriendly ship environment. Yet no one system should be optimized at the expense of the ship's overall performance and missions. Of course, *not* being able to optimize goes against the grain of an engineer. By education and training, if not by nature, the engineer has a burning desire to demonstrate his professional ability to provide optimum designs. It is a matter of pride to produce the very best possible. Normally, in the engineering world the one real restriction is cost. Within the bounds of economy the engineer is given free rein — and is in fact expected — to design his best. Not so with shipboard antenna systems. The name of the game is compromise. Even if he were given the ideal situation of complete freedom to select sites anywhere on the ship, the engineer faces an alien environment. Instead he is presented with a raft of constraints forcing him to compromise at the very outset. Giving here, giving there, his proposed design must inevitably deteriorate from optimum to fair, to satisfactory — maybe even to merely acceptable. Why? Because the overall performance capabilities of the ship must not be degraded below a certain tolerence level. Ship operators, for example, who must navigate and communicate twenty-four hours a day must also learn to appreciate the fact that communications and navigation performance are less than optimum because these capabilities were traded off during the combat system design process in an attempt to ensure the fulfillment of more critical anti-aircraft, anti-submarine, or surface warfare *fighting* missions — if and when those priority needs arise. After all, the purpose of a warship, first and foremost, is to fight. The purpose of the naval shipboard antenna engineer, under these circumstances, is to do his best — for the ship's sake.

To

Guglielmo Marconi

Whose Vision Guides Us Now Upon the Seas

Chapter 1

Historical Background

1-0 EARLY WIRELESS MILESTONES

It can be said fairly, without equivocation, that the United States Navy is the founder of our nation's radio communications. With the tremendous success of present-day commerical broadcasting, and the current low esteem of things military by our public, it is easy to ignore or lose sight of this important fact. To understand the U. S. Navy's essential role in the establishment of radio electronics we must retrace the sequence of far-reaching events that occurred in the last quarter of the 19th century, laying the foundation of wireless transmission:

I. In 1873, the British physicist James Clerk Maxwell published his classical treatise on the theory of electromagnetic radiation, proving mathematically that electromagnetic waves travel through space with a speed precisely equal to that of light. Unfortunately, he never saw its practical use. He died six years later at the age of 48.[1]

II. In 1886, 29-year old German physicist Heinrich Hertz, experimentally proving Maxwell's theorectical equations, demonstrated that high-frequency oscillations produce an effect some distance away from the source and that this phenomenon was the result of electromagnetic waves.[2] Curiously, neither did Hertz live to see the practical use of his work. Eight years after his initial discoveries, he was dead, at the age of 37. It is significant to note that Hertz's experimental apparatus was comprised almost entirely of *antenna* devices. The "transmitter" was no more than a capacitor discharging through a small spark gap at the center of a dipole made from short lengths of heavy wire. The "receiver" was even more a rudimentary antenna — a mere loop of wire with its own gap.

III. In 1890, the French professor Edouard Branly developed the "coherer" — a receiver to detect electromagnetic energy much more efficiently than Hertz's spark gap.[3]

IV. In 1895, a year after Hertz's death, a young Italian experimenter named Guglielmo Marconi first put electromagnetic theory into practical application. Near Bologna, Italy, Marconi succeeded in communicating by wireless, thus heralding a new and phenomenally successful enterprise. He was then 21 years old. By the next year Marconi was transmitting Morse code messages nearly 2 miles. Moving to England to obtain patents on his devices, he demonstrated reception over 8 miles. In 1897, he exhibited the use of wireless between ship and shore; in 1899 he communicated across the English Channel and used wireless for the first time on naval maneuvers with equipment installed on the British Navy ships HMS *Alexandria, Europa,* and *Juno,* exchanging messages up to 75 miles. Then, on December 12, 1901, using an antenna hung from a kite and a low frequency (313 kHz) at 10 kW of power, he made the first transatlantic wireless communication, sending the letter *S* over 1700 miles from England to St. John's, Newfoundland.[4] Marconi, by the way, lived until 1937, witnessing the entire growth and modern success of radio communications.

V. In 1900 Reginald A. Fessenden, an American, made the first transmission of voice by wireless.[5]

During this same period, a young U.S. Naval Officer, Lieutenant Bradley A. Fiske, became interested in the possibilities offered by communications without wires. Knowing that fog was the most dreaded of all weather conditions by mariners, he was convinced that the Navy was in need of some form of signaling which could be used during fog and darkness as well as bright sunlight. Fiske conducted numerous experiments of wireless telegraphy between vessels by interrupted current (signals) in large coils of insulated cable wrapped around their hulls.[6]

We can readily appreciate, therefore, the international contributions in the pioneering years to the theory and experimentation of wireless communications. British, French, German, Italian, American, and others — notably Alexander Popov of Russia and Nikola Tesla, an American born in Yugoslavia — contributed to these early efforts of the 19th century. However, the immediate *application* was almost totally sea-related. It was in fact the navies and maritime industries of the world that recognized the significance, and need, of wireless communication, nurturing it along in its infancy and assuring its eventual success. The folks on land, after all, were quite smugly satisfied with 50 years of highly developed, well-established wire telegraphy. (In 1866, thirty-five years before Marconi's wireless message across the sea, a transatlantic telegraph cable had been laid by Cyrus Field.)[7] Moreover, telephone networks were being rapidly installed and put into use. These wire services seemed quite satisfactory, thank you!

Happily, Marconi had the great foresight to see beyond such technological parochialism. Instead of attempting to compete with the transatlantic wire telegraph he sought to apply wireless where there was no practical alternative, namely to ships on the open seas.[8] Indeed, the U.S. Navy was interested.

1-1 INITIAL NAVAL EXPERIMENTS

By the dawn of the 20th century, the Navy had begun to realize that wireless communications had reached a stage of practicality such that its use might have a great impact on tactics and strategy. Long distance communications, impossible before now, could increase enormously the effectiveness of naval operations at sea. Writing in historical retrospect, two world wars and 60 years later, Fleet Admiral Chester W. Nimitz stressed, "Until the present century a ship's isolation was complete once she navigated out of port and sailed over the horizon. For centuries the ocean was both the great water highway of the world and a silent barrier. Radio raised the curtain to a vast and universal change. With the installation and perfection of the miracle of instantaneous wireless telegraph, a radio-equipped ship would never again be totally cut off from land."[9]

Purchasing the best equipment available, the Navy proceeded with initial installations on its ships and ashore — an ambitious program that resulted in the Navy becoming the principal sponsor of radio electronics in this country and a pioneer in its development.[10] On November 2, 1899, the first official naval message was transmitted from the *USS New York* to the naval shore station at Navesink, New Jersey.[11] By 1900, Marconi was able to report that the U.S. Navy easily communicated up to 36 miles between the battleships *New York* and *Massachusetts* in a demonstration to British authorities during maneuvers.[12]

Now fully committed, the Navy acquired wireless sets manufactured by various firms, with the result that, before the end of 1902, equipments from two American and four European companies were undergoing naval evaluation. Test stations were set up at the Washington Navy Yard and the U.S. Naval Academy at Annapolis. Two ships were made available, the *USS Prairie* and the *USS Topeka,* and testing was conducted during the fall of 1902 and spring of 1903. Communication between the *Prairie* and Annapolis was maintained out to a distance of 90 miles and between the two ships up to 62 miles. With the completion of these initial tests and procurement of equipment early in 1904, the U.S. Navy's wireless communications system was launched.[13] Stations were established at 23 key points on the east and west coasts of the United States, in Panama, Cuba, Puerto Rico, Hawaii, the Philippines, Guam, and the Marianas[14] — thereby providing the first global radio broadcasting network in the world. Admiral George Dewey, pleading for naval control of wireless, stated then, "To the Navy wireless telegraphy is absolutely essential. The principal use of wireless telegraph is now, and long will be, at sea — between ship and ship, or ship and shore. On shore other means of communications always exist, often better. But from ships at sea out of sight of flags or lights, and beyond the sound of guns, the electric wave, projected through space, invisible and inaudible, can alone convey the distant message."[15]

Meanwhile, other sea-going activities were quick to perceive the usefulness of wireless services. As early as 1898, the Dublin (Ireland) *Daily Express*, impressed with Marconi's claims, seized the opportunity to "scoop" their competitors in the news business. Its owners invited Marconi to do wireless reports of the Kingston Regatta of July 1898 from the steamer *Flying Huntress*, the first ship equipped with a wireless system for commerical purposes. Marconi stated that his reporting of the races was entirely satisfactory in every respect, over 700 messages having been sent and received during the regatta.[16]

1-2 PUBLIC INTEREST IN SAFETY

Less than a year later a dramatic incident brought worldwide attention to the value of radio for maritime safety. In a dense fog at 4 o'clock in the morning of 3 March 1899, the East Goodwin Sands Lightship was rammed by the steamer *R. F. Mathews* outward bound from London. The lightship was one of four light vessels marking the Goodwin Sands, and, fortunately, it happened to be the one with a wireless installed. It immediately flashed an emergency message and tugs were sent out to tow the lightship in, saving the entire crew.[17]

L. S. Howeth, in his definitive history of radio electronics in the U.S. Navy, describes the next event which greatly aroused public interest:

> On 22 January 1909, the palatial 15,000 ton White Star liner *S. S. Republic* departed New York, bound for Gilbralter and the Mediterranean, with some 440 passengers and a large quantity of supplies for our 'Great White Fleet' homeward bound from its triumphant world cruise. The fleet's passage through the Mediterranean had coincided with the earthquake at Messina and large quantities of its supplies had been expended for relief. Running into thick fog, in the early morning hours of 23 January, the *Republic* collided with the Italian *S. S. Florida*, crowded with 830 immigrants, most of whom were refugees from the Messina catastrophe. The *Republic's* sole radio operator escaped death by the merest chance, the *Florida's* sharp bow having cut into the bulkhead only a few feet from where he slept. His radio apparatus remained undamaged, but the inrushing waters shorted the ship's generators, forcing him to use the emergency storage batteries. He transmitted the message: '*Republic* rammed by unknown steamship, 26 miles southwest of Nantucket. Badly in need of assistance.' This was received by the Marconi station at Siasconsett, Massachusetts. A U.S. revenue cutter was quickly dispatched to the *Republic's* assistance. By daylight an assortment of vessels was gathering at the scene of the collision. About 1,650 persons were transferred from the two ships, with but six lives reported lost from the *Republic*. Radio played a major role in limiting the loss of life and created such a favorable impression upon the public that it [radio], like life preserves, became considered a necessity.
>
> Within a fortnight of this incident there was considerable clamor . . . that legislation should be enacted requiring all ocean going passenger steamers to be fitted with radio equipment.[18]

Thus by 1910 the United States had begun requiring wireless equipment and a skilled operator on all vessels carrying 50 or more passengers between ports 200 or more miles apart.[19] (It was about this time, 1910, that the original word "wireless," as relating to the propagation of elecromagnetic waves, was superseded by the word "radio." At the same time the electron vacuum tube was being experimented with, giving rise to the new term "radio-electronics.")[20]

However, it was the 1912 *Titanic* disaster that, in the public's eye, decisively changed radio from a scientific curiosity to a vital necessity. Again quoting from Howeth:

> In the early hours of 15 April 1912, the White Star liner *Titanic*, then on her maiden voyage, struck an iceberg and sank. The world was stunned at the sudden and tragic end of the newest, largest, most luxurious, and 'unsinkable' 46,000 ton, $12 million masterpiece of shipbuilding. The most astounding part of the calamity was that on a clear, calm, beautiful, starlit night, with scarcely a ripple on the surface of the sea, and with nearly 3 hours in which to abandon ship under ideal conditions, over 1,517 human beings, including a long list of prominent figures in the arts, public life, and business world, met death in the icy waters of the North Atlantic. Tragically, there were ships within range of the *Titanic* that did not know of her plight because they either were not equipped with radio or employed a single operator (off duty during the quiet night hours) . . . Great as was the loss of life, without radio there might not have been a single survivor.[21]

A few ships with radio did finally receive the message and converged on the area to pick up survivors. For three days and three nights a 21 year-old radio operator named David Sarnoff stayed at his post, without sleep and with but very little food, atop Wanamaker's department store in New York City, keeping in direct contact with rescue operations and passing on the information to the news-hungry public.[22]

1-3 DEVELOPMENT OF NAVAL RADIO

Understandably it was the importance and uniqueness of radio communications to sea operations that "gave the Navy dominance over other users and placed it in a strong position to obtain funding from Congress, thus enabling the Navy to provide the principal support in fostering the growth of a radio industry in this country." It must be emphasized that this early support was "especially important during the critical period following the installation of equipment on commercial vessels to provide safety at sea. After this initial activity little financial return could be expected from other radio manufacturing — a situation that would continue for nearly two decades until the advent of commerical broadcasting during the 1920's."[23]

As electronics technology advanced, many experimental broadcasts using vacuum tube transmitters took place between the United States and Europe, South America, and the Pacific. In one of the more interesting ventures the American Telephone and Telegraph Company (AT&T), using the Navy's antenna facility at Arlington, Virginia, broadcast phonograph music and live speech on a frequency of 50 kHz at about 3 kW across the Atlantic to the Eiffel Tower. A formal demonstration was arranged and on 23 October 1915, a number of observers in Paris listened to these radio telephone transmissions from Arlington, 3600 miles away.[24]

World War I encouraged the bourgeoning use of radio and again it was the Navy to the fore. In 1916, a year before America entered the conflict, the U.S. Navy began working on a system for voice contact between ships at sea and ship-to-shore.[25] Concurrently, the President authorized the Navy to take control of the nation's radio communications facilities, to operate them for the duration of the war, and to handle all radio traffic with foreign countries.[26] Furthermore, Congress gave serious consideration to a plan that would permit the Navy to operate broadcast stations in competition with private firms even in peacetime. The plan never went into effect.[27] Still, the Navy did become "the spokesman for the Government in an effort which resulted in the first legislation to control radio use in the U.S., and it originated the plan which became the basis for the international control of the assignment of frequency channels in the radio spectrum accepted by the nations of the world. After the war the Navy played a major part of the formation and establishement of the organization that became the present Federal Communications Commission."[28] Moreover, the Navy spurred on a healthy interest in research and development. With just pride Admiral Nimitz wrote, "Originally vacuum tubes had a life expectancy of some 70 hours and cost fifty dollars each, limitations which made their widespread use economically unfeasible. However, shortly after World War I the Navy solicited bids on a 5000 hour tube costing no more than five dollars. The successful bidder met these revolutionary specifications, and, in so doing, the Navy had sired a major contribution to the wildfire growth of commerical radio in this country, and set our electronics industry off to a booming start."[29]

As the country entered the "Roaring Twenties" there was a rapid rise of public interest in regularly scheduled entertainment broadcasting. By this time the superheterodyne radio receiver had been developed and marketed for home use. General Electric bought

American Marconi and formed the Radio Corporation of America. With WWJ Detroit and KDKA Pittsburgh on the air, commerical broadcasting was born.[30] At this time, too, the Navy, as part of its continuing investigation of propagation characteristics and experimental work, started broadcasting music and talks by distinguished scientists and educators on 858 kHz from its Naval Aircraft Radio Laboratory in Anacostia. In fact, in 1922, the station established a number of radio "firsts."[31]

I. First radio broadcast by a Congressman (Rep. J. L. Cable of Ohio);

II. First radio broadcast by a Senator (Sen. Henry Cabot Lodge of Massachusetts);

III. First radio broadcast by a Supreme Court Judge (Chief Justice White),

IV. First radio broadcast by a President of the United States (Dedication of the Lincoln Memorial on 30 May 1922 by President Warren G. Harding);

V. First radio broadcast of a session of Congress (President Harding delivering annual address to joint session on 8 December 1922);

VI. First radio broadcasts of regular weather forecasts.

With the linking of WEAF in New York and WNAC in Boston in 1923, the first radio broadcasting network was formed[32] and the future of the radio-electronics industry was solidly ensured. As Howeth so aptly noted, radio and its future offspring owe an everlasting debt to the Navy for its aid and contribution in the advance of the art.[33]

It had been the need for communications that provided the principal impetus for "wireless" transmission — communications at sea, where hard-wire hook-up was impossible. Along the road of radio development other benefits to the military and mankind were being carefully observed: direction-finding, navigation, detection; all significant and crucial steps in the evolution of that most important new utility of radio frequency energy kindled by the approaching winds of World War II — radar.[34] As radio electronics applications mushroomed, more and more antennas of nearly every size and shape were required to efficiently transmit and receive the desired information. For Navy ships, the design of topside antenna systems became a new art.

REFERENCES

1. Bern Dibner, *Ten Founding Fathers of the Electrical Science* (Norwalk, Connecticut: Burndy Engineering Co., Inc., 1954), p. 45.
2. Andy Santoni, "Radio," *Electronic Design*, 1 September 1977, p. 58.
3. Louis A. Gebhard, *Evolution of Naval Radio-Electronics and Contributions of the Naval Research Laboratory* (Washington, D.C.: Naval Research Laboratory, Report 8300, 1979), p. 6
4. Santoni, *op.cit.*, p. 59.
5. *Ibid.*
6. L. S. Howeth, *History of Communications-Electronics in the United States Navy* (Washington, D.C.: U.S. Government Printing Office, 1963). pp. 16-17.
7. Morris Grossman, "Data Communications," *Electronic Design*, 1 September 1977, p. 87.
8. Santoni, *op.cit.*, p. 59.
9. Howeth, *op.cit.*, p. xii
10. Gebhard, *op.cit.*, p. 1.
11. *Ibid.*, p. 2.
12. Morris Grossman, "The Communications Era (1879-1905)," *Electronic Design*, 16 February 1976, p. 94.
13. Gebhard, *op.cit.*, p. 5.
14. *Ibid.*
15. Howeth, *op.cit.*, p. 74.
16. *Ibid.*, p. 21.
17. *Ibid.*
18. *Ibid.*, p. 154.
19. Santoni, *op.cit.*, pp. 59-60.
20. Gebhard, *op.cit.*, p. iii.
21. Howeth, *op.cit.*, p. 160.
22. Santoni, *op.cit.*, p. 60.
23. Gebhard, *op.cit.*, p. 1.

24. Santoni, *op.cit.*, pp. 61-62.
25. *Ibid.*
26. Gebhard, *op.cit.*, p. 1.
27. Santoni, *op.cit.*, p. 62.
28. Gebhard, *op.cit.*, p. 1.
29. Howeth, *op.cit.*, p. xiii.
30. Santoni, *op.cit.*, p. 59.
31. Gebhard, *op.cit.*, pp. 22-24.
32. Santoni, *op.cit.*, p. 59.
33. Howeth, *op.cit.*, p. 77.
34. Gebhard, *op.cit.*, p. 24.

Chapter 2

Communications Antennas

2-0 INTRODUCTION

It should come as no great surprise to learn that, since the initial use of radio was for long distance wireless communications at low frequencies, the first shipboard antennas were all made of haphazard lengths of wire strung as high as possible above the ship's topside. Evidently the thinking was that the longer and higher the wire, the better the results. With only the most rudimentary tuning incorporated, the transmitter frequency was largely controlled by the resultant wavelength characteristics of the antenna. Consider the following report of the first U.S. Navy investigations of wireless, in October 1899, aboard the *USS New York* and *USS Massachusetts*:

> The aerials employed were, in each case a rubber insulated single wire, made of seven copper strands, each 1/25th-inch in diameter. Three ebonite rods, each 18 inches long and one inch in diameter, suspended in series, formed the aerial insulator at the top of the wire. The temporary wooden masts attached to the topmast of each battleship has an 18-foot sprit hoisted to the top by halliards bent on, at about one-third of its length. The aerial was attached to the longer end while the shorter one was held steady by a down haul keeping the wire away from lightning conductors on the masts and from metal rigging. The antenna was, therefore, more or less of the vertical type, being at an acute angle with the mast, and extending from the top of the sprit to the after gunroom, where it was attached to the wireless apparatus. The upper ends were 130 feet and 140 feet, respectively, above decks of the *New York* and the *Massachusetts*. The detector was the coherer, with each of the telegraph instruments being equipped with a Morse inkwriter, so that messages received were recorded on tape. Some sources state that a plain antenna circuit was used; i.e., that the coherer was directly in series between antenna and ground; others that a coupling transformer was employed. The transmitted wave was dependent entirely on length of antenna, and the signal

> emitted was very broad. The receiver was slightly more refined, as an induction coil was used to connect the antenna circuit to the detector. This coil was not tuned but was intended merely to raise the voltage which operated the coherer. Frequencies transmitted by the two larger ships and the shore station were about the same. Since all three stations had antennas approximately the same height they were tuned to each other as closely as was necessary.[1]

Additonally, in that same autumn of 1899, wireless news reporting was being experimented with at the Americas Cup Races. Lieutenant G.W. Denfield, USN, was stationed aboard one of the Marconi-equipped ships as a naval observer. Denfield wrote concerning the antennas:

> They were suspended carefully from masts and carefully insulated from them. These antennas were of insulated wire, about 100 to 130 feet long. In leading them into the rooms containing the apparatus, great care was exercised to ensure that the insulation was satisfactory. During transmission of messages the antenna was connected to one of the knobs of the secondary winding of the transmitter and the other knob was connected to earth. These knobs were about eight-tenths of an inch apart. The telegraph key was in series with the primary winding of the coil. When the telegraph key was closed, sparks between the knobs made a noise similar to striking several matches. When receiving messages, the antenna was disconnected from the transmitter and secured to one terminal of the primary winding of the transformer, while the other end of this winding was grounded. From the transmitter, the antenna was rapidly charged and discharged, thereby producing an electrical effect which acted on the wire in a receiving antenna in a manner which reduces resistance in the receiver which was either in series with the second wire or in a circuit with one of the transformer coils, the other coil being in series with the antenna. When resistance of the receiver circuit was sufficiently reduced, the local battery energized it, and the dot or dash was recorded.[2]

As the Navy became fully convinced of the necessity for wireless aboard ship, instructions were issued to prepare the masts to accommodate antennas. On 4 January 1902, the Bureau of Equipment requested that all ships under construction and those to be built in the future be provided with masts suitable for the use of wireless telegraph apparatus.[3] This heralding of a new science was not met with whole-hearted enthusiasm by all navy men, however. George H. Clark, the Navy's first civilian radio expert recorded, doubtless with some relish, "The older officers mistrusted, did not understand, and made fun of wireless. I recall the remarks of one officer who, seeing a flat-top antenna not yet drawn up taut, remarked testily, 'How can they expect to tune a thing like that when its strings are loose.' "[4]

Despite the obvious need for wireless at sea, the growth of this new technology for naval use was not rapid. Along with the extremely limited knowledge of early radio — and lack of operational training — most naval officers sincerely believed that the large topside antennas would be quickly damaged in battle by gunfire blasts from their own weapons, or by enemy fire, so that "it was nothing short of folly to depend on such a (communication) system."[5] Nevertheless, with Navy support and sponsoring as outlined earlier in Chapter One, radio communication became essential and widespread throughout the fleet. With it came the requirement for designing and installing antennas for shipboard use. To be as efficient as possible at the low frequencies employed, these antennas were constructed of long wires running from mast to mast, and mast to bow or stern.

By the early 1920s, the most common shipboard antennas were the flat-top, the T, the inverted L, and inclined wire[6] — antenna types that remained in naval service with some modification for another half-century. Illustrations of early inclined and flat-top wire antennas for U.S. battleships in the 1918-1920 era are given by Figures 2-1, 2-2, and 2-3. Despite appearing primitive at first glance, one cannot help but admire upon closer inspection the careful and precisely detailed design evidenced by these old plans.

2-1 LF/MF/HF COMMUNICATIONS ANTENNAS

2-1.1 Wire-Rope Antennas — General

In view of the foregoing introduction, it is only fitting that we begin our discussion with the subject of wire antennas. For the first two decades after the turn of the century, the Navy used a variety of radio equipment operating in the LF and MF (low and medium) frequency range from about 50 to 1500 kHz. This provided maximum communication distance by surface wave propagation as a function of transmitter power and receiver sensitivity. Eventually, most naval radio operations began to converge around 750 kHz.[7] However, by the mid-1920s, the Navy was forced to relinquish all operating frequencies falling within the 55-1500 kHz span to the new commercial broadcasting interests — the now-familiar "AM Radio" band. Continuing to use some LF and MF frequencies below 500 kHz for fleet operations, the Navy at the same time moved rapidly to exploit the promising new technology of HF (shortwave) communications. Whether LF, MF, or HF, shipboard antennas (except for direction-finding "radio compass" loop antennas) remained variations of the old straight-wire types.

LF/MF shipboard wire antennas, because of the extreme length required for efficient performance, remain virtually unchanged to this day (where they can be found). Figures 2-4 and 2-5 depict World War II configurations, an inclined wire running from foremast to bow on the *USS Curtiss*, and multi-element horizontal wires running between towers, from stem to stern, on the old aircraft carrier *Essex*, an arrangement considered absolutely intolerable by today's standards. Figures 2-6 through 2-9 illustrate various postwar forms of the common flat-top, mainstay of low and medium frequency communications for nearly three quarters of a century. Later, as high frequency techniques were perfected, and satellite radio was inaugurated fleetwide, thus affording solidly reliable ship-to-ship and ship-to-shore communications, the Navy gradually abandoned use of the lower frequencies for surface ships. By the early 1970s, a single MF frequency, 500 kHz, had been retained for receive-only of international distress messages. For reception of emergency signals, a single MF wire is used, typical of that in Figure 2-10. Nowadays even these MF wire types are often replaced by a 35-foot twin V located high on a mast or stack. Nevertheless, in our modern Navy the wire-rope antenna remains an important cornerstone of shipboard communications, a key factor of the extensively used, worldwide, HF system.

Modern Wire-Rope Antennas

Even in the pioneer days of ship antenna design, it was recognized that the horizontal wire portion of the flat-top, the T, or the inverted L is not the actual radiator or receptor per se, but is in fact a deliberately added capacitance to provide increased electrical height, and therefore, greater overall *effective* height. Thus, where vertical length is severely limited aboard ship, the antenna's performance is thereby improved by increasing the antenna efficiency with capacitive top-loading.[8] The true antenna is the mast, or the tall superstructure, or long "rat-tail" downleads, in combination with the horizontal or inclined wires strung to, and between, the top of the tall structures. Accordingly, the intended purpose of this composite wire antenna is to transmit or receive *vertically polarized* electromagnetic waves. It should be understood, moreover, that this compound antenna is non-directional to vertical wavefronts. This fact is of importance to naval low-band HF communications. Not only does vertically polarized propagation attenuate far less rapidly along the surface (or, put another way, transmit greater field strength at a greater distance over sea water), but it allows omnidirectional coverage (ideally) independent of the ship's position — characteristics not available by use of horizontally polarized shipboard antennas. Therefore, even though shipboard wire-rope antennas are frequently referred to as "long-wire antennas," they are not actually long-wire in the classic sense. True long-wire antennas are devised to be directional — the opposite of what shipboard HF antennas must be.

When it became fully understood that high frequency wire antennas could be fashioned from tall structures existing on deck, the evolutionary design process — using masts, yardarms, outriggers, and even stacks — began as a serious, scientific endeavor. Through experimentation and modeling, wire antennas were custom-made for individual ship class topsides. By 1951, a broadband inclined fan-type antenna had been proposed for shipboard use so that more than one transmitter could be used with a single antenna in order to reduce the total number of antennas cluttering the ship's topside.[9] This concept has progressed to the point that today wire-rope fan antennas are standardized for the lower portions of the HF range, particularly the low band (2-6 MHz) and, infrequently, the mid-band (4-12 MHz). Used chiefly for transmitting (but becoming more and more applied to transceiving), these antennas are designed to be as omnidirectional as possible, taking into account the distorting effects of structural interference on deck. Through the use of matching networks at the input terminals of the antenna, an impedance characteristic allowing a maximum VSWR of 3:1 is maintained over the 2-6 MHz, 4-12 MHz, and 10-13 MHz frequency bandwidths.

Figure 2-11 is a drawing of a typical wire-rope fan installation showing the major parts used in construction prior to 1984. Figure 2-12 is a photograph of an actual ship installation. The standard wire used is 5/16-inch diameter, Type I, Class 2 Construction 1, 6×19 phosphor bronze material conforming to Federal Specifications RR-W-410. Wire rope is made by twisting a number of individual wires into a single strand and then twisting these strands together over either a fiber of wire core. For example, the standard 6×19 wire-rope contains 6 strands with 19 wires in each strand. Phosphor bronze wire is used because of its corrosion-resistant and non-magnetic properties.

For several years vinyl jacketing or stranded wire rope was specified for use on shipboard HF wire-type antennas. The requirement for a protective covering arose in an effort to reduce precipitation static, a special noise caused by charged particles traveling through the air at high velocity and impinging upon the antenna wires. This phenomenon is experienced particularly in sleet, snow, sand, and dust storms, and is associated with receive type antennas, or in the receive mode of transceive antennas. It was thought that a protective covering, such as vinyl, would provide a degree of isolation between the impacting particles and the wire. However, no method was ever perfected to seal the end points of the vinyl jacket; that is, the point at which the antenna wire terminates in a clamp, connector plate, or end fitting. These termination points required elaborate procedures for wrapping, coating, and sealing with moisture and corrosion resistant materials. The procedures were time-consuming, costly, and ineffective after short periods of sea service. Normal ship motion, producing continuous flexures of the wire antenna elements, soon caused the coating material on the end fitting to crack and separate, allowing moisture to enter. Once the water crept down the line, it was captured between the vinyl jacket and the wire. The moisture, being a solution of salts and exhaust contaminates in constant contact with the metal, accelerated corrosion and the ultimate deterioration of the antenna wire. Thus resulted the situation in which a covering meant to provide protection against corrosion became in fact an agent hastening corrosion. It should be noted, moreover, that phosphor bronze wire is itself corrosion resistant and should not require any added protective coating.

With the onset of the worldwide oil crisis, polyvinyl chloride (pvc) material, being a petroleum byproduct, became increasingly difficult to obtain and ever more expensive. Insofar as there never was any clearly proven benefit of vinyl covering for wire transmit antennas, the protective jacket requirement was eliminated for naval shipboard use except in the case of receive and transceive.[10] Since wire antennas are rarely used for receive-only purposes (having been replaced by whip types) vinyl jacketing has been retained only for transceive wire fans.

Referring again to Figure 2-11 the individual sections of wire rope antennas are tensioned during installation for proper dress and to eliminate excessive slack that might create interference with other sections of the antenna, the ship's structure or rigging, and other systems. Additional stresses are placed on wire rope antennas by heavy winds, ice accumulation, and whipping of the wire by ship motion in a seaway. The ultimate result of these hazards is failure of the antenna. To prevent this, safety links are installed. These links are usually designed to have a breaking strength of approximately one quarter of the breaking strength of the wire rope used for the antenna. A safety pendant is provided in the safety link so that if the link should break the pendant will carry the load and prevent the antenna from falling to the deck. The length of the safety pendant, made of the same material as the antenna, should be such that when fully extended, the antenna will not foul adjacent sections of the antenna, the ship's rigging or superstructure. The safety pendant should be held to the safety link with a light seizing to prevent whipping due to wind or the ship's motion. Figure 2-13 shows the proper method of constructing safety links for use with 5/16-inch wire rope antennas. Safety links are installed in each section of wire rigged to yardarms or structures, normally at the highest point of connection of the antenna.[11] An alternative concept of a typical safety link is illustrated in Figure 2-14.[12]

Through the use of careful modeling on an antenna range, a variety of wire fans is tailored to each ship class. The most common type is the 2-6 MHz three-element twin fan, as in Figure 2-15 and 2-16. (Figure 2-16 also shows a few wire antennas which are no longer commonplace.) Examples that are more unusual are shown in Figure 2-17 (a 2-30 MHz full range single fan requiring a base tuner), Figure 2-18 (a 2-6 MHz yoked twin fan — i.e., a double twin draped fore and aft and connected together below decks), Figure 2-19 (showing a single section of yoked twin fan), Figure 2-20 (a multi-element, five-wire, 4-12 MHz single fan), and Figure 2-21 (a multi-element, fifteen-wire, 2-6 MHz yoked twin fan, and a multi-element, fifteen wire 6-18 MHz yoked twin fan.) The latter is one of a kind, unlikely to occur again.

Wire rope antennas of the type described above are constructed on the ships by riggers of the installing activity using design drawings developed from the model study for the ship class. These antennas normally are not given a standard Navy nomenclature. Following, however, are examples of the rare instances when they have:

ANTENNA AS-2801/SRT

a. TYPE/USE — The AS-2801/SRT is an LF/MF wire rope antenna designed for use onboard LHA amphibious assault ships. It consists of two wire rope sections, one for the starboard antenna and the other for the port antenna. The antennas are strung between the yardarms of two masts. The starboard antenna is fed from a tuner on the after mast and the port antenna from a tuner on the forward mast. This antenna may be used with any transmitting equipment capable of operating within its frequency range and power handling limitations.
b. PHYSICAL CONFIGURATION — [See Figure 2-22].
c. DIMENSIONS — (1) Length: 120 feet each section; (2) Antenna is assembled from 5/16-inch phosphor bronze wire rope; (3) Weight: 80 lbs. each section.
d. FREQUENCY RANGE — 300 to 1500 kHz.

e. INPUT IMPEDANCE/VSWR — 50 ohms to tuner.
f. RF POWER RATING — 1.0 kW average.
g. POLARIZATION — Vertical.
h. TYPE OF FEED — Unbalanced wire rope from tuner.
i. INPUT CONNECTOR — Bolt to feedthrough insulator stud.
j. INSTALLATION REQUIREMENTS — The wire rope sections are attached to padeyes on the yardarms. The lead to each tuner is made from a TEE fitting to a strain insulator and jumper to the tuner. A safety link is used at the end opposite the feed point of each horizontal section.
k. SPECIAL CONSIDERATIONS — After installation the antenna must be stressed as specified in the installation requirements. The area in the vicinity of the couplers and feed points must be kept clear of interference from ship's rigging.
l. REFERENCE — CHU Associates Drawing Number 31730.

ANTENNA AS-2803/SRC

a. TYPE/USE — The 2803/SRC antenna is an HF twin fan wire-rope antenna initially designed for installation on LHA amphibious assault ships. The antenna consists of dual triangular wire-rope fans and a matching unit used to transform the antenna input impedance to the impedance of the transmission line. The radiating elements of the antenna are manufactured from phosphor bronze wire rope. This antenna may be used with any HF transmitting or receiving equipment designed to operate within its frequency range and RF power handling limitations.
b. PHYSICAL CONFIGURATION — [See Figure 2-23(a)].
c. DIMENSIONS — Antenna: (1) Height: 53 feet; (2) Width: at yardarm (8 feet - 4 inches); (3) Weight: 340 lbs. Matching Network: (1) Height: 21 $^{3}/_{16}$ inches O.A.; (2) Width: 37⅛ inches (including entrance insulator); (3) Depth: 28⅛ inches O.A.
d. FREQUENCY RANGE — 2 to 6 MHz.
e. INPUT IMPEDANCE/VSWR — 50 ohms, matching network supplied to limit VSWR to not greater than 3:1 over frequency range.
f. RF POWER RATING — 20 kW average.
g. POLARIZATION — Vertical.
h. TYPE OF FEED — Unbalanced, coaxial cable.
i. INPUT CONNECTOR — Matching network 1⅝-inch EIA flange.
j. INSTALLATION REQUIREMENTS — A typical installation of a twin fan antenna is shown in Figure 2-23(a). The exact configuration of each individual installation is dependent on the ship's structure.
k. SPECIAL CONSIDERATIONS — This antenna contains numerous mechanical and electrical connections. Extreme care must be exercised to insure that all metal-to-metal connections are clean and tight and that all brazed connections have been made properly.
l. REFERENCE — CHU Associates Drawing Number 31958.

ANTENNA AS-2804/SRR

a. TYPE/USE — The AS-2804/SRR antenna is a conformal fan HF receiving antenna initially designed for use onboard LHA amphibious assault ships. It consists of a three-wire, single fan installed between the ship's deck and a yardarm. The upper ends of the fan elements are mechanically and electrically grounded at the yardarm terminations. This type of antenna makes use of the ship's structure as part of the HF radiator. To obtain the desired frequency coverage the dimensions of the fan, fan spread, and distance from ship's structure at top and bottom must be determined, usually through a model study. After installation these dimensions must be maintained to obtain satisfactory operation. This antenna may be used with receiving equipment capable of operating within its frequency range.
b. PHYSICAL CONFIGURATION — [See Figure 2-23(b)].
c. DIMENSIONS — (1) Height: 44 feet; (2) Width: 14 feet (top spread); (3) Weight: 30 lbs. (antenna only).
d. FREQUENCY RANGE — 2 to 4 MHz.
e. INPUT IMPEDANCE — 50 ohms.
f. RF POWER RATING — Not applicable, receive only.
g. POLARIZATION — Vertical.
h. TYPE OF FEED — Unbalanced. Wire-rope from matching unit feed-through insulator to lower end of fan.
i. INPUT CONNECTOR — Bolt connection to fan connector plate.
j. INSTALLATION REQUIREMENTS — The insulator at the lower end of the fan may be secured to padeyes on the deck. The upper ends of the fan are terminated at the yardarm. Standard antenna fittings and insulators are used in constructing the antenna.
k. SPECIAL CONSIDERATIONS — The area in the vicinity of the base of the antenna should be kept clear of interference for at least three feet. After installation the individual wires must be stressed to specifications.
l. REFERENCE — CHU Associates Drawing Number 31760.

Lightweight Broadband Communications Fan Antenna (LWCA)

In 1981, the Navy initiated a development effort to modernize the traditional wire-rope fan antenna, including investigation of new materials and construction techniques that could improve reliability, durability, maintainability, electromagnetic compatibility, and overall performance of the antenna. Of particular interest was the need to reduce topside weight, improve the combat survivability of Command and Control Communications by increasing the shock resistance of the primary shipboard broadband HF antenna, and to lessen the potential for generating electromagentic interference by existing HF antennas. This effort resulted in the lightweight broadband HF communications antennas (LWCA) being established as the new standard wire-rope antenna for Navy shipboard applications. As such, it is the single most important change in the overall design of U.S. Navy wire-rope antennas to occur in the last fifty years, and is currently being adapted for use fleetwide.

Overall dimensions and electrical configuration of the LWCA are identical to those of the old-style antenna which it replaces. Antenna hardware, however, has undergone significant changes as seen in Figure 2-24(a). Wire used for construction of the LWCA is $^3/_{16}$-inch diameter 7×19 phosphor bronze material, with six of the strands twisted around the seventh to form a wire strand core. Installation is simplified and cable fittings are more readily applied. The wire-rope is jacketed with a $^1/_{16}$-inch thick vinyl cover.

Insulators used for construction of the LWCA are fabricated of Delrin acetal resin, a thermoplastic polymer. The primary advantages of Delrin over ceramic compounds are its superior toughness and lighter weight. All electrical connections are encapsulated in Delrin, providing a seal for the end points of the vinyl jacket of the antenna wire, and eliminating the need for wrapping and coating wire terminations, as was the case with the old-style antenna. Two unique features of the LWCA are the use of shock isolating insulators to attach the antenna wires to the yardarm, and incorporation of a single shear pin at the apex attachment point, thereby eliminating the need for safety links in the antenna (see Figure 2-24(b)). Shock and environment loads on the antenna wires cause the shock insulators to extend as much as eight inches to absorb sudden transient increases. Antenna stress exceeding 1100 pounds causes the apex shear pin to snap, releasing the apex end of the antenna from the padeye to which it is attached. A retainer line connected between the apex insulator and supporting structure prevents uncontrolled whipping of the antenna after the shear pin breaks. Improvements in hardware design and materials have resulted in a significant weight reduction (approximately 65%) over a comparable fan antenna of the older style, and a product much easier to maintain in good operating condition by ship personnel.

The LWCA is preassembled in accordance with design drawings developed for each ship class. A typical installation is pictured in Figure 2-24(c). Navy nomenclature has been assigned by ship class to all LWCA configurations. For example, that applied to CG 47 Class ships is the AS-3727(V) given below.

ANTENNA GROUP OE-367(V)/SRC (AS-3727(V) LWCA Wire Fan)

a. TYPE/USE — Antenna group OE-367(V)/SRC is an HF broadband lightweight communications antenna (LWCA) designed for installation on CG 47 Class ships. Because of topside structural differences between CG 47 lead ship and follow-on ships, the antenna group is produced with two variants: (V)1 applies to CG 47 and CG 48; (V)2 applies to CG 49 and later ships of the class.
The OE-367(V)1/SRC group consists of dual triangular multi-element wire-rope antennas, AS-3727(V)1/SRC (starboard) and AS-3727(V)2/SRC (port) connected together to form a single omnidirectional transceive antenna capable of supporting up to eight HF transmit circuits in the 2-6 MHz frequency range, as well as HF receive circuits in the 2-12 MHz range. The fans are connected to a passive matching network in a drip-proof enclosure to transform the antenna input impedance to the impedance of a coaxial transmission line. Four ground wire assemblies, MX-10632/SRC (2) and MX-10633/SRC(2) are grounded to, and connected between, the foremast and stack to prevent parasitic interaction of these vertical structural members. Antenna Group OE-367(V)2/SRC is similar in makeup, differing only in physical dimensions of the starboard and port antennas, AS-3727(V)3/SRC and AS-3727(V)4/SRC.

b. PHYSICAL CONFIGURATION — [See Figure 2-24(d)].

c. DIMENSIONS — (1) Height: 45 feet; (2) Width: at yardarm (50 feet); between apex insulators (26 feet); (3) Weight: 176 pounds (wire assembly only; not including matching network box and components).

d. FREQUENCY RANGE — 2 to 6 MHz (Transmit); 2 to 12 MHz (Receive).

e. INPUT IMPEDANCE/VSWR — 50 ohms at 3:1 VSWR over frequency range of 2 to 6 MHz (matched).

f. RF POWER RATING — 4 kW average.

g. POLARIZATION — Vertical.

h. TYPE OF FEED — Unbalanced, 50 ohm coaxial cable.

i. INPUT CONNECTOR — Matching Network 1⅝-inch flange.

j. INSTALLATION REQUIREMENTS — A typical installation of the OE-367(V)/SRC antenna group is shown in Figure (2-24(c)). The AS-3727(V)/SRC antennas are delivered as assembled units. The installing activity must attach the shock insulators to the yardarm padeyes, then connect the lower ends of the antenna wires to the apex insulator. The feed wires must also be connected to the feed adapter at the matching network input insulator. Ground wire assemblies must be connected between the appropriate padeyes on the superstructure, and terminated at grounding studs located near the padeyes.

k. SPECIAL CONSIDERATIONS — Care must be exercised by the installing activity to insure proper corrosion protection instructions are followed when installing the apex insulators and feedwires.

l. REFERENCES — (1) Space and Naval Warfare Systems Command Drawing No. 28687-0137561. (2) Eldyne, Inc. Technical Manual No. 60595-0927A.

2-1.2 HF Monopole Antennas

The most commonly used and simplest of all naval shipboard communications antennas is the plain monopole. Thin, semi-rigid, supported by an insulator at its base, this antenna — generally known as the "whip" — can be found on virtually every ship from small craft to large combatant, and is used for both transmit and receive. The self-supporting feature of the whip makes it particularly useful where space is at a premium and in locations not suitable for other types of general purpose communications antennas. The monopole most extensively used is the single element, vertical 35-foot aluminum or fiberglass whip; a type simple and economical to construct, install, and maintain. Figures 2-25 and 2-26 show a series of 35-foot whips installed on naval ships.

Monopole antennas are narrowband and, being so electrically short below 5 MHz, are quite inefficient. With quasi-minimum atmospheric noise the limiting signal-to-noise factor at these frequencies, the inefficiency is of little consequence for the receive case. In fact, the maximum receiving antenna deficiency, without excessive degradation of effective receiving sensitivity, is desirable aboard ship to reduce the interference effect of local transmissions.[13] Said differently, a lossy antenna affords receiver protection. However, for the transmit case, the situation is quite different. To cover the entire 2-30 MHz range the narrowband monopole must be provided with a base tuner to allow maximum power transfer (by optimum impedance matching) between transmitter and antenna.

The main disadvantage of the tuned monopole is that a limited number can be installed onboard ship with sufficient spatial separation to prevent excessive mutual coupling interference between adjacent base tuners. A minimum of 40 feet is required, for example, between transmit whips using the standard Navy 2-30 MHz AN/URA-38 base coupler.[14]

There are other, less common, HF single whips and monopoles in naval service. These special types include the self-tunable helical monopole, and the very thin five-foot probe whip with an integral RF tuner. Descriptions of these seldom-used, limited-range antennas are given below under AS-2875, AS-2876, and AS-1857.

Whip Type Monopole Antennas

For many years the NT-66046 and NT-66047 were the standard whip antennas used aboard Navy ships. These antennas are sectional non-telescoping aluminum poles mounted on feedthrough base insulators by means of an aluminum adapter piece. The NT-66046 is 28 feet and the NT-66047 is 35 feet high.

ANTENNA NT-66046

a. TYPE/USE — The NT-66046 antenna is a four-section 28-foot aluminum whip used for HF transmitting, and LF/MF/HF receiving. The four sections are tapered from 2 1/2 inches at the bottom to one inch at the top section, and are connected together by use of screw-type couplings. The NT-66046 whip may be mounted to either an IL-18/U or IL-19/U base insulator by means of a cast aluminum adapter piece.
b. PHYSICAL CONFIGURATION — [See Figure 2-27].
c. DIMENSIONS — (1) Height: 336 inches (including base insulator); (2) Diameter (max): 2 1/2 inches (less insulator); (3) Weight: 55 lbs.
d. FREQUENCY RANGE — Receiving: 500 kHz to 30 MHz. Transmitting: 2 to 30 MHz (depending on the antenna coupler or matching network).
e. INPUT IMPEDANCE/VSWR — Used for receiving in a 50-ohm system, the VSWR may vary from 2000:1 at 2 MHz to approximately 1:1 near the antenna quarter-wave frequency. Used for transmitting, the antenna must be connected to a base tuner for matching to the transmitter with a VSWR of 3:1 or less.
f. RF POWER RATING — Antenna not rated. RF power limitation is dictated by the heat rating and flash-over voltages of the base insulators.
g. POLARIZATION — Vertical when mounted on a true ground plane. Elliptical depending on mounting, frequency and direction in the vertical and horizontal planes when mounted onboard ship.
h. TYPE OF FEED — Unbalanced.
i. INPUT CONNECTOR — None. Input to antenna is made to base insulator terminal.
j. INSTALLATION REQUIREMENTS — The NT-66046 is mounted to the base by a cast aluminum adapter piece. It is secured to the insulator mating surface by four 5/8-11 bolts on a six-inch diameter bolt circle.
k. SPECIAL CONSIDERATIONS — The location of the antenna should be selected to provide as clear an area as possible in the vicinity of the antenna with a minimum of 35 feet from other antennas. This requirement is necessary to reduce the effect of the ship's structures on the antenna radiation pattern, and the mutual coupling between antennas.
l. REFERENCE — NAVSEA Drawing RE-2691719.

ANTENNA NT-66047

a. TYPE/USE — The NT-66047 antenna is a non-telescopic, five-section, 35-foot aluminum whip. It was formerly the standard Navy shipboard whip for transmitting and receiving in the LF,MF, and HF ranges. The five sections taper from three inches O.D. at the base to one inch O.D. at the top, and are connected together by means of screw-type couplings. The whip is mounted on either an IL-18/U or IL-19/U base insulator by means of a cast aluminum adapter piece.

b. PHYSICAL CONFIGURATION — [See Figure 2-28].
c. DIMENSIONS — (1) Height: 420 inches (including base insulator); (2) Diameter (max): 3 inches (excluding base insulator); (3) Weight: 70 lbs.
d. FREQUENCY RANGE — Receiving: 14 kHz to 32 MHz. Transmitting: 2 to 32 MHz, depending on the antenna coupler or matching network used.
e. INPUT IMPEDANCE/VSWR — Used for receiving in a 50-ohm system, the VSWR may vary from 2000:1 at 2 MHz to 1:1 near the antenna quarter-wave frequency. Used for transmitting, the antenna must be connected to a base tuner for matching to the transmitter with a VSWR of 3:1 or less.
f. RF POWER RATING — Antenna not rated. RF power limitation is dictated by the heat rating and flashover voltages of the base insulator.
g. POLARIZATION — Vertical when mounted on a true ground plane. Elliptical depending on mounting, frequency, and direction in the vertical and horizontal planes when mounted aboard ship.
h. TYPE OF FEED — Unbalanced.
i. INPUT CONNECTOR — None. The input to antenna is made to base insulator terminal.
j. INSTALLATION REQUIREMENTS — The NT-66047 is mounted on the base insulator by means of a cast aluminum adapter piece. It is secured to the insulator mounting surface by four 5/8-11 inch bolts on a 6-inch diameter bolt circle.
k. SPECIAL CONSIDERATIONS — The location of this antenna should be selected to provide as clear an area as possible in the vincinty of the antenna and a minimum distance of 35 feet from other antennas. These requirements are necessary to reduce the effect of the ship's structures on the impedance, radiation pattern, and the mutual coupling between adjacent antennas.
l. REFERENCES — NAVSEA Drawings RE-C2692050, RE-F2691719.

AS-2537/SR and AS-2537A/SR Whip Antennas

The AS-2537/SR and AS-2537A/SR are 35-foot fiberglass whip antennas designed to replace the NT-66047. These antennas are equal electrically, but have minor mechanical differences. They are manufactured in two sections that are screwed or bolted together. The base insulator is an integral part of the antenna. The base bolt mounting circle is identical to feedthrough insulator Type NL222B30-139, used in certain cases with NT-66047 antennas, and, therefore, may use the same foundation after removal of the insulator. However, when replacing NT-66047 antennas using feedthrough insulators NL222B31-203 or NL222B32-240, it will be necessary to provide an adapter plate to match the bolt circle of the insulator drilled to match the base of the antenna. RF feed to this type of antenna is made to a stud above the base since this connection is above the foundation rather than below as with the feedthrough types. Therefore, the enclosure for the coupling or matching networks also should be located above the mounting platform.

ANTENNA AS-2537/SR

a. TYPE/USE — The AS-2537/SR is a two-section 35-foot fiberglass whip antenna consisting of an 18-foot top section bolted on the 17-foot lower section. A pin and socket arrangement is used to connect the radiating elements of the two sections which are embedded in fiberglass and a thermosetting epoxy resin.
b. PHYSICAL CONFIGURATION — [See Figure 2-29] (note especially the type of base).
c. DIMENSIONS — (1) Height: 438 inches; (2) Base Daimeter: 12 inches; (3) Weight: 130 lbs.
d. FREQUENCY RANGE — Receiving: 14 kHz to 32 MHz; Transmitting: 2 to 32 MHz, depending on the antenna coupler or matching network used in the system.
e. INPUT IMPEDANCE/VSWR — Used for receiving in a 50 ohm system, the VSWR may vary from 2000:1 at 2 MHz to 1:1 near the antenna quarter-wave frequency. Used for transmitting, the antenna must be connected to a base tuner for matching to the transmitter with a VSWR of 3:1 or less.
f. RF POWER RATING — 5 kW average.
g. POLARIZATION — Vertical when mounted on a true ground plane. Elliptical, depending on type of mounting, frequency, and direction in the vertical and horizontal when mounted onboard ship.
h. TYPE OF FEED — Unbalanced.
i. INPUT CONNECTOR — $^3/_8$-16 × 1$^1/_8$ inch bolt.
j. INSTALLATION REQUIREMENTS — The AS-2537/SR bottom section has eight $^{11}/_{16}$-inch bolt holes on a 10$^3/_4$-inch bolt circle for installation. The bolt holes and bolt circle are the same as the mounting flange on the IL-19/U base insulator used with the NT-66047 antenna. The AS-2537/SR can be installed on the existing base used as a direct replacement. If the antenna is to be installed at a location other than where a base insulator was used, a suitable mounting plate drilled to match the antenna mounting bolt holes is required.
k. SPECIAL CONSIDERATIONS — The location selected should provide as clear an area as possible in the vicinity of the antenna. A minimum of 35 feet should be maintained from other antennas. This requirement is necessary to reduce the effect of the ship's structure on the antenna radiation pattern and mutual coupling between antennas.

l. REFERENCES — (1) NAVSEA Drawing RE-F-2698386. (2) Similar type on commercial vessels is Mackay Marine ITT MR-106-11A and MR-106-12A.

ANTENNA AS-2537A/SR

a. TYPE/USE — The AS-2537A/SR is a two-section 35-foot fiberglass whip antenna consisting of an 18-foot top section that screws into a similar bottom section; the bottom section has an integral mounting base and an attachment bolt for the antenna feed line. The radiating elements are six beryllium copper tapes embedded in fiberglass and thermosetting epoxy resin.

b. PHYSICAL CONFIGURATION — [See Figure 2-30].

c. DIMENSIONS — (1) Height: 432 inches; (2) Base Diameter: 12 inches; (3) Weight: 120 lbs.

d. FREQUENCY RANGE — Receiving: 14 kHz to 32 MHz. Transmitting: 2 to 32 MHz, depending on the antenna coupler or matching network used in the system.

e. INPUT IMPEDANCE/VSWR — Used for receiving in a 50-ohm system, the VSWR may vary from 2000:1 at 2 MHz to 1:1 near the antenna quarter-wave frequency. Used for transmitting, the antenna must be connected to a base tuner for matching to the transmitter with a VSWR of 3:1 or less.

f. RF POWER RATING — 5 kW.

g. POLARIZATION — Verticial when mounted on a true ground plane. Elliptical, depending on type of mounting, frequency, and direction in the vertical and horizontal planes when mounted onboard ship.

h. TYPE OF FEED — Unbalanced.

i. INPUT CONNECTOR — $^3/_8$-16 × 1 inch cap screw.

j. INSTALLATION REQUIREMENTS — The AS-2537A/SR bottom section has eight 11/16-inch bolt holes and a $10^3/_4$ inch bolt circle for installation. The bolt holes and bolt circle size are the same as the mounting flange on the IL-19/U base insulator used with the NT-66047 antenna. The AS-2537A/SR can be installed on the existing base, used as a direct replacement. If the antenna is to be installed at a location other than where a base insulator was used, a suitable mounting plate drilled to match the antenna mounting bolt, is required.

k. SPECIAL CONSIDERATIONS — The location should be selected to provide as clear an area as possible in the vicinity of the antenna. A minimum distance of 35 feet should be maintained from other antennas. This requirement is necessary to reduce the effect of the ship's structure on the antenna radiation pattern and mutual coupling between antennas.

l. REFERENCES — (1) Technical Manual, Antenna AS-2537A/SR, NAVELEX 0967-LP-423-9010. (2) NAVSEA Drawing RE-F2698386. (3) Valcom Inc., Ltd., Model V-132 Antenna.

ANTENNA AS-2807/SRC

a. TYPE/USE — The AS-2807/SRC antenna is a general purpose omnidirectional 35-foot monopole consisting of three tapering 6061T6 aluminum sections supported by a fiberglass base insulator on a square mounting base. Constructed of riveted aluminum panels, the antenna measures 12×12 inches at the base, tapering to 1 inch square at the top. The fiberglass base insulator is inserted inside the bottom section of the antenna and the aluminum or steel mounting base. RF connection is made to a stud on the bottom section. Twelve 9/16-inch bolt holes fasten the mounting base to the ship's structure. The enclosure for the coupling or matching networks should be located above the mounting platform close to the antenna connection. Unconventional in that it is a square, tapered monopole, this antenna can hardly be considered a "whip."

b. PHYSICAL CONFIGURATION — [See Figure 2-31].

c. DIMENSIONS — (1) Height: 435 inches; (2) Width: 12 inches at base; (3) Depth: 12 inches at base; (4) Weight: 150 lbs.

d. FREQUENCY RANGE — Receiving: 14 kHz to 32 MHz. Transmitting: 2 to 32 MHz depending on the antenna coupler or matching network used.

e. INPUT IMPEDANCE/VSWR — Used for receiving, the VSWR of this antenna in a 50-ohm system may vary from 2000:1 at low frequencies to approximately 1:1 near the antenna quarter-wave frequency. Used for transmitting, the antenna must be connected to the base tuner for matching to the transmitter with a VSWR of 3:1 or less.

f. RF POWER RATING — 5 kW.

g. POLARIZATION — Vertical when mounted on a true ground plane. Elliptical, depending on mounting, frequency, and direction in the vertical and horizontal planes when mounted onboard ship.

h. TYPE OF FEED — Unbalanced.

i. INPUT CONNECTOR — Bolt connection at base of lower section.

j. INSTALLATION REQUIREMENTS — The antenna may be installed on the ship's deck or structure by means of a 19-inch square base plate drilled for twelve 9/16-inch bolt holes. The base plate has an $11^{3}/_{16}$ inch square ×7 inches high adapter piece that fits outside the antenna fiberglass insulator. The antenna insulator is then bolted to the base adapter piece.

k. SPECIAL CONSIDERATIONS — This antenna may be installed on base tilting mounts where required for aircraft operations. The antenna base may be fastened directly to the tilting mount using the regular antenna base.

l. REFERENCE — CHU Associates Drawing Number 31968.

ANTENNAS AT-1011/U AND AT-1047/U

a. TYPE/USE — The AT-1011/U and AT-1047/U are self-supporting, very thin, "fishing pole" type fiberglass antennas. Due to their high flexibility and the continual maintenance required to keep corrosion-free tight couplings between the numerous sections, these antennas are not widely used aboard ship. Nevertheless they are frequently employed on small craft for both transmit and receive in the MF/HF ranges. Moreover, because of its very light weight the AT-1011 has been supplied for use with the AN/URC-35 Radio set on major ships — not as a permanently installed antenna, but for emergency use in the event of losing the main antenna system.

The AT-1011 consists of eight 4-foot sections and the AT-1047 is made up of four small diameter 4-foot sections. Both antenna configurations use the same 20-inch base section which contains the base insulator, mounting flange, and input connection.

b. PHYSICAL CONFIGURATION — [See Figure 2-32].
c. DIMENSIONS — AT-1011/U: (1) Height: 436 inches O.A.; (2) Weight: 19 lbs. AT-1047/U: (1) Height: 228 inches O.A.; (2) Weight 15 lbs.
d. FREQUENCY RANGE — 2 to 30 MHz, depending on the antenna coupler or matching network used in the system.
e. INPUT IMPEDANCE — Used for receiving in a 50-ohm system, the VSWR may vary from 2000:1 at 2 MHz to 1:1 near the antenna quarter-wave frequency. Used for transmitting, the antenna must be connected to a base tuner for matching to the transmitter with a VSWR of 3:1 or less.
f. RF POWER RATING — 2 kW.
g. POLARIZATION — Vertical.
h. TYPE OF FEED — Unbalanced.
i. INPUT CONNECTOR — Strap or clamp to upper portion of base section.
j. INSTALLATION REQUIREMENTS — The base section plate has six 13/32 inch bolt holes equally spaced on a $7^3/_4$ inch bolt circle for mounting the antenna. In an emergency the antenna may be attached to a flat surface by "C" clamps.
k. SPECIAL CONSIDERATIONS — These antennas are designed primarily for mobile operations and have a high degree of flexibility compared to standard 35-foot whip antennas. Onboard ship the whipping of the antenna in a seaway may require the use of guys or standoff-insulated bracing to prevent the antenna from coming into contact with the ship's structure. These antennas are not recommended for permanent installation on ships other than small craft.
l. REFERENCE — Shakespeare Communications Products Catalog.

Tuner Whip Antennas

A few antenna groups such as the Navy's AN/SRA-43 have whip antennas with integral tuner units. These antennas are physically small and are often mounted either horizontally or vertically on yardarms and other parts of the ship's structure not suitable for larger antennas. A typical example, once used heavily by the Navy, is the five-foot AS-1857 whip described below:

ANTENNA AS-1857

a. TYPE/USE — The AS-1857 is an integral part of the TN-438/SRA-43 RF tuner. The antenna, tuner, and antenna control unit, C-6828/SRA-43, comprise the antenna group AN/SRA-43 which is used for low-band HF receive. A 60-inch whip antenna, the AS-1857 is mounted on a base feed-through insulator attached to the RF tuner housing. The antenna element is a copper rod encased in fiberglass. The RF tuner is remotely controlled by the C-6828/SRA-43 control unit. Since the tuner is the mounting base for the antenna the following specifications and requirements include both units.

b. PHYSICAL CONFIGURATION — [See Figure 2-33].
c. DIMENSIONS — (1) Height: 66-3/8 inches O.A.; (2) Width: 21 inches (tuner); (3) Depth: 13 inches (tuner); (4) Weight: 32.5 lbs total.
d. FREQUENCY RANGE — 2 to 8 MHz.
e. INPUT IMPEDANCE — 50 ohms, tuner input.
f. RF POWER RATING — Not applicable, receive only.
g. POLARIZATION — Vertical. (Antenna must be mounted high on mast or superstructure to be fully effective.)
h. TYPE OF FEED — Unbalanced, coaxial cable (tuner).
i. PRIMARY POWER REQUIRED — Tuner power supplied from control unit.
j. OPERATING TEMPERATURE — Minus 28°C to +65°C.
k. INPUT CONNECTOR —To tuner: (1) RF-Type N UG-58A/U; (2) Control Cable-Type MS.
l. INSTALLATION REQUIREMENTS — The AS-1857 whip antenna may be mounted either horizontally or vertically with the tuner attached to a suitable mounting plate by four 3/8-inch bolts (bolt holes centered on 9-inch by 20 1/4-inch centers). Because of the small size of the tuner and antenna, the unit is often installed on yardarms, mast platforms, stacks, and other locations not suitable for larger antennas. Care should be exercised in selecting locations that will not cause interference with signal halyards and the ship's rigging. Support is necessary when mounted horizontally, to prevent fractures, and subsequent failure of the whip, caused by excessive vibration. Details of a suitable fiberglass support widely used by the Navy are shown in Figure 2-34.
m. SPECIAL CONSIDERATIONS — The distance between the RF tuner and the remote control unit should not exceed 150 feet. When the tuner is mounted on masts or structures higher than 25 feet, fixed capacitors C3 and C4 should be shunted to improve efficiency from 3.5 to 8 MHz.
n. REFERENCES — (1) Technical Manual, Antenna Group AN/SRA-43, NAVSEA 0967-269-4010. (2) NAVSEA Drawing RE-D2696416.

Helical Monopole Antennas

Not all HF monopoles used on naval surface ships are whip-types. Tunable helical monopoles are also used, though not extensively. Of uniform diameter (i.e., non-tapered), these antennas resemble a true pole.

ANTENNA AS-2875/SRC

a. TYPE/USE — Antenna AS-2875/SRC is a tunable, omnidirectional, helical monopole used for transmitting or receiving. It operates in the quarter wavelength mode over the 2-6 MHz frequency band with an efficiency greater than 25%, and maximum tuned input VSWR less than 2:1 in 50 ohms. It is also capable of operating from 6-30 MHz in higher order 3/4 wavelength mode.

   The antenna consists of an encased helical radiating element and an aluminum base. Servo circuits, through either semi-automatic or fully automatic control, adjust the amount of active (resonant) helix, changing its electrical length to tune the antenna. The radiating element is enclosed with a sealed cylindrical fiberglass radome 10 1/4 inches in diameter and 15 feet tall. The radome is attached to the antenna base which contains the tuning servo-drive motor, associated tuning circuits, and the RF input and control cable connectors. A hinged mounting plate permits the radome and antenna to be manually tilted 90 degrees for servicing.

b. PHYSICAL CONFIGURATION — [See Figure 2-35].

c. DIMENSIONS — (1) Height: 177 inches; (2) Diameter: $10^1/_4$ inches; (3) Weight: 355 lbs.
d. FREQUENCY RANGE — 2 to 6 MHz (also 6-30 MHz in $^3/_4$ wavelength mode).
e. INPUT IMPEDANCE/VSWR — 2:1 maximum in 50 ohms.
f. RF POWER RATING — 1 kW average.
g. POLARIZATION — Vertical.
h. TYPE OF FEED — Unbalanced, coaxial cable.
i. PRIMARY POWER REQUIRED — Control voltages supplied by control group.
j. OPERATING TEMPERATURE — Minus 54°C to +65°C.
k. INPUT CONNECTOR — Type HN receptacle.
l. INSTALLATION REQUIREMENTS — The base mounting flange has fourteen $^3/_{16}$-inch diameter bolt holes for $^3/_4$-inch bolts around its periphery. Cable connections for RF input, servo power input, and antenna data are located on the base casting. Two $^3/_8$-16-inch tapped holes are located on the base mounting flange for attaching bond straps.
m. SPECIAL CONSIDERATIONS — The antenna is secured to a hinged plate that permits it to be manually tilted 90 degrees. This hinged plate is released from the base by a quick-release pin and the antenna is lowered by operating a pivot mechanism consisting of a rack and pinion actuator. A charging valve and pressure indicator, for use in purging and charging the antenna with dry air or nitrogen, are located on the base. The antenna is normally pressurized to 2 psig. The coaxial and control cables to the base of the antenna should have sufficient slack to permit lowering the antenna without stretching the cables.
n. REFERENCE — CHU Associates Drawing Number 31992.

ANTENNA AS-2876/SRC

a. TYPE/USE — Antenna AS-2876/SRC is a helical monopole very similar to the description and operating characteristics of the AS-2875, except that it operates in the basic quarter-wavelength mode over the 6-30 MHz band. As such, it is shorter and lighter — only $97^1/_4$ inches tall and 260 pounds. All other features are identical to the AS-2875.
b. PHYSICAL CONFIGURATION — [See Figure 2-36].

2-1.3 Broadband HF Monopole Antennas

Since the very early days of naval radio, there has been a continuous effort to reduce the number of communications antennas aboard ship. Before the days of sophisticated filtering and receivers with high selectivity and good front-end protection, attempts to transmit on more than one or two circuits with brute force (high power, broad bandwidth) energy resulted in serious interference being picked up by adjacent antennas. Within the bounds of a ship's topside, little relief could be realized at these low frequencies (long wavelengths) by maximizing the separation between transmitting and receiving antennas. Yet, as constant improvements were made to equipment and antennas over the years, allowing narrower bandwidth transmission with remarkably higher circuit quality, the predictable outcome was a proportionate increase in requirements for *more* communications channels. The end result was, of course, a demand which would soar dramatically during World War II operations, resulting in a proliferation of antennas, which only caused greater interference and generated the need for an even higher degree of electronic sophistication. The obvious way to stem such a runaway problem was to find a method utilizing multiple transmission and reception from only a few antennas. If a means could be devised to connect several transmitters or receivers to one antenna then the total number of communications antennas could be significantly reduced. Such "multicoupling," however, requires *broadband* antennas.

Theoretical calculations and measurements predicted that as the length-to-diameter ratio of cylindrical antennas is reduced, the variation of impedance with frequency decreases greatly. In other words, a thick antenna has a much broader frequency response than a thin one. Shipboard space and weight limitations preclude the installation of large cylindrical stub antennas of sufficient broadband response to accommodate Navy transmitters.

In 1959, the Naval Research Laboratory developed the conical monocone HF broadband antenna, consisting of four sloping-wire radiating elements fed from a tall pole mast and draped symmetrically down to the deck.[15] Covering a 5:1 frequency range with a VSWR not exceeding 3:1 throughout, this was the first shipboard application of a broadband HF antenna. The only problem was that it was too large for general shipboard use. However, it was found that a simple 35-foot whip could be made to respond like a thick cylinder by the process of trussing or caging. Thus, a broadening of the antenna impedance over a wide bandwidth is accomplished by physically fattening the antenna to increase its effective diameter. The trussed whip, therefore, has evolved to fulfill the broadband requirements for shipboard multicoupling at HF. As will be discussed later, in addition to trussed whip monopoles, other broadband HF antennas used in naval service are the twin monopole, the discone, and the discage monopole — all designed to cover 3:1 frequency bandwidths with a maximum VSWR of 3:1 over the band.

At first, the NT-66047 aluminum 35-foot whip was easily converted to a broadband antenna by simply adding the necessary appendages and trussing wires. With no specific methods required, however, different shipyards and design activities each devised its own type of trussing, using a variety of materials, so that several configurations of the trussed whip materialized over the years. There were the 35-foot three-wire open-tip (trussed about 75% of total length) "Christmas trees," shown by Figures 2-37, 2-38, and 2-39. Note that in Figures 2-37 and 2-39 the lower ends of the trussing wires are connected to a rigid spreader assembly as part of the antenna, while in Figure 2-38 the wires are anchored directly to the ship structure. Also in naval use was a three-wire closed-tip 35-foot trussed whip, where the trussing wires extended the entire whip length from the top down.

In addition to the three-wire types, there were a number of four-wire trussed whips. Figures 2-40, 2-41, and 2-42 depict three different versions of the four-wire open-tip type. Figures 2-43 and 2-44 illustrate simple four-wire closed-tip types, while Figures 2-45 and 2-46 show an elaborately trussed four-wire closed-tip trussed whip. An unusual six-wire trussed monopole, guyed near the midsection for stability, is shown in Figure 2-47. Note how the spreader assembly positions differ in the various types from Figure 2-37 through 2-47, so that the antenna shapes vary from "Christmas tree" to diamond to obelisk. As if these were not enough, other kinds of monopole broadbanding are illustrated by the cylindrical cage type of Figure 2-48 and the multi-element monopole of Figure 2-49.

In 1974, the Navy decided to put an end to the multiplicity of trussed whip types and material. As a result of the study done by the Naval Electronics Laboratory Center in San Diego (now Naval Ocean Systems Center) a 35-foot three-wire open-tip trussed monopole was standardized for 2-30 MHz transmit and a 15-foot four-wire open-tip trussed monopole for 10-30 MHz receive was recommended, using a support pole of type 5086-H32 aluminum, and a $^5/_{16}$-inch phosphor-bronze wire-rope with associated hardware for trussing material.[16] Issuing the plans and details accompanying Reference (16) to all shipyard design activities, the Navy immediately implemented these standardized trussed monopoles for all new and upgraded installations. Henceforth, all transmitting trussed whips would appear similar to those in Figure 2-39 — except that the upper spreader, as well as bottom spreader, is now made of rigid materials. The standardized 15-foot four-wire receive antenna is like that of Figure 2-40.

Prior to adopting the two standardized trussed monopoles discussed above, there were two specialized broadband monopoles that did receive Navy nomenclature, although they are not widely used. They are the AS-2805 and AS-2806 described as follows:

ANTENNA AS-2805/SRC

a. TYPE/USE — Antenna AS-2805/SRC is a 35-foot trussed monopole general purpose antenna for use with military communications equipment.
The antenna consists of an aluminum radiating element, a fiberglass base insulator, a truss assembly, and a matching network. The aluminum radiating element, fabricated in three sections, has a cross-section which tapers from an 11-inch square at the base to a 1-inch square at the top. A vibration damper is installed at the tip.
The base insulator consists of four $1/4$ inch-thick fiberglass plates attached to the sides of the radiating element. A reinforcing angle section is installed along each corner.
The truss assembly consists of four $1/8$ inch-diameter flexible wire ropes used to broaden the electrical properties of the antenna. The wires are attached between the top of the second section of the whip and the base spreaders.
The matching network is contained in an aluminum box, 12 inches square, and a porcelain feedthrough insulator installed in one side connects to the whip. A type N coaxial receptacle provides connection to associated equipment.
A square base-mounting foundation is required for installation of the antenna. For fixed installations, the foundation can be of steel or aluminum. For installations requiring a tilting foundation, antenna mount MT-4480/SRC may be used.
This antenna can be installed and operated in a vertical, tilted or horizontal position as in aircraft carrier deck edge installations.
AS-2805/SRC has been tested and qualified to meet MIL-S-901C, Grade A Shock; MIL-STD-167, Type I Vibration; and all Service Conditions of MIL-E-16400F, Class 2 Equipment.
b. PHYSICAL CONFIGURATION — [See Figure 2-50].
c. DIMENSIONS — (1) Height: 435 inches; (2) Width: 96$1/2$ inches; (3) Depth: 96$1/2$ inches; (4) Weight: 200 lbs.
d. FREQUENCY RANGE — 4 to 12 MHz.
e. INPUT IMPEDANCE/VSWR — 50 ohms. Matching networks are required to limit VSWR to not greater than 3:1 in transmit mode and not greater than 5:1 in receive mode.
f. RF POWER RATING — 1 kW average.
g. POLARIZATION — Vertical when mounted on a true ground plane. Ellipitical depending on mounting, frequency, and direction in the vertical and horizontal planes when mounted onboard ship.
h. TYPE OF FEED — Unbalanced. Wire-rope feedline from matching unit feedthrough insulator to base of antenna.
i. INPUT CONNECTOR — Type N receptacle.
j. INSTALLATION REQUIREMENTS — The antenna may be installed on the ship's deck or structure by means of an adapter piece fitted with angles drilled for 5/16-inch mounting bolts. The truss wires must be stressed to specifications after initial installation.

k. SPECIAL CONSIDERATIONS — This antenna may be installed on a base tilting mount where required for aircraft operations. In such cases it may be mounted directly to the tilting mount using the regular antenna base.
l. REFERENCE — CHU Associates Drawing Number 31844.

ANTENNA AS-2806/SRC

a. TYPE/USE — Antenna AS-2806/SRC is a 13.5-foot trussed monopole general purpose antenna for use with military communications equipment.
The antenna consists of an aluminum radiating element, a fiberglass base insulator, a truss assembly, and a matching network. The aluminum radiating element has a cross-section which tapers from an 11-inch square at the base to a $7^1/_2$-inch square at the top.
The base insulator consists of four $^1/_4$ inch-thick fiberglass plates attached to the sides of the radiating element. A reinforcing angle section is installed along each corner.
A square base-mounting foundation is required for installation of the antenna. For fixed installations, the foundation can be of steel or aluminum. For installations requiring a tilting foundation, antenna mount MT-4480/SRC may be used.
The truss assembly has four $^1/_8$ inch-diameter flexible wire-ropes used to broadband the electrical properties of the antenna.
The fixed-tuned impedance matching network is contained in an aluminum box 12 inches square. A porcelain feedthrough insulator installed in one side connects to the antenna, and a type N coaxial receptacle provides connection to associated equipment.
This antenna has been tested and qualified to meet MIL-S-901C, Grade A Shock; MIL-STD-167, Type I Vibration; and all Service Conditions of MIL-E-16400F, Class 2 Equipment. It can be installed and operated in a vertical, tilted or horizontal position as in aircraft carrier deck edge installations.
b. PHYSICAL CONFIGURATION — [See Figure 2-51].
c. DIMENSIONS — (1) Height: $162^3/_4$ inches; (2) Width: $96^1/_2$ inches; (3) Depth: $96^1/_2$ inches; (4) Weight: 160 lbs.
d. FREQUENCY RANGE — 10 to 30 MHz.
e. INPUT IMPEDANCE/VSWR — 50 ohms. Matching networks are required to limit VSWR to not greater than 3:1 in transmit mode or not greater than 5:1 in receive mode.
f. RF POWER RATING — 1 kW average.
g. POLARIZATION — Vertical when mounted on a true ground plane. Elliptical depending on mounting, frequency, and direction in the vertical and horizontal planes when mounted onboard ship.
h. TYPE OF FEED — Unbalanced. Wire-rope feedline from matching unit feedthrough insulator to base of antenna.
i. INPUT CONNECTOR — Type N receptacle.

j. INSTALLATION REQUIREMENTS — The AS-2806 may be installed on the ship's deck or structure by means of an adapter piece fitted to the antenna below the spreader. This piece is fitted with angles drilled for $^5/_{16}$-inch mounting bolts. The truss wires must be stressed to specifications after initial installation.

k. SPECIAL CONSIDERATIONS — This antenna may be installed on a base tilting mount using the regular antenna base.

l. REFERENCE — CHU Associates Drawing Number 31845.

Twin Whip Antennas

An alternative method of broadbanding monopole antennas by increasing the effective diameter is to operate plain 35-foot whips in pairs as a single unit. Twin whip antennas so connected as a single antenna for better performance are widely used aboard ship. In twin configuration the individual antennas are usually mounted on a common platform that may be either stationary or tilting. If the distance between the antennas is relatively close, they may be connected with a crossbar which is then fed at the center point from a matching network using a flexible strap. Alternatively, a wire-cable from each antenna to the matching network may be used, provided the wires do not place undue strain on the matching network insulator or come in contact with the ship's structure during antenna motion in a seaway. Figure 2-52(a) is a picture of a twin whip in a fixed vertical installation, 35-feet high, used for 4 to 12 MHz communications. A Navy-nomenclatured version of this type antenna is the OE-214/U, illustrated in Figure 2-52(b), using AS-4014 aluminum whips. (The AS-4014 is electrically and physically identical to the AS-2807 described earlier except for a different type of vibration damper.) Figure 2-53 shows a twin whip in a vertical position, but mounted with a tilting mechanism so that the antenna may be lowered to a horizontal position during air operations. Figure 2-54 shows a twin whip mounted at an angle in a permanent tilt,a method frequently used for 4 to 12 MHz HF communications.

Another technique used to achieve broadbanding with monopoles is the twin trussed-whip arrangement shown in Figure 2-55. Finally, Figures 2-56(a) and (b) show a "stubby" twin whip made of two 12-foot general purpose AS-4038 and AS-4039 aluminum whips (the lower portion of an AS-2087). Used for high-band 10-30 MHz HF communication, this antenna has been given the designation AS-4037. A similar 10-30 MHz broadband antenna is the OE-213/U twin whip dual-array, depicted in Figure 2-56(c), comprised of two pairs of 16-foot AS-4015 type aluminum whips.

2-1.4 Conical HF Antennas

The Navy, over the past 25 years, has developed and put into use a variety of conical-type broadband antennas for mid- and high-range HF communications. As in the case of the trussed monopoles, the purpose of conical shaping is to increase the effective diameter and thereby broaden the frequency response of the antenna. However, because there are many more wires involved, the conical antennas are more caged than they are trussed, so they cannot be classified strictly as trussed monopoles. Moreover, the feed system of conical antennas, rather than the simple base connection of the trussed whip, is more complex and critical to proper performance. The three principal types of conical HF antennas currently in widespread shipboard use are the discage, the discone, and the inverted cone.

The HF Discage Antenna

In 1957, the Naval Research Laboratory designed what is believed to be the first comparatively short, fat, broadband omni-directional high frequency antenna.[17] Christened the "conical monopole" and intended for shore installations, it quickly came into wide use at military communications sites throughout the world. Realizing the possibility of developing smaller versions for installation aboard ship (at higher portions of the HF range and narrower bandwidth), the Navy soon adopted the conical monopole for shipboard use. By adding a disk to the top of the conical monopole, thus cleverly merging the principles of the discone and cage-type antennas, a two-band antenna was born — actually two antennas in one — the discone/cage, or, more simply, the "discage" antenna. The upper cone, from top disk to midsection, is the high-band 10-30 MHz discone, while the entire structure from top to bottom, fed separately at the base, is the mid-band 4-12 MHz conical monopole cage.

Figures 2-57 through 2-63 show differing views and types of discages in naval service. Note that the most common location is on the ship's bow, to provide as much separation from other ship structures as possible. Built individually by shipyards so that the mounting, support pole, discone feed asembly, and spreader diameter and location differ, these antennas all use eight wires spaced symmetrically to form the cage.

The discage being specified now for new installations and to replace worn out older types, is the 16-wire AS-2802/SRC, as seen in Figure 2-64. It is currently the only nomenclatured discage in Navy inventory.

ANTENNA AS-2802/SRC

a. TYPE/USE — Antenna AS-2802/SRC is a 32-foot, broadband, general purpose transmitting/receiving discage for use with military communications equipment. It provides vertically polarized, omnidirectional azimuth radiation coverage of the 4-12 MHz and 10-30 MHz frequency bands for inputs up to 12 kilowatts average power.

The radiating structure consists of two truncated wire-rope cones, connected base to base and supported by a grounded central mast. A twelve-spoke disk assembly with matching network, attached to the top of the mast, is used in conjunction with the upper cone as a broadband discone radiator for 10 to 30 MHz operation.

For operation at the lower frequencies, the entire structure is driven as a broadband monopole. RF input for 4 to 12 MHz operation is at a deck mounted matching network which connects to the lower cone near the base of the mast.

The central mast is supplied by the installing activity.

This antenna has been qualified to meet MIL-S-901C, Grade A Shock; MIL-STD-167, Type I Vibration; and all Service Conditions of MIL-E-16400F, Class 2 Equipment.

b. PHYSICAL CONFIGURATION — [See Figure 2-65].

c. DIMENSIONS — Antenna: (1) Height: 32 feet-9 inches O.A.,: (2) Maximum Diameter: 21 feet, (cone); (3) Maximum Diameter: 18 feet, (disk); (4) Weight: 730 lbs. Matching Network: (1) Height: 21-3/16 inches O.A.; (2) Width: 37 1/8 inches (including entrance insulator); (3) Depth: 28 1/8 inches O.A.; (4) Weight: 114 lbs.

d. FREQUENCY RANGE — 4 to 12 and 10 to 30 MHz.

e. INPUT IMPEDANCE/VSWR — 50 ohms. Matching networks supplied to limit VSWR to not greater than 3:1 over frequency range.

f. RF POWER RATING — 12 kW average.

g. POLARIZATION — Vertical.

h. TYPE OF FEED — Unbalanced, coaxial cable.

i. INPUT CONNECTORS — 1 5/8-inch EIA flange.

j. INSTALLATION REQUIREMENTS — The base of the mast section is welded directly to the ship's deck. Eight gussets, equally spaced around the circumference of the mast, are required for strengthening. Eight insulated guys, also equally spaced around the circumference, are required from the lower spreader to the ship's deck. After the antenna has been set in position and assembled, the wire-rope sections must be stressed equally by the eight turnbuckles supplied with the rope sections. Various mechanical fittings require torque adjustments in accordance with the manufacturer's specifications during final installation.

k. SPECIAL CONSIDERATIONS — The antenna should be located in an area as free as possible from surrounding structures and other antennas. Ship's structures in the vicinity of antennas cause radiation pattern distortion and consequently a loss of omnidirectional characteristics. Close proximity of other antennas increases the degree of mutual coupling between the systems. The matching network should be located close to the base of the antenna in order to keep the connection from the matching unit to the antenna as short as possible.

l. REFERENCES — (1) CHU Associates Drawing Number 31941. (2) NAVSEA Technical Manual 0967-474-7010.

The HF Discone Antenna

The discone antenna is in essence one half of a biconical dipole with the missing half replaced by a disk.[18] The antenna is then mounted vertically to provide broadband, omnidirectional radiation and reception. At HF these antennas become quite large. A special variation of the discone developed specifically for shipboard high-power transmission in the 10-30 MHz range is the so-called "bow stub" antenna. Three versions of this rugged antenna are illustrated in Figures 2-66, 2-67, and 2-68. Note that in figure 2-67 the trussing wires are anchored to the deck, while in Figure 2-68 the wires are attached to a ring assembly at the antenna base.

The only nomenclatured bow stub in Navy service is the AS-2865/SRC described below:

ANTENNA AS-2865/SRC

a. TYPE/USE — The AS-2865/SRC is an omnidirectional broadband antenna designed for deck installation on the ship's bow. This type is generally classified as a bow stub antenna. It consists of a central mast and top radiator of aluminum. The top disk consists of twelve equally spaced elements with a resultant diameter of 14 feet. The antenna may be used with any transmitting or receiving equipment designed to operate within its frequency range and RF power handling capabilities.
b. PHYSICAL CONFIGURATION — [See Figure 2-69].
c. DIMENSIONS — (1) Height: 12 feet-2$^3/_4$ inches; (2) Diameter: 14 feet (top radiator); (3) Diameter: 12 inches (mast upper); (4) Diameter: 16 inches (mast lower); (5) Weight: 1325 lbs. (with foundation).
d. FREQUENCY RANGE — 10 to 30 MHz.
e. INPUT IMPEDANCE/VSWR — 50 ohms, VSWR not greater than 3:1.
f. RF POWER RATING — 10 kW average.
g. POLARIZATION — Vertical.
h. TYPE OF FEED — Unbalanced, coaxial cable, connected to top (disk center).
i. INPUT CONNECTOR — 1$^5/_8$-inch EIA flange.
j. INSTALLATION REQUIREMENTS — The support mast base flange is 24 inches in diameter with twelve equally spaced bolt holes on a 20-inch bolt circle drilled to accommodate 1-inch diameter bolts. The antenna is mounted to a mast foundation welded to the ship's deck.
k. SPECIAL CONSIDERATIONS — The foundation for the antenna must be made to suit the camber and sheer of the ship's deck so that the antenna mating surface is level. The faying surface between the foundation and

antenna base must be treated to prevent electrolytic action between dissimilar metals.

l. REEFERENCES — (1) CHU Associates Drawing Number 32221. (2) NAVSEA Technical Manual 0967-475-0010.

The HF Inverted Cone Antenna

Another version of the conical antenna used for HF shipboard communication is the inverted cone. Originally designed for mounting atop gun turrets, this antenna is used for receiving only. Figures 2-70 and 2-71 depict the inverted cone installed on five-inch gun mounts. The only Navy nomenclatured inverted cone is the AS-2866/SRR described in detail below:

ANTENNA AS-2866/SRR

a. TYPE/USE — The AS-2866/SRR is an omnidirectional inverted cone HF communications receiving antenna, designed for installation atop five-inch gun turrets. The antenna elements are manufactured from aluminum and are supported by a sealed fiberglass base. This antenna may be used with any receiving system operating within its frequency range.

b. PHYSICAL CONFIGURATION — [See Figure 2-72].

c. DIMENSIONS — (1) Height: 74 inches; (2) Diameter: radiator — 120 inches max.; (3) Diameter: base — 26$^3/_4$ inches; (4) Weight: 200 lbs.

d. FREQUENCY RANGE — 10 to 30 MHz.

e. INPUT IMPEDANCE/VSWR — 50 ohms. VSWR not greater than 5:1 when installed on a five-inch gun mount.

f. RF POWER RATING — Not applicable, receive only.

g. POLARIZATION — Vertical.

h. TYPE OF FEED — Unbalanced, coaxial cable.

i. INPUT CONNECTOR — Type N, female.

j. INSTALLATION REQUIREMENTS — The antenna base mounting flange has eight 11/16-inch bolt holes equally spaced on a 24-inch diameter circle for mounting the antenna. A foundation with matching bolt holes, made to suit the configuration of the turret top so that the antenna mounting flange is level, must be provided by the installing activity. The turret top must be penetrated to accommodate the coaxial transmission line.

k. SPECIAL CONSIDERATIONS — The coaxial cable transmission line must be sufficiently slack and arranged in such a manner that it will not be damaged by turret rotation.

l. REFERENCE — CHU Associates Drawing 32370.

2-1.5 Special LF/MF/HF Shipboard Communications Antennas

The communications antennas described in the previous paragraphs of Section 2-1, the wire-rope fans, whips, trussed monopoles, cones, and cages, are all found in fairly wide use throughout the fleet. In addition to these there are a few special purpose types that are found less often in shipboard use at low, medium, and high frequencies. The category includes electrically small antennas, HF loops, log-periodic, and vertical marine types which are presented below.

The Minaject AN/SRA-51 Antenna Group

The AN/SRA-51 is a miniaturized high-Q receiving antenna and preselector used aboard ship to minimize coupled energy interference by rejecting undesired signals outside its passband and presenting an extremely small parasitic aperture to other antennas in the area. The basic antenna preselector is designed to operate from 2 to 16 MHz in three remotely tuned bands: 2-4 MHz, 4-8 MHz, and 8-16 MHz. Each band contains a four-resonator, narrowband, tunable band-pass filter to provide better than 75 dB rejection at 5 percent frequency spacing. The top-loaded AS-2071 antenna is non-parasitic in that portion of the HF band off frequency to which it is tuned and is electrically small, being only 59 inches in height. A coating of low loss insulating material is used to reduce precipitation static. To compensate for system losses, a low noise RF amplifier is used, followed by another filter for additional isolation and impedance match to the receiver transmission line. Figure 2-73 shows two AS-2071 minaject antennas installed. Following is a technical description:

ANTENNA AS-2071/SRA-51

a. TYPE/USE — The AS-2071/SRA-51 is an integral part of the CU-1691/SRA-51 antenna coupler. It is an electrically short, physically small, top-hat monopole antenna. The AS-2071/SRA-51, CU-1691/SRA-51, and C-7340/SRA-51 Antenna Coupler Control comprise the AS-2071 antenna group. The antenna group is used to improve the performance of the ship's receiving system by permitting normal operation of receivers in the proximity of strong off-frequency RF voltages from nearby HF transmitters, radar equipment, and extraneous signals, and by reducing the parasitic effects from closely coupled shipboard antennas. The antenna group may be used with any receiving equipment capable of operating within its frequency range. Since the coupler is the mounting base for the antenna, the following specifications and requirements cover both units.
b. PHYSICAL CONFIGURATION — [See Figure 2-74].
c. DIMENSIONS — (1) AS-2071/SRA-51 Height: 59 inches; Width: 14 inches (top-hat); Weight: 2 lbs.; (2) CU-1691/SRA-51 Height: 35 inches; Width: 19-$^{13}/_{16}$ inches; Depth: 16-$^{7}/_{16}$ inches; Weight: 115 lbs.
d. FREQUENCY RANGE — 2.0 to 16.0 MHz.
e. INPUT IMPEDANCE — 50 ohms nominal (output impedance of coupler).
f. RF POWER RATING — Not applicable, receive only.
g. POLARIZATION — Vertical.
h. TYPE OF FEED — Unbalanced, coaxial cable.

i. PRIMARY POWER REQUIRED — Coupler tuning power supplied by coupler control unit.
j. OPERATING TEMPERATURE — Antenna and coupler -28° to +65°C.
k. INPUT CONNECTOR — Antenna base insulator is integral to coupler unit. Coupler output Type N, UG-30D/U.
l. INSTALLATION REQUIREMENTS — The antenna coupler should be deck mounted to facilitate maintenance. It may be mounted on masts or other areas of the ship's superstructure, but maintenance access may require removal and lowering to deck levels. Placing the unit within a nominal 15 feet of metal structures should be avoided. The minimum allowable distance between the coupler unit antenna and the antenna of a one kilowatt transmitter is 20 feet. The coupler is mounted to a suitable foundation by eight 1/2-inch bolts.
m. SPECIAL CONSIDERATIONS — A source of dry air is required to purge and pressurize the coupler unit each time the interior is exposed to the environment. A humidity indicator is located within the unit and it must be replaced each time the cabinet is opened.
n. REFERENCES — (1) Technical Manual, Antenna Group AN/SRA-51, NAVSEA 0967-299-4010. (2) NAVSEA Drawing RE-2697045.

LF/MF Tuner Whip AT-924/SR (AN/SRA-17)

In the past it was not feasible on some ships to install long-wire LF/MF receiving antennas, particularly if there were a large number of circuits required. Electrically small, very thin tuner whips were found to be effective alternatives when installed high on the mast. Thus the 10-foot AT-924/SRA-17 became quite prevalent as the standard LF/MF tuner whip unit for many years. However, modern coupler equipment allows the combined use of HF wire-rope fan antennas for LF/MF reception. The AN/SRA-17 has, therefore, been gradually phased out of service and is now found only on older ships. Figure 2-75 shows an upper mast cluster of AT-924/SRA-17 antennas, described below:

ANTENNA AT-924A/SR and AT-924B/SR

a. TYPE/USE — The AT-924A/SR antenna is an integral part of the TN-334A/SRA-17C RF tuner. The antenna, tuner, and antenna control unit C-2536/SRA-17 comprise antenna coupler group AN/SRA-17B. Antenna coupler group AN/SRA-17C consists of the AT-924B/SR antenna, TN-418/SRA-17C RF tuner and C-6193/SRA-17 antenna control unit. The AT-924A/SR and AT-924B/SR have equal overall dimensions and are mounted on the same type of base insulator, but the configuration and mounting dimensions differ between the TN-334A/SRA-17 and TN-418/SRA-17. Both tuners may be remotely controlled by their respective control units. The antenna coupler groups are physically small but efficient receiving antenna systems for use on small craft where long wires are not available or on ships where large, closely spaced antennas would interact. These antenna groups are designed for use with any receiving equipment tunable over the VLF/LF/MF frequency spectrum. Since the tuner units form the mounting base for the antennas, the following specifications and requirements cover both units.

b. PHYSICAL CONFIGURATION — [See Figure 2-76].

c. DIMENSIONS — Antenna Height: 129 1/4 inches; Antenna Weight: 5 1/2 lbs. RF Tuner TN-334A/SRA-17: (1) Height: 9 1/2 inches; (2) Width: 12 inches; (3) Depth: 12 inches; (4) Weight: 37 1/2 lbs.; RF Tuner TN-418/SRA-17C: (1) Height: 16 5/8 inches; (2) Width: 12 inches; (3) Depth: 12 inches; Weight: 36 lbs.

d. FREQUENCY RANGE — 14 to 600 kHz.

e. INPUT IMPEDANCE — 72 ohms (tuner input).

f. RF POWER RATING — Not applicable, receive only.

g. POLARIZATION — Vertical.

h. TYPE OF FEED — Unbalanced, coaxial cable.

i. PRIMARY POWER REQUIRED — All power required for remote control circuitry is supplied by control unit.

j. OPERATING TEMPERATURE — Minus 28°C to +65°C.

k. INPUT CONNECTOR — (1) TN-334A/SRA-17 — Watertight fittings for control and coaxial cables. (2) TN-418/SRA-17C — Type N, UG-58A/U for coaxial cable and MS-3102R-24-20P for control cable.

l. INSTALLATION REQUIREMENTS — The RF tuner units may be mounted vertically or horizontally. When mounted horizontally, the support method given in Figure 2-34 must be applied. Bolt holes are provided on the tuner housing for securing the units to suitable pads on the ship's structure.

m. SPECIAL CONSIDERATIONS — To obtain maximum operating performance the antenna should be mounted near the top of the mast, king post, stack or other vertical member of the superstructure. The distance between the tuner and its associated receiver should not exceed 150 feet for satisfactory operation.

n. REFERENCES — (1) Technical Manual, Antenna Coupler Groups, AN/SRA-17B and AN/SRA-17C, NAVSEA 94127. (2) NAVSEA Drawing RE-F2695039 and RE-C2692097.

HF Miniloop Antenna

In the unending quest to find an antenna exhibiting improved performance in a high-interference environment, unusual types are now and then encountered. One such antenna that seemed to hold real promise, causing great interest by the Navy for a period of time in the 1970s, is the so-called Miniloop HF antenna. Being a vertical H-field (magnetic) antenna, rather than the commonly-used E-field (electric), and being tuned (high-Q) and directional, the Miniloop possesses the inherent characteristic of isolation from both man-made noise and undesired radiated energy from other shipboard E-field antennas — a very important advantage indeed. These features, it was claimed, provide the Miniloop with as much as 30 dB decoupling, making it relatively unresponsive to interference from local transmitting antennas. Moreover, because it is a balanced antenna, the Mililoop does not produce electrical currents in its pole-mast support structure, so that it greatly reduces potential RF burn hazards, and is little influenced by nearby metallic objects.

Despite these decided advantages, the Miniloop suffers from some rather severe drawbacks which hinder its application to shipboard HF communications. First and foremost, it is a tuned antenna and, therefore, since it is not broadband, cannot be multicoupled for general purpose shipboard reception. As a receive antenna it must be dedicated to a single receiver at a time, undesirable for ship use. And as a transmit antenna it has to be tuned and matched singly with a dedicated transmitter, just as in the case of the 35-foot tuned whip monopole system. Unfortunately, however, the Miniloop cannot reach the entire 2 to 30 MHz frequency range of the 35-foot tuned whip. In fact, the widest range Miniloop model, the MLA-2B, is capable of covering from only 3 to 25 MHz, a bit less than the full range required for naval HF communications.

Another disadvantage, evident in Figure 2-77, is that, in spite of its rather ironic name, the Miniloop is by no means small. It requires much more unenclosed volume topside than a 35-foot tuned whip. Furthermore, to achieve a similitude of omnidirectivity in its azimuthal pattern (a loop antenna is inherently bi-directional, giving a "figure-eight" radiation pattern at low elevation angles), the Miniloop must be rotatable — significantly increasing the required installation space.

Nevertheless, it was hoped that the Miniloop would be of useful service aboard ship, not a replacement, but a complement to the existing HF antennas. Thus it was given thorough testing, at sea and in naval laboratories,[19,20,21] and was found to have a serious drifting problem attributed to the tuning capacitor's sensitivity to widely fluctuating temperatures. These sharp variations in temperature were caused by the capacitor heating up when transmitter power was applied and the harsh, rapidly changing, atmosphere at sea. The drifting/tuning problem was well on its way to being resolved (through use of an automatic frequency compensation device and improved methods of sealing against water seepage), when the Miniloop being tested at the naval laboratory failed to meet the shock and vibration specifications.[22] Since that time, interest in the Miniloop has waned to the point where it is no longer found on U.S. Navy ships. A background summary of the Miniloop has been included here because of the intense enthusiasm it once evoked and because it still sees some maritime service on Coast Guard and commercial vessels. A technical description of the Model MLA-1 lower range Miniloop antenna is given below.

ANTENNA MLA-1 MINILOOP

a. TYPE/USE — The MLA-1 is a tuned vertical loop antenna system designed for high-frequency transmitting and receiving communications. Meant to be particularly useful in a high noise environment because of its sharp selectivity (high-Q) and directional characteristics, it consists of a large, tuned, single-turn, elliptically shaped main loop fabricated of four-inch diameter tubing and a small untuned, single-turn, circular feed loop inductively coupled. The main loop is tuned remotely by means of a variable vacuum capacitor located at the top of the loop. The antenna is mounted on a six-foot vertical mast and may be rotated in azimuth either manually or by a remotely controlled drive motor mechanism inserted between the loop and mast top.
b. PHYSICAL CONFIGURATION — [See Figure 2-78].
c. DIMENSIONS — Height: $11^1/_2$ feet; Size: Loop proper is elliptical, 5 feet high by 7 feet across (max) with a conductor diameter of approximately 4 inches; Weight: 215 lbs. (including mast and base).
d. FREQUENCY RANGE — (1) 2.25 to 16.5 MHz (model MLA-1/D); (2) 1.80 to 14.5 MHz (model MLA-1/E).
e. INPUT IMPEDANCE/VSWR — 50 ohms/2:1 maximum VSWR.
f. RF POWER RATING — 1kW rms, continuous.
g. POLARIZATION — Vertical at 0° (on horizon) - horizontal overhead. (Pattern is a figure-eight at 0° elevation angle, but becomes essentially omnidirectional above 15° elevation angle.)
h. TYPE OF FEED — Coaxial Cable RG-9B/U or equivalent.
i. PRIMARY POWER REQUIRED — (1) Tuning Unit: 90 to 130 vac; 50-60 Hz; 2 amps maximum. (2) Rotator: 90 to 130 vac; 50-60 Hz; 1 amp maximum.
j. INPUT CONNECTOR — Type N female.
k. INSTALLATION REQUIREMENTS — The antenna may be installed in any clear area on deck, but, if rotatable, sufficient additional clearance must be allowed. Mounting is accomplished by bolting the flanged base support to the deck.
l. REFERENCES — (1) NAVELEX Technical Manual 0967-461-2010. (2) Antenna Research Associates, Inc. Catalog.

Rotatable Log-Periodic HF Antennas

So far, in our treatment of shipboard antennas, all but the Miniloop have been omnidirectional and vertically polarized. Even the Miniloop was bi-directional. However, there is one type of shipboard HF communications antenna that is completely different from all others in that it is neither omnidirectional nor vertically polarized. It is the horizontally polarized, unidirectional, log-periodic antenna. Its chief advantage is that, being in the class of so-called frequency independent antennas, it is very broadband, covering a 4:1 frequency range. Being directional, it has high gain compared to the vertical monopoles and cones. Being very large, it can handle high transmitter power. Thus, it is meant for reliable, long distance, high power, multicoupled, HF naval communications. Yet it has some rather serious drawbacks: because it is so large horizontally, it requires very large horizontal installation space which is available on ship only at the top of a truly sturdy mast or tower, and, because it is directional, it must be rotatable — requiring a good deal of primary power and difficult-to-reach maintenance. Therefore, it finds only limited service on a few large, stable ships. Figure 2-79 shows a tower-mounted shipboard rotatable log-periodic (RLPA). The only Navy nomenclatured RLPA, the AS-2874/SRC, is described in detail below:

ANTENNA AS-2874/SRC

a. TYPE/USE — Antenna AS-2874 is a high power, HF, rotatable, log-periodic dipole array for shipboard use. The antenna consists of a main boom, ten radiating elements, and a matching section to provide 6 to 10 dB gain over an isotrope. The longer radiating elements have been foreshortened to reduce the array turning circle diameter to 58 feet. These elements are resonated by series reactances and top loading sections to maintain high radiation efficiency. The central beam support, a rigid truss section, is the transmission line for feeding the elements. It is completely enclosed for weatherproofing. Each element is provided with an internal vibration damper and is connected to adjacent elements of each tip to minimize vibration. Lightning protection is provided by a high capacity dc path across the feedpoint. The antenna will operate in winds up to 100 knots (or 60 knots with $^1/_2$-inch icing).
The antenna is supported on a stub mast and pedestal. The pedestal contains the antenna drive motor, azimuth synchro generator, and rotary joint. Control of the antenna azimuth is accomplished from a C-9264/SRC antenna control unit.

b. PHYSICAL CONFIGURATION — [See Figure 2-80].

c. DIMENSIONS — (1) Length: 45 feet; (2) Width: $39^1/_2$ feet; (3) Turning Diameter: 58 feet; (4) Weight: 1530 lbs.

d. FREQUENCY RANGE — 7.5 to 30 MHz.

e. INPUT IMPEDANCE/VSWR — 50 ohms. VSWR not greater than 2.5:1 over frequency range.

f. RF POWER RATING — 40 kW average.

g. POLARIZATION — Horizontal.

h. GAIN — 6 to 10 dB over isotropic.

i. HALF POWER BEAMWIDTH — (1) Azimuth 60° average. (2) Vertical 130° average.

j. FRONT-TO-BACK RATIO — 15 dB average (7 dB minimum).

k. MATERIAL — High-strength aluminum alloy and fiberglass.
l. TYPE OF FEED — Unbalanced, coaxial cable.
m. PRIMARY POWER REQUIRED — Antenna drive motor and synchro reference voltages are supplied from control unit. 110 vac and 30 amps.
n. OPERATING TEMPERATURE — Minus 20°F to +150°F.
o. INPUT CONNECTOR — 1 5/8 inch EIA flange.
p. INSTALLATION REQUIREMENTS — The base of the pedestal casting has three mounting pads 120° apart for installing the antenna. Each pad has one 1 1/16-inch bolt hole for a one-inch bolt and two 25/32-inch bolt holes for 3/4-inch bolts. The bolt holes are on a 33-inch diameter bolt circle. A minimum swing circle of 60 feet is required to permit 360° azimuth coverage. The antenna pedestal bench mark must be aligned with the ship's centerline, heading forward. The transmission line, pedestal, stub mast, and antenna feed section may be pressurized with dry air or nitrogen to 5 psi.
q. SPECIAL CONSIDERATIONS — This antenna is designed for use on LHA, LCC, and similar types of ships having sufficient space to support the installation requirements. It should be located as high and in as clear an area as possible. It should not be located in the proximity of communications receiving antennas.
r. REFERENCE — CHU Associates Drawing Number 6406.

MF/HF Vertical Marine Antenna MR 106-38A

A rather unusual antenna developed for shipboard use at the request of the Maritime Administration to cover a wide frequency range is the 40-foot, top-loaded, vertical monopole MR 106-38A. This commercial antenna is included here because it is sometimes used for the main communications on Military Sealift Command ships as specified by naval systems design engineers.

ANTENNA MR 106-38A

a. TYPE/USE — The MR 106-38A is a vertical monopole distinguished by its large, geometric top-hat. The two-section pole is made of heavy wall, high tensile strength aluminum alloy tubing and the top-hat is fabricated from stainless steel. Designed especially for the shipboard environment, all aluminum parts are protected from corrosion by an iridite chromate finish and weather-resistant paint. The antenna will withstand 120 mph winds and 2 inches of ice loading.
b. PHYSICAL CONFIGURATION — [See Figure 2-81].
c. DIMENSIONS — (1) Height: 40 feet; (2) Diameter: 13 feet (top-hat); 4 inches (pole); (3) Weight: 1060 lbs. (less gussets).
d. FREQUENCY RANGE — 410 kHz through 24 MHz.
e. RF POWER RATING — 1 kW PEP.
f. POLARIZATION — Vertical.
g. TYPE OF FEED — Copper single-wire lead-in.

h. INSTALLATION REQUIREMENTS — The antenna is mounted on a vertical hinge base. Except for the gussets to be supplied by the customer or installing activity, all mounting hardware is provided with the antenna, including vertical hinge mount MR 487-13.
i. SPECIAL CONSIDERATIONS — When selecting a site aboard ship, care should be exercised in choosing a location which provides the shortest possible lead length between the transmitter and the base of the antenna. Lead lengths greater than 10 feet can reduce antenna efficiency, particularly if run close to metallic surfaces. It is also important that the antenna be located in as clear an area as possible to provide omnidirectional radiation. It is recommended that it be located at least 15 feet away from interfering objects to allow for maximum deflections under extreme conditions.
j. REFERENCE — Mackay Marine ITT Corporation Catalog.

2-2 VHF COMMUNICATIONS ANTENNAS

In the middle 1920s, at the time that the Navy was required to surrender the lower frequencies to commercial radio broadcasting, the very high frequencies from 30 MHz and up were assigned by international agreement to amateurs and experimenters.[23,24] Many of the experiments were being conducted in earnest by naval radio engineers and scientists. Whereas HF would serve the needs of long-distance radio communications admirably, there was still an urgent need for short-range tactical communications in the fleet, free from enemy interception. Thus, by 1929, operational research of its first VHF equipment convinced the Navy of the usefulness of this new medium for limited-range shipboard communication.[25] The potential advantages were quite promising: the higher frequencies inherently would provide additional radio channels (increased capacity) and the limited ranges would require only very low transmission power and, therefore, smaller, lighter equipment. The latter was of particular value to developers of airborne radio. Furthermore, the long trailing wire and larger, fixed-wire HF antennas on high-speed aircraft caused a prohibitive drag. Short, efficient VHF antennas could be aerodynamically shaped to allow greatly increased speed and maneuverability, an especially important feature for the new generation of fighter planes then on the drawing board.

Realizing these potential benefits, the Navy made plans to adopt VHF communications for ship-to-shore amphibious landing operations, ship-to-air flight operations, ship-to-ship tactical maneuvers, convoy duty, and air-to-air command and control. Subsequently, the Sixth International Radio Conference, meeting in Cairo in 1938, allocated the new VHF band between 30 and 300 MHz,[26] and the Navy made its first large buy of VHF equipment: the TBS, a rugged, dependable set that would serve the fleet for nearly 20 years. During World War II, the 10-watt, 10-channel AN/ARC-1 radio (115-156 MHz) became the Navy's principal ship-to-air communications equipment.

With the war's end and the return to commercial enterprise, however, the VHF band was soon absorbed by television, FM radio broadcasting, civilian air traffic control, radioastronomy, and amateur radio. The Navy moved up, almost entirely, to the lower part of the UHF band (225-400 MHz) where the advantages of increased channel capacity, lower transmission power, and smaller equipment design were even more favorable. Nevertheless, the Navy today still retains a few portions of the VHF band from 30 through 156 MHz for amphibious and land-mobile operations. Consequently, a few shipboard VHF antennas are in service, chiefly on amphibious class ships. These antennas take the form mainly of whips and dipoles, although there are some special types such as cages and log-periodics also in use.

2-2.1 VHF Whip Antennas

The only VHF whip commonly used in naval service is the 10-foot fiberglass AS-1729/VRC. It is employed mostly on such amphibious class ships as LPDs, LPHs, LHAs, and LSTs in the manner shown by Figures 2-82 and 2-83.

ANTENNA AS-1729/VRC

a. TYPE/USE — The AS-1729/VRC is a center fed, remotely step-tuned whip antenna heavily used on large ships equipped with AN/VRC-46s for amphibious operations. It consists of two fiberglass covered sections (AT-1095/VRC and AS-1730/VRC), a base spring assembly, and a base switching assembly (MX-6707/VRC). The base switching assembly contains ten LC tuned circuits to match the transmission line to the antenna over the operating range of the AN/VRC-46. Selection of the individual tuned circuits is automatic from the channel selector switch on the transceiver, but they may also be selected manually by a switch on the base of the assembly.

b. PHYSICAL CONFIGURATION — [See Figure 2-84].

c. DIMENSIONS — (1) Height: 120 inches; (2) Width: 5½ inches (base diameter); (3) Weight: 9 lbs.

d. FREQUENCY RANGE — 30 to 76 MHz.

e. INPUT IMPEDANCE/VSWR — 50 ohms nominal, VSWR not greater than 3:1 over frequency range.

f. RF POWER RATING - 70 watts.

g. POLARIZATION — Vertical.

h. TYPE OF FEED — Unbalanced, coaxial cable.

i. PRIMARY POWER REQUIRED — Switching voltage for base tuned circuit selection (22-30 Vdc) supplied from transceiver.

j. OPERATING TEMPERATURE — Minus 32°C to +150°C.
k. INPUT CONNECTOR — Type C, modified UG-631/U.
l. INSTALLATION REQUIREMENT — The base assembly has four bolt holes for mounting the antenna to its support structure. The base may be mounted through a metal deckplate, bulkhead, or on an L-shaped bracket attached to the ship's superstructure or mast. If the AS-1729/VRC is installed on other than a deck mounting, it will be necessary to provide a ground plane. The ground plane should consist of four horizontal, equally spaced metal rods, 54 inches in length, mounted out from the base of the antenna and electrically connected to the ship [see Figure 2-83].
m. SPECIAL CONSIDERATIONS — The AS-1729/VRC antenna was designed for use on land or amphibious vehicles. The spring base was provided to furnish the flexibility for vehicular operation where contact between the antenna and overhead obstructions could be expected. Since this requirement does not exist onboard ship and corrosion has been experienced due to dissimilar metals used in the manufacture of the spring mount, a rigid adapter piece has been designed for replacement of the spring section. Details for the manufacture and installation of this adapter piece are contained in NAVSEA 0967-177-3030 Shipboard Antenna Systems, Antenna Couplers, Section G.
n. REFERENCES — (1) NAVSEA Drawing RE-D2696712. (2) Technical Manual NAVSEA 0967-LP-375-1090.
NOTE — Another small, flexible, whip type VHF antenna is the very thin five-foot AS-3015 used for 156-162 MHz bridge-to-bridge communications. However, this antenna is currently being replaced by the more rugged AS-2809 vertical dipole described in the next section.

2-2.2 VHF Dipole Antennas

The class of antennas most used for shipboard VHF communications is the vertical dipole. A variety of types are used, varying from the simple fixed-frequency version to broadband biconicals, typified by the following descriptions:

ANTENNA NT-66095

a. TYPE/USE — The NT-66095 is a fixed half-wave dipole. The lower element is grounded to the support housing and a concentric two-step matching section mates to a 50-ohm transmission line so that the antenna may be used with any equipment designed to operate within its frequency range and RF power handling lmitations. Long a standard Navy VHF dipole, the NT-66095 is usually installed at the end of a yardarm, as shown in Figure 2-85, for omnidirectivity.
b. PHYSICAL CONFIGURATION — [See Figure 2-86].
c. DIMENSIONS — (1) Height: 46½ inches; (2) Width: 24¼ inches; (3) Depth: 4½ inches.
d. FREQUENCY RANGE — Fixed dipole, operating frequency 135 MHz.
e. INPUT IMPEDANCE/VSWR — 50 ohms. VSWR will vary depending on the

percentage that the operating frequency is removed from the resonant frequency of the antenna.

f. RF POWER RATING — 250 watts average.
g. POLARIZATION — Vertical.
h. TYPE OF FEED — Unbalanced, coaxial cable.
i. INPUT CONNECTOR — Type N.
j. INSTALLATION REQUIREMENTS — The location of the NT-66095 antenna aboard ship should be selected to provide maximum unobstructed coverage within the physical constraints of the ship's structure. Propagation in this portion of the spectrum is generally line-of-sight. Therefore, antennas that are masked by ship structures produce limited coverage in azimuth and elevation.
k. SPECIAL CONSIDERATIONS — The antenna is not a broadband type and as such has maximum operating efficiency at its resonant frequency.
l. REFERENCES — (1) Technical Manual, Antenna Assembly NT-66095, 0967-LP-089-5010. (2) NAVSEA Drawing RE-D2691809.

ANTENNA AS-2231/SRA-60(V)

a. TYPE/USE — The AS-2231/SRA-60(V) antenna, a broadband omnidirectional biconical dipole, is a component of the AN/SRA-60(V) antenna coupler group. The antenna consists of a center support post and two cylindrical cage type radiating elements. A teflon insulator for isolating the two sections is located on top of the support column. The antenna transmission line is connected to the upper radiating element at the insulator. For mechanical rigidity and vibration dampening the vertical members of the upper and lower cones are connected by a fiberglass insulator. Figures 2-87 and 2-88 show two methods of installing the AS-2231 antenna.
b. PHYSICAL CONFIGURATION — [See Figure 2-89].
c. DIMENSIONS — (1) Height: 156 inches O.A.; (2) Diameter: 48 inches O.A.; (3) Weight: 409 lbs.
d. FREQUENCY RANGE — 30 to 76 MHz.
e. INPUT IMPEDANCE/VSWR — 50 ohms nominal, VSWR not greater than 3:1 over frequency range.
f. RF POWER RATING — 400 watts average, 1600 watts PEP.
g. POLARIZATION — Vertical.
h. TYPE OF FEED — Unbalanced, coaxial cable.
i. INPUT CONNECTOR — Type N.
j. INSTALLATION REQUIREMENTS — The base flange of the antenna center support post has eight $1^1/_{32}$-inch bolt holes located on a 14-inch bolt circle for installing the antenna onboard ship. The antenna is designed for vertical mounting on a deck, deck house top, or a pedestal of any desired height. The transmission line insulation should be stripped from the outer conductor to allow bonding to the antenna cable clamps

located on the antenna support column. The outer conductor must be grounded directly to where the transmission line meets the deck.

k. SPECIAL CONSIDERATIONS — The location aboard ship for the AS-2231/SRA-60(V) should be selected to provide maximum omnidirectional coverage within the physical constraints of the ship's topside. If it is necessary to locate the antenna near large structures, a minimum separation of 35 feet should be maintained in order to reduce the effects on antenna impedance. Propagation within the frequency range of the antenna is generally line-of-sight. Therefore, antennas that are masked by surrounding structures may produce limited coverage in azimuth and elevation.

l. REFERENCES — (1) Technical Manual, Antenna Coupler Group AN/SRA-60(V), NAVELEX 0967-421-4010. (2) NAVSEA Drawing RE-E2698643.

ANTENNA AS-2808/SRC

a. TYPE/USE — The AS-2808/SRC antenna is an omnidirectional broadband, vertical, center-fed dipole. All major components of the radiating and support structures are aluminum, with the antenna feed sections protected and weatherproofed by laminated fiberglass housings. The antenna may be used with any VHF transmitting or receiving equipment designed to operate within its frequency range and RF power handling limitations.

b. PHYSICAL CONFIGURATIONS — [See Figure 2-90].

c. DIMENSIONS — (1) Height: 110½ inches; (2) Radiator: 6-inch diameter; (3) Mounting Flange: 8½-inch diameter; (4) Weight: 48 lbs.

d. FREQUENCY RANGE — 30 to 100 MHz.

e. INPUT IMPEDANCE/VSWR — 50 ohms nominal. VSWR not greater than 3:1 above 45 MHz.

f. RF POWER RATING — 250 watts.

g. POLARIZATION — Vertical.

h. TYPE OF FEED — Unbalanced, coaxial cable.

i. INPUT CONNECTOR — Type N, UG-58A/U.

j. INSTALLATION REQUIREMENTS — The antenna assembly mounting flange has eight bolt holes for ½-inch bolts on a 7¼-inch bolt circle for use in mounting the antenna. The antenna may be installed on a suitable mounting plate having matching bolt holes and a center hole to accommodate the antenna transmission line.

k. SPECIAL CONSIDERATIONS — The location for the AS-2808/SRC should be selected to provide maximum omnidirectional coverage within the physical constraints of the ship's topside. VHF propagation in the frequency range of the antenna is generally line-of-sight. Therefore, coverage in azimuth or elevation may be limited if the antenna is masked.

l. REFERENCES — (1) CHU Associates Model CA-3095H Antenna. (2) Technical Manual, Antenna AS-2808/SRC, NAVSEA 0967-475-3010.

ANTENNA AS-2809/SRC

a. TYPE/USE — The AS-2809/SRC antenna is a base mounted omnidirectional broadband vertical center-fed dipole. All major components of the radiating and support structures are aluminum, with the central feed section and radiating elements protected by a laminated fiberglass cover. The AS-2809/SRC may be used with any VHF transmitting or receiving equipment designed to operate within its frequency range and RF power handling limitations.
b. PHYSICAL CONFIGURATION — [See Figure 2-91].
c. DIMENSIONS — (1) Height: 47 inches; (2) Diameter: 2½ inches; (3) Mounting Flange: 4½-inch diameter; (4) Weight: 7½ lbs.
d. FREQUENCY RANGE — 115 to 162 MHz.
e. INPUT IMPEDANCE/VSWR — 50 ohms nominal, VSWR not greater than 2:1.
f. RF POWER RATING — 100 watts.
g. POLARIZATION — Vertical.
h. TYPE OF FEED — Unbalanced, coaxial cable.
i. INPUT CONNECTOR — Type N, UG-58A/U.
j. INSTALLATION REQUIREMENTS —The antenna assembly mounting flange has four 9/32-inch holes for ¼-inch bolts on a 3⅞-inch diameter bolt circle for use in mounting the antenna. The antenna may be installed on a suitable mounting plate having matching bolt holes and a center hole to accommodate the antenna transmission line.
k. SPECIAL CONSIDERATIONS — The location for the AS-2809/SRC should be selected to provide maximum omnidirectional coverage within the physical constraint of the ship's topside. VHF propagation in the frequency range of this antenna is generally line-of-sight. Therefore, antennas that are masked by surrounding structures produce limited coverage in azimuth or elevation.
l. REFERENCES — (1) CHU Associates Model CA-3124 Antenna. (2) Technical Manual, Antenna AS-2809/SRC, NAVSEA 0967-475-4010.

ANTENNA AS-2811/SRC

a. TYPE/USE — The AS-2811 antenna is an omnidirectional, very broadband, vertical biconical dipole covering nearly all of the VHF band, and up to the lower part of the UHF band. All major components of the radiating and support structures are aluminum. The central feed section is protected and weatherproofed by a laminated fiberglass cover. The AS-2811/SRC may be used with any VHF/UHF transmitting or receiving equipment designed to operate within its frequency range and power handling limitations.
b. PHYSICAL CONFIGURATION — [See Figure 2-92].
c. DIMENSIONS — (1) Height: 38½ inches; (2) Diameter: 34½ inches; (3) Weight: 21 lbs.
d. FREQUENCY RANGE — 90 to 500 MHz.

e. INPUT IMPEDANCE/VSWR — 50 ohms nominal. VSWR not greater than 2.5:1.
f. RF POWER RATING — 250 watts.
g. POLARIZATION — Vertical.
h. TYPE OF FEED — Unbalanced, coaxial cable.
i. INPUT CONNECTOR — Type N, UG-58A/U.
j. INSTALLATION REQUIREMENTS — The antenna assembly mounting flange has eight holes for ⅜-inch bolts on a 4¾-inch diameter bolt circle. The overall diameter of the mounting flange is 5¾ inches. The antenna may be installed on a suitable mounting plate having matching bolt holes and a center hole to accommodate the antenna transmission line.
k. SPECIAL CONSIDERATIONS — The location of the AS-2811/SRC should be selected to provide maximum omnidirectional coverage within the physical constraints of the ship's topside. Propagation within the frequency range of this antenna is generally line-of-sight. Therefore, antennas that are masked by surrounding structures may produce limited coverage in azimuth or elevation.
l. REFERENCES — (1) CHU Associates Model CA-3096 Antenna. (2) Technical Manual, Antenna AS-2811/SRC, NAVSEA 0967-475-6010.

ANTENNA AS-3226/URC

a. TYPE/USE — The AS-3226 is a base-mounted, broadband vertical dipole currently being introduced into the fleet as a replacement for the older AS-1729 VHF whips. It is a rugged, lightweight 12-foot cylindrical pole sealed against the marine environment by a fiberglass housing that provides an omnidirectional azimuthal radiation pattern.
b. PHYSICAL CONFIGURATION — [See Figure 2-93(a)].
c. DIMENSIONS — (1) Height: 144 inches; (2) Diameter: 4 inches; (3) Weight: 39 pounds.
d. FREQUENCY RANGE — 30 to 76 MHz.
e. INPUT IMPEDANCE/VSWR — 50 ohms, VSWR not greater than 3.5:1 maximum.
f. RF POWER RATING — 800 watts.
g. POLARIZATION — Vertical.
h. TYPE OF FEED — Unbalanced, RF coaxial cable.
i. INPUT CONNECTOR — Type N receptacle.
j. INSTALLATION REQUIREMENTS — Six ½-inch bolts on a 6.75-inch circle through an 8-inch diameter aluminum base flange.
k. REFERENCE — CHU Associates Model CA-1123 Antenna.

ANTENNA AS-3227/URC

a. TYPE/USE — The AS-3227 is a low power version of the base-mounted broadband vertical dipole for shipboard use. A lightweight, thin 12-foot pole, it is protected from the marine environment by a fiberglass covering, and provides omnidirectional radiation in the horizontal plane.
b. PHYSICAL CONFIGURATION — [See Figure 2-93(b)].

c. DIMENSIONS — (1) Height: 144 inches; (2) Diameter: 3 inches (tapering to 2 inches); (3) Weight: 30 pounds.
d. FREQUENCY RANGE — 30 to 76 MHz.
e. INPUT IMPEDANCE/VSWR — 50 ohms, VSWR not greater than 3.5:1 maximum.
f. RF POWER RATING — 100 watts.
g. POLARIZATION — Vertical.
h. TYPE OF FEED — Unbalanced, RF coaxial cable.
i. INPUT CONNECTOR — Type N receptacle.
j. INSTALLATION REQUIREMENTS — Six ½-inch bolts on a 5.75-inch circle through a 7.5-inch diameter aluminum base flange.
k. REFERENCE — CHU Associates Model CA-1124 Antenna.

2-2.3 Special VHF Shipboard Communications Antennas

In addition to the whips and dipoles used for VHF communications, there are a few special types too, such as discages and log-periodics. Representative examples are described below:

ANTENNA AS-2867/SRR

a. TYPE/USE — The AS-2867/SRR is a VHF/UHF broadband omnidirectional discage receiving antenna. It is a dual-frequency antenna. The lower range uses the entire structure as a broadband monopole and the upper range uses a discone radiator. The two coaxial tranmission line connectors are located in the base of the antenna. A metal groundplane at least eight feet in diameter or a set of four ground radials each 96¼ inches long is required. The antenna may be used with any receiving equipment designed to operate within its frequency range. Figure 2-94(a) shows the AS-2867 properly installed with groundplane.
b. PHYSICAL CONFIGURATION — [See Figure 2-94(b)].
c. DIMENSIONS — (1) Height: 65¾ inches; (2) Diameter: Antenna 20 inches, Antenna with groundplane 96 inches, Antenna with radials 157½ inches; (3) Weight: Antenna 50 lbs, Antenna with ground radials 65 lbs.
d. FREQUENCY RANGE — (1) 30 to 150 MHz (entire structure). (2) 150 to 1000 MHz (upper discone).
e. INPUT IMPEDANCE/VSWR — 50 ohms. VSWR 30 to 33 MHz not greater than 8:1 and 33 to 1000 MHz not greater than 4:1.
f. RF POWER RATING — Not applicable, receive only.
g. POLARIZATION — Vertical.
h. TYPE OF FEED — Unbalanced, coaxial cable.
i. INPUT CONNECTOR — Type N, female (2 each).
j. INSTALLATION REQUIREMENTS — The antenna mounting flange has eight 7/16-inch bolt holes equally spaced on a 9⅞-inch diameter bolt circle. The antenna may be mounted to bulkheads, stacks, masts or on an antenna pedestal. A groundplane made of expanded metal is required for installation where the antenna is attached to the ship's structure. A

minimum distance of 48 inches is required between the vertical centerline of the antenna and the ship's structure in the same plane. The ground radials should be used whenever the antenna is mounted on a pedestal and there is sufficient space to accommodate the radials. The antenna foundation must have a hole for the coaxial transmission lines.

k. SPECIAL CONSIDERATIONS — The location of the AS-2867/SRR should be selected to furnish maximum coverage within the physical constraints of the ship's topside. Propagation in the frequency range of this antenna is generally line-of-sight and antennas that are masked by surrounding structures may produce limited coverage in azimuth or elevation.

l. REFERENCE — CHU Associates, Drawing Number 32271, Antenna CA-3138.

ANTENNA AS-2814/SRC

a. TYPE/USE — The AS-2814/SRC is a broadband, unidirectional, log-periodic dipole array. The antenna consists of a main boom and eight radiating elements. It may be installed for operation with either vertically or horizontally polarized signals and may be used with any VHF transmitting or receiving equipment capable of operating within its frequency range and RF power handling capabilities. Two AS-2814s installed vertically, port and starboard, are shown in Figure 2-95.

b. PHYSICAL CONFIGURATION — [See Figure 2-96].

c. DIMENSIONS — (1) Height: 203 3/16 inches; (2) Width: 116¾ inches; (3) Depth: 7 inches; (4) Weight: 40 lbs.

d. FREQUENCY RANGE — 30 to 76 MHz.

e. INPUT IMPEDANCE/VSWR — 50 ohms, VSWR not greater than 2.5:1 over frequency range.

f. RF POWER RATING — 100 watts.

g. POLARIZATION — Horizontal or vertical.

h. TYPE OF FEED — Unbalanced, coaxial cable.

i. INPUT CONNECTOR — Type N.

j. INSTALLATION REQUIREMENTS — The antenna is supplied with a mounting flange for installation onboard ship. This aluminum flange is 6½ inches by 6½ inches with twelve holes for ⅜-inch bolts equally spaced on a 5½-inch by 5½-inch square pattern. Vinyl jacketed mylar rope is supplied for use between the flexible radiating elements to reduce vibration.

k. SPECIAL CONSIDERATIONS — The location of the antenna should be selected to provide optimum directional coverage within the physical constraints of the ship's structure. Two AS-2814/SRC antennas are often used, on opposite sides of the ship, to obtain maximum coverage.

l. REFERENCE — CHU Associates Model CA-3126 Antenna.

2-3 UHF COMMUNICATIONS ANTENNAS

Recognizing the advantages attendant with even higher frequencies (namely the increased communications channel capacity; higher quality and reliability; short-range line-of-sight circuits resistant to enemy interception; and smaller, lighter, low-power equipment), by World War II the Navy had fully adopted ultra-high frequency (UHF) radio. Although technically embracing a portion of the upper VHF band, the range of frequencies from 225 to 400 MHz used by the military is commonly referred to as the UHF band. Today it is the principal band of frequencies used for U.S. and allied forces tactical communications.[27]

The benefit for the antenna systems engineer of UHF communications is immediately apparent: the shorter wavelengths at these frequencies allow small antennas, simpler in design, making it much easier to find suitable installation sites on the ship's topside — usually high on mast platforms or yardarms.

Most shipboard UHF line-of-sight radio antennas are variations of the dipole. However, a few specialized types such as the stub-on-counterpoise, the discage, and conical spiral or log-periodic sector arrays are used as well. As is the case with all naval shipboard fixed communications antennas, the objective is to achieve azimuthal omnidirectivity independent of the ship's heading.

2-3.1 UHF Dipole Antennas

The simplest and perhaps most durable shipboard UHF antenna in terms of years of service is the AT-150 dipole, still recommended after 40 years of naval UHF radio. The antennas normally are installed on the ends of yardarms. However, as seen by Figure 2-97, in situations such as aircraft carriers where a great number of UHF circuits are in use, the AT-150s may end up in rather unusual locations.

UHF dipoles of more complex design have evolved over the years. Finding wide use aboard naval ships, these include the "wrap-around array," the biconical dipole, and the vertical cylindrical dipole.

ANTENNA AT-150/SRC

a. TYPE/USE — The AT-150/SRC is a broadband, coaxial dipole antenna used for transmitting and receiving. It consists of vertical radiating elements and a balun arrangement that electrically balances the antenna to ground. The antenna may be used with any UHF transmitting or receiving equipment designed to operate within its frequency range and power handling limitations.
b. PHYSICAL CONFIGURATION — [See Figure 2-98].
c. DIMENSIONS — (1) Height: 17⅜ inches; (2) Depth: 3¼ inches; (3) Width: 22 5/16 inches; (4) Weight: 4.5 lbs.
d. FREQUENCY RANGE — 220 to 400 MHz.

e. INPUT IMPEDANCE/VSWR — 50 ohms nominal. VSWR not greater than 2:1 over its frequency range.
f. RF POWER RATING — 200 watts average.
g. POLARIZATION — Vertical.
h. TYPE OF FEED — Unbalanced.
i. INPUT CONNECTOR — Type N special mate with UG-21B/U.
j. INSTALLATION REQUIREMENTS — The antenna must be mounted with the radiating element in the vertical plane. It is supplied with a mounting base and clamp that fits around the horizontal portion of the antenna. This mounting has four 7/16-inch bolt holes on 4½ by 3¾ bolt centers for attachment to a suitable mounting pad on the ship's mast, yardarm, or similar structure. These antennas are often mounted on stub masts above the radar platforms and on the outboard ends of yardarms.
k. SPECIAL CONSIDERATIONS — The location for mounting the AT-150/SRC antenna onboard ship should be selected to provide maximum unobstructed coverage within the physical constraints of the ship's topside. UHF propagation is generally line-of-sight. Therefore, antennas that are masked by surrounding structures produce limited coverage in azimuth or elevation. The antenna is designed to be sufficiently airtight to maintain a dielectrically stable atmosphere within the antenna. The presence of moisture may be detected by the removal of the pipe plug located on the connector flange. The antenna should not be located in stack gas drafts or exhausts where the temperatures exceed the operating temperature limitations.
l. REFERENCES — (1) Technical Manual, Antenna AT-150/SRC NAVSEA 0967-LP-060-5010. (2) NAVSEA Drawing RE-F2680668.

ANTENNA AS-1735/SRC

a. TYPE/USE — The AS-1735/SRC UHF antenna consists of four broadband radiating dipole elements mounted equidistant from the central mast and equally spaced from each other. The individual antennas are fed in-phase to furnish omnidirectional coverage over the frequency bandwidth of the antenna. The aluminum framework supporting the radiating elements contains integral coaxial line matching transformers and feedlines. This framework also serves to ground the radiators to the mast, provides a balun feed system and forms isolation traps which limit current flow in the mast. Effective interconnection of each radiating element to provide the proper impedance match to the transmission line is accomplished by first combining two elements of a pair and then combining the two pairs into a single input. The antenna and all mounting hardware are made of aluminum. Normal siting is near the top of the mast as in Figure 2-99.
b. PHYSICAL CONFIGURATION — [See Figure 2-100].
c. DIMENSIONS — (1) Height: 27 inches; (2) Diameter 29½ inches; (3) Weight: 31 lbs. antenna, less adapters. The weight of the mounting adapter is dependent on the size of the mast and does not exceed 32 lbs. for mast diameters to 15 inches.

d. FREQUENCY RANGE — 225 to 400 MHz.
e. INPUT IMPEDANCE/VSWR — 50 ohms. VSWR not greater than 1.8:1 over frequency range.
f. RF POWER RATING — 1 kW.
g. POLARIZATION — Vertical.
h. TYPE OF FEED — Unbalanced, coaxial cable.
i. INPUT CONNECTOR — Type LC, UG-352/U.
j. INSTALLATION REQUIREMENTS — The antenna is secured to the support mast by mounting flanges and adaptors. Adaptors are supplied to accommodate mast diameters to 12 inches. A special yoke is required for mast diameters to 15 inches.
k. SPECIAL CONSIDERATIONS — The antenna should be mounted as high as possible on the mast to provide maximum unobstructed coverage within the physical constraints of the ship. Care should be exercised to eliminate fouling of the antenna by flag halyards and ship's rigging. Caution signs should be provided to indicate that the antenna is not to be used as a support by personnel working on the mast.
l. REFERENCES — (1) Technical Manual, Antenna AS-1735A/SRC, NAVSEA 0967-303-7310. (2) NAVSEA Drawing RE-B2698607.

ANTENNA AS-2812/SRC

a. TYPE/USE — The AS-2812/SRC antenna is a small, broadband, omnidirectional vertical biconical dipole intended for shipboard and similar applications where rugged, reliable design is necessary. All major components of the radiating and support structures are aluminum. The central feed section is protected and weatherproofed by a laminated fiberglass cover. The AS-2812/SRC may be used with any UHF transmitting or receiving equipment designed to operate within its frequency range and RF power handling limitations. An example of a shipboard installation is shown in Figure 2-101.
b. PHYSICAL CONFIGURATION — [See Figure 2-102].
c. DIMENSIONS — (1) Height: 24 inches; (2) Diameter: 15¾ inches; (3) Weight: 6 lbs.
d. FREQUENCY RANGE — 200 to 1300 MHz.
e. INPUT IMPEDANCE/VSWR — 50 ohms nominal. VSWR not greater than 2.5:1.
f. RF POWER RATING — 250 watts.
g. POLARIZATION — Vertical.
h. TYPE OF FEED — Unbalanced, coaxial cable.
i. INPUT CONNECTOR — Type N, UG-58A/U.
j. INSTALLATION REQUIREMENTS — The antenna assembly is supplied with a mounting pad for vertical mounting. This pad has four holes for 5/16-inch bolts on a three-inch square pattern. The antenna may be installed on a suitable mounting plate having matching bolt holes.

k. SPECIAL CONSIDERATIONS — The locations of the AS-2812/SRC should be selected to provide maximum omnidirectional coverage within the physical constraints of the ship's topside. Propagation within the frequency range of the antenna is generally line-of-sight. Therefore, antennas that are masked by surrounding structures may produce limited coverage in azimuth or elevation.

l. REFERENCES — (1) CHU Associates Model CA-3080 Antenna. (2) Technical Manual, Antenna AS-2812/SRC, NAVSEA 0967-475-7010.

ANTENNA AS-1018/URC

a. TYPE/USE — The AS-1018/URC antenna is a broadband, omnidirectional UHF communications antenna. It is a collinear array of two cylindrical, parallel-fed dipole elements. The nonresonant dipoles provide broadband coverage and a horizontal pattern that is omnidirectional within ±1.0 dB. The lower half-power point of the vertically polarized radiation pattern falls on or slightly below the horizon. Pattern tilt is accomplished by feeding the dipoles in such a manner that the upper dipole lags the lower dipole in phase by a few degrees. The upper dipole is altered to provide both horizontal and vertical components, thus producing the upward propagation required to fill the cone of silence overhead when communicating with aircraft. The dipole elements and internal transmission line are enclosed in a polyester fiberglass mast. This antenna may be used with any UHF transmitting or receiving set designed to operate within its frequency range and RF power handling limitations. It is typically installed on yardarms as shown in Figure 2-103.

b. PHYSICAL CONFIGURATION — [See Figure 2-104].

c. DIMENSIONS — (1) Height: 72⅜ inches; (2) Diameter: 9¼ inches; (3) Weight: 50 lbs.

d. FREQUENCY RANGE — 225 to 400 MHz.

e. INPUT IMPEDANCE/VSWR — 50 ohms nominal. VSWR not greater than 2:1 over frequency range.

f. RF POWER RATING — 2 kW average.

g. POLARIZATION — Vertical.

h. TYPE OF FEED — Unbalanced, coaxial cable.

i. INPUT CONNECTOR — Type N, UG-352B/U.

j. INSTALLATION REQUIREMENTS — The antenna assembly mounting flange has eight ⅞-inch diameter bolts holes on an 11¾-inch bolt circle. The antenna may be installed on the ship's structure by a suitable mounting plate having matching bolt holes and a center hole to accommodate the antenna transmission line.

k. SPECIAL CONSIDERATIONS — The location of the AS-1018/URC antenna onboard ship should be selected to provide maximum unobstructed coverage within the physical constraints of the ship's topside. UHF propagation is generally line-of-sight and antennas that are masked by surrounding structures produce limited coverage in azimuth or elevation. At the time of installation, the antenna should be purged to remove any moisture or damp air entrapped within the enclosure. Purging should be accomplished in accordance with the procedures outlined in the equipment technical manual.

l. REFERENCES — (1) Technical Manual, Antenna AS-1018/URC, NAVSEA 0967-971-2010. (2) CHU Associates Model CA-4000 Series Antenna.

ANTENNA AS-2810/SRC

a. TYPE/USE — The AS-2810/SRC antenna is a low-power, lightweight, broadband, omnidirectional vertical center-fed dipole. All major components of the radiating and support structures are aluminum with the central feed section and radiating elements enclosed in a laminated epoxy fiberglass radome. An electrical isolating section is incorporated in the feed system to provide balance feed to the radiating elements and decoupling from the feed line. The AS-2810/SRC may be used with any UHF transmitting or receiving equipment designed to operate within its frequency range and RF power handling limitations.

b. PHYSICAL CONFIGURATION — [See Figure 2-105].

c. DIMENSIONS — (1) Height: 29 inches; (2) Diameter: 2⅞ inches (antenna); (3) Diameter: 5¼ inches (base flange); (4) Weight: 6 lbs.

d. FREQUENCY RANGE — 225 to 400 MHz.

e. INPUT IMPEDANCE/VSWR — 50 ohms. VSWR not greater than 2.5:1 over frequency range.

f. RF POWER RATING — 100 watts.

g. POLARIZATION — Vertical.

h. TYPE OF FEED — Unbalanced, coaxial cable.

i. INPUT CONNECTOR — Type N.

j. INSTALLATION REQUIREMENTS — The antenna base flange has four 11/32-inch bolt holes equally spaced on a 4⅝-inch bolt circle for use in mounting the antenna. The antenna may be installed on a suitable mounting plate having matching bolt holes and a center hole to accommodate the antenna transmission line.

k. SPECIAL CONSIDERATIONS — The location for the AS-2810/SRC should be selected to provide maximum omnidirectional coverage within the physical constraints of the ship's topside. UHF propagation in the frequency range of this antenna is generally line-of-sight. Therefore, antennas that are masked by surrounding structures produce limited coverage in azimuth or elevation.

l. REFERENCES (1) CHU Associates Drawing Number 31815, Antenna Model CA-3123. (2) Technical Manual, Antenna AS-2810/SRC, NAVSEA 0967-475-5010.

ANTENNA AS-2877/SRC

a. TYPE/USE — The AS-2877/SRC is a double-channel, broadband, omnidirectional UHF antenna consisting of a two-element dipole array (channel A) and a single dipole (channel B), enclosed and sealed in a cylindrical fiberglass housing.

The dipoles are configured to give a minimum isolation of 25 dB between channels. A waveguide runs from the base plate through the center to the top of the antenna shroud, thus providing facilities to install an ECM antenna, such as the AS-1175/SLR, on top when desired. The AS-2877/SRC may be used with any UHF transmitting or receiving equipment capable of operating within its frequency range and RF power handling capabilities.

b. PHYSICAL CONFIGURATION — [See Figure 2-106].
c. DIMENSIONS — (1) Height: 120 inches; (2) Diameter: 13½ inches; (3) Weight: 64 lbs.
d. FREQUENCY RANGE — 225 to 400 MHz.
e. INPUT IMPEDANCE/VSWR — 50 ohms. VSWR not greater than 1.8:1 over frequency range.
f. RF POWER RATING — 1 kW PEP.
g. POLARIZATION — Vertical.
h. TYPE OF FEED — Unbalanced, coaxial cable.
i. INPUT CONNECTOR — (1) UHF antennas, type HN. (2) ECM antenna, waveguide.
j. INSTALLATION REQUIREMENTS — The antenna assembly mounting flange has eight 11/16-inch diameter bolt holes equally spaced on an 11¾-inch bolt circle for use in mounting the antenna. The antenna may be installed on a suitable mounting plate having matching bolt holes and a center hole to accommodate the antenna transmission lines and waveguide.
k. SPECIAL CONSIDERATIONS — The location for the AS-2877/SRC should be selected to provide maximum omnidirectional coverage within the physical constraints of the ship's topside. UHF propagation in the frequency range of the antenna is generally line-of-sight. Therefore, antennas that are masked by surrounding structures produce limited coverage in azimuth or elevation.
l. REFERENCES — CHU Associates Drawing Number 6401, Antenna Model CA-1080A.

NOTE — Other multiple-function vertical dipole collinear arrays similar in outward appearance to the AS-2877 exist, such as the AS-3020 (four independent dipoles: two UHF, one VHF, and one IFF) and the AS-3021 (two independent UHF dipole antennas). However, they are in principle extensions of the AS-1018 and AS-2877, and are used only in a limited capacity (e.g., DD-963 class ships). Reference for the AS-3020 is NAVSEA Technical Manual 0967-LP-568-1010; for the AS-3021 it is NAVSEA 0967-LP-568-0010.

2-3.2 Special UHF Antennas

Dipole antennas are by no means the only type used for shipboard UHF communications. One of the most extensively used UHF antennas in the fleet is the AS-390, an enduring example of rugged simplicity, and (where a very great number of ship-to-air circuits are required) the high-gain circularly polarized AN/SRA-62 sector arrays are used. Finally, there is the worldwide FLEETSATCOM UHF system described in the next section (Section 2-4).

ANTENNA ASSEMBLY AS-390/SRC

a. TYPE/USE — The AS-390/SRC antenna is an unbalanced broadband coaxial stub used for transmitting and receiving. It consists of a radiator and groundplane (or counterpoise) of eight elements bent downward 37° from the horizontal. The lower ends of the elements form points of a circle 23 inches in diameter. The lower section of the radiator assembly contains a stub for adjusting the input impedance of the antenna. This is determined at the time of manufacture and locked during final assembly of the antenna. This antenna may be used with any UHF transmitting or receiving equipment designed to operate within its frequency range and RF power handling limitations. It is normally installed on masts or yardarms as in Figure 2-107, but occasionally may be mounted elsewhere as shown in Figure 2-108.
b. PHYSICAL CONFIGURATION — [See Figure 2-109].
c. DIMENSIONS — (1) Height: 18 inches; (2) Diameter: 23 inches; (3) Weight: 3½ lbs.
d. FREQUENCY RANGE — 220 to 400 MHz.
e. INPUT IMPEDANCE/VSWR — 50 ohms nominal. VSWR not greater than 2:1 over frequency range.
f. RF POWER RATING — 200 watts average.
g. POLARIZATION — Vertical.
h. TYPE OF FEED - Unbalanced, coaxial cable.
i. INPUT CONNECTORS — Type N, special, mates with UG-21B/U.
j. INSTALLATION REQUIREMENTS — The base of the assembly is provided with a threaded ¾-inch, 14 IPS pipe, 1½ inches long, for use in mounting the antenna. The antenna may be installed using a matching size pipe and pipe coupling welded to the ship's structure.
k. SPECIAL CONSIDERATIONS — The location for the AS-390/SRC antenna should be selected to provide maximum unobstructed omnidirectional coverage within the physical constraints of the ship's topside. UHF propagation is generally line-of-sight and antennas that are masked by surrounding structures produce limited coverage in azimuth or elevation. The antenna is designed to be sufficiently air-tight to maintain a dielectrically stable atmosphere within the antenna. The presence of moisture can be checked by the removal of a pipe plug located on the underside of the counterpoise plate. The antenna should not be located in stack gas drafts or exhausts where the temperatures exceed the operating limitations.
l. REFERENCES — (1) Technical Manual, Antenna Assembly AS-390/SRC, NAVSEA 0967-LP-060-5010. (2) NAVSEA Drawing RE-F2691919; RE-D2691804; RE-F2688195.

ANTENNAS AS-1496/U, AS-1497/U (PHASOR-90)

a. TYPE/USE — The AS-1496/U and AS-1497/U are used with the PHASOR-90 UHF system. The antennas are arrays consisting of two circularly polarized conical spirals spaced one wavelength apart at 300 MHz. The AS-1496/U has a right-hand spiral, while the AS-1497/U has a left-hand spiral; otherwise the two antennas are identical. The two conical spirals are mounted on a small groundplane or mounting plate. Each conical spiral is fed at the apex of the cone through a balun and has a nominal impedance of 100 ohms. When two conical spirals are connected in parallel a nominal impedance of 50 ohms is presented to the transmission line. A full system consists of two AS-1496/U and two AS-1497/U antennas, and a power divider. The power divider is a low loss transformer which divides the power from a single 50-ohm source into four separate 50-ohm outputs to the antennas.

   The PHASOR-90 system was designed for use aboard aircraft carriers for ship-to-aircraft communications. Figures 2-110 and 2-111 depict carrier installations of the paired antennas. Note that this system is considered obsolete and is being replaced by the AN/SRA-62 sector array.

b. PHYSICAL CONFIGURATION — [See Figure 2-112].

c. DIMENSIONS — Single Array: (1) Height: 69½ inches; (2) Width: 31⅜ inches; (3) Depth: 40⅜ inches, exclusive of ground plane; (4) Weight: 100 lbs.

d. FREQUENCY RANGE — 225 to 400 MHz.

e. INPUT IMPEDANCE/VSWR — 50 ohms nominal. VSWR not greater than 3:1.

f. RF POWER RATING — 200 watts.

g. POLARIZATION — Circular.

h. TYPE OF FEED — Unbalanced, coaxial cable.

i. INPUT CONNECTOR — Type N, UG-29B/U.

j. INSTALLATION REQUIREMENTS — The antenna arrays are installed by bolting each unit to the ship's structure through bolt holes on the groundplane. They are usually installed on the carrier's island, located in such a manner as to provide unobstructed coverage in all quadrants. The arrays on adjacent sides of the ship's island must be of opposite sense; i.e., if the fore and aft arrays are left-hand circular the port and starboard must be right-hand circular. The arrays should be free from obstruction to the antenna near-field within 20 degrees of the horizon in the respective quadrant. All lengths of coaxial cable between the power divider and the four antennas should be very nearly equal.

k. SPECIAL CONSIDERATIONS — Installation of additional arrays require that they be separated as far as possible to provide isolation between systems. Additional isolation can be achieved by placing oppositely sensed arrays next to each other.

l. REFERENCE — Instruction book for PHASOR-90 UHF Antenna System, NAVSEA 95682.

ANTENNA AS-2493/SRA-62

a. TYPE/USE — The AS-2493/SRA-62 is a component of the AN/SRA-62 sector array UHF antenna system. This system is composed of four AS-2493/SRA-62 UHF sector antennas and one RF-247/SRA-62 power divider, which allows all four sector antennas to be fed from a single 50-ohm coaxial transmission line. The system is used to provide high-gain omnidirectional coverage for ship-to-air communications. Each sector antenna is made up of two cross-log periodic antennas tilted together at a 14° angle to maintain a near constant aperture with frequency, and a polarization network to produce a circularly polarized signal. The sector antennas are placed on the ship's superstructure roughly equally spaced fore and aft, and port and starboard. The outputs of the two individual antennas of each sector are combined in a two-way power divider, and then the outputs of each of the four sectors are connected to a four-way power divider to obtain a single high-gain output for the entire array. The system has been especially designed to withstand the shock, vibration, and corrosive environment aboard ship. Figure 2-113 shows a set of two sectors each nested together, oppositely polarized to achieve isolation.

b. PHYSICAL CONFIGURATION — [See Figure 2-114].

c. DIMENSIONS — [Dimensions for one mounted sector antenna]. (1) Height: 52 inches; (2) Width: 20 inches; (3) Depth: 50¾ inches; (4) Weight: 68 lbs.

d. FREQUENCY RANGE — 225 to 400 MHz.

e. INPUT IMPEDANCE/VSWR — 50 ohms. VSWR not greater than 2:1 over frequency range.

f. RF POWER RATING — 500 watts average, 2000 watts peak.

g. POLARIZATION — Right-hand circular and left-hand circular, alternating among the four sector antennas.

h. TYPE OF FEED — Unbalanced, coaxial cable.

i. INPUT CONNECTOR — Type N.

j. INSTALLATION REQUIREMENTS — The antennas may be mounted directly to the ship's structure or suitable platforms with mounting brackets drilled to accommodate ⅜-inch bolts for this purpose. The antennas should be located fore, aft, port, and starboard roughly in a square around the superstructure. The polarization of the antennas must be arranged so that there is a right-hand polarized antenna port and starboard, and a left-hand polarized antenna fore and aft. Distance between the antennas is not critical; but the cables connecting the sector antennas to the power divider should be carefully cut so that the lengths are equal to within ±¼ inch.

k. SPECIAL CONSIDERATIONS — If it is necessary to increase channel capacity, additional sector antennas may be mounted on the ship's superstructure. The antennas on the same side of the superstructure should be separated as far as possible to provide isolation between systems. To further increase isolation adjacent antennas should be oppositely polarized.

l. REFERENCE — Technical Manual, Sector Array UHF Antenna System, AN/SRA-62, NAVSEA 0967-374-4010.

2-4 SATELLITE COMMUNICATIONS ANTENNAS

In 1951, the Navy began conducting a series of experiments to investigate the feasibility of reflecting radio signals from the surface of the moon as an alternative means of long range point-to-point propagation. The conventional method of worldwide HF communications via skywave refraction, in addition to being limited in channel capacity, had proven to be very much affected by the daily, seasonal, and geographical vagaries of the ionosphere. Thus in 1959, after nearly a decade of tests, the Navy established the world's first rudimentary satellite system from Washington, D.C. to Hawaii, using the earth's moon as the satellite relay. On 28 January 1960, this UHF Communications Moon Relay, or CMR, was publicly demonstrated with both voice and facsimile. That same year the Navy Research Lab and NASA, in a joint effort, performed communications experiments using a passive 100-foot diameter plastic spherical satellite called *ECHO I.* Almost two years later, on 15 December 1961, the first ship-to-shore CMR SATCOM link was established using the *USS Oxford.*

Spurred by these preliminary successes, interest in satellite communications accelerated. In 1963, The *USNS Kingsport* became the world's first ship to transmit and receive voice messages via an active satellite. The following year, the Geneva Radio Conference approved designation of the 8 GHz region of SHF (super-high frequency), for the Defense Satellite Communications System uplink, and 7 GHz for the downlink. By 1969, following rapid developments, the Navy, using six ships in the fleet, had established worldwide satellite communications.[28]

As the 1970s dawned, optimism and enthusiasm about satellite communications was so great that momentum was building to virtually replace HF in the fleet with SATCOM. Touting the promise of the Fleet Satellite Communications System (FLTSATCOM) "to provide a highly reliable, timely, secure and flexible medium for communications to a degree not achievable by other media," the Navy announced its plan "to reduce selectively the conventional HF broadcast after the FLTSATCOM system becomes operable. It is not anticipated that all HF facilities will be deactivated. Some HF/MF gear will be retained for polar region coverage, FLTSATCOM backup, and intra-task force tactical communications ... [but] FLTSATCOM will take over primary ship-to-shore communications."[29]

Nevertheless, despite all the advantages and the initial excitment over SATCOM, toward the end of the '70s caution began to surface about too great a commitment to SATCOM. A nagging concern began to creep in about the vulnerability of satellites, how easily command and control might be lost with a few strategic "kills." It was realized that perhaps a second look should be taken at the relative immunity of HF. Consequently, there exists at the present time a balanced mixture of SATCOM and HF capabilities in naval use.

Satellite communications did introduce new complexities in the design of shipboard antennas. The direct line-of-sight between antenna and satellite requires the antenna to "see" from horizon to overhead and 360° in azimuth, i.e., total hemispherical coverage. This is fulfilled in the case of transceive antennas through the use of rotatable, high-gain antennas often installed in pairs aboard ship to attain full coverage, irrespective of blockage from masts, stacks, and other deck objects. The ship platform itself imposes even more stringent requirements. In spite of constant vibrating, pitching, rolling, and yawing, the SATCOM antenna's narrow radiation beam must be pointed accurately from any position on the high seas so that the beacon signal emitted by the satellite for acquisition and tracking can be picked up. Once this beacon signal is received, tracking is automatic. Such complexities are an order of magnitude greater than those encountered in the design of the general purpose, fixed, communications antennas previously described in this chapter.

There are two distinctly different naval shipboard SATCOM systems currently in use. One, the AN/WSC-3 is a UHF system, and the other is the AN/WSC-6 SHF system. The shipboard antennas associated with these two systems will be described in separate sections following.

2-4.1 UHF Satellite Communications Antennas

At the present time there are three UHF SATCOM transceive antenna types developed for naval shipboard service. The oldest of these is the AS-2410 crossed-dipole array shown in Figure 2-115. The second type is an improved version of the crossed-dipole array, the AS-3018 (Figure 2-116). Finally, the latest type, entirely different from the first two, is the so-called "oil drum," or "wash tub" AS-3018A shown in Figure 2-117. All are intended for use with the AN/WSC-3 (and the AN/WSC-5 in some special cases) UHF Fleet Satellite Communications System.

ANTENNA AS-2410/WSC-1(V) and AS-3018/WSC-1(V)

a. TYPE/USE — The AS-2410 and AS-3018 UHF SATCOM antennas are similar in function, characteristics, and appearance. The AS-2410 is part of antenna group OE-82A/WSC-1, while the AS-3018 is part of antenna group OE-82B/WSC-1. (The OE-82 antenna groups are comprised of such ancillary equipment as amplifier-filters, antenna control units, and sensing/switching systems in addition to the antennas themselves.) These antennas consist of four crossed-dipole elements mounted above the groundplane with a 90-degree hybrid, two combining networks, and terminating load arranged so that the array radiates a right-hand, circularly polarized wave. The antenna assembly is attached atop a pedestal that permits it to be rotated through 360 degrees in azimuth and from near horizon to approximately 20 degrees beyond zenith (i.e., +2° to 110° elevation angle). The upper compartment of the pedestal contains the elevation drive motor and its associated synchro data package. Two limit switches in the upper compartment disconnect drive motor power as the elevation end stops are approached so that drive motor damage will not occur. A variable cam assembly is provided near the bottom of the upper pedestal compartment that permits establishment of azimuth zones (sectors) where radiation from the antenna may be prohibited. The zones are adjustable in increments of approximately 7 degrees azimuth and can easily be adjusted to conform to a wide variety of ship superstructure configurations. A switch associated with this variable cam assembly detects the radiation-blocked sector and provides an electrical signal to the antenna control. The access opening in the upper pedestal compartment is made weather-tight by a removable, gasketed cover plate that is secured in place by 16 captive hexhead capscrews.

The lower compartment of the antenna pedestal contains the azimuth drive motor and its associated synchro data package. Also included in this compartment are the multipin connectors that terminate the control and response circuits from other units of the antenna group. The connections are made within the pedestal, and the RF cable from the amplifier-filter is connected to an external ⅞-inch EIA flanged fitting near the safety switch. One set of contacts on the safety switch control the dc power to the antenna pedestal and another set of contacts on this switch interlocks the sector switch in the companion antenna for signaling purposes. A slipring assembly in the lower compartment of the antenna pedestal interconnects the electical circuits of the lower and upper compartments. A sound-powered phone circuit is terminated in the lower compartment permitting maintenance conversations with personnel at the antenna control and at the companion antenna. Two access openings in the lower compartment are equipped with weather-tight, removable, gasketed covers, each secured with 14 captive hexhead capscrews.

The antennas are used to transmit and receive simultaneously in the UHF range, provide the necessary azimuth and elevation controls to direct the antenna at the desired satellite, and maintain tracking during ship movement. Normally, dual antennas are used to prevent masking of the antenna system by the ship's superstructure. If one antenna's coverage is blocked, the control system automatically switches operation to the other antenna.

b. PHYSICAL CONFIGURATIONS — [See Figure 2-118].
c. DIMENSIONS — (1) Height: 50 inches; (2) Width: 50 inches; (3) Depth: 27½ inches; (4) Weight: 325 pounds.
d. FREQUENCY RANGE — 240 to 318 MHz.
e. INPUT IMPEDANCE/VSWR — 50 ohms, VSWR not greater than 1.5:1 for 240 to 290 MHz and not greater than 1.2:1 for 290 to 350 MHz as measured at the input to the antenna pedestal.
f. RF POWER RATING — 150 watts CW, 250 watts PEP.
g. POLARIZATION — Right-hand circular.
h. GAIN — 12dB minimum.
i. TYPE OF FEED — Unbalanced, coaxial cable.
j. PRIMARY POWER REQUIRED — For antenna drive motors, antenna position, and control: 440 Vac, three-phase, three-wire, 50/60 Hz, supplied from ship's power source or from radio set group, at 2500 volt-amperes maximum, 0.92 power factor.
Ship power: 115 Vac, one-phase, 400 Hz (nominal), 250 volt-amperes (approx.) 1.0 power factor.
Own ships course information: 115 Vac, one-phase, 400 Hz (nominal), 250 volt-amperes (approx.) 1.0 power factor.
k. OPERATING TEMPERATURE — Minus 28°C to +65°C.
l. INPUT CONNECTOR — 7/8-inch EIA flange.
m. INSTALLATION REQUIREMENTS — The antenna pedestal mounting base has 12 equally spaced ¾-inch holes on a 16½-inch bolt circle for use in the installation of the antenna. The overall diameter of the base plate is 17½ inches. The antenna may be deck mounted or installed on a suitable mounting base. Two antennas are installed with each system to give maximum view areas of the satellite within the physical constraints of the ship's structure and fittings. The location of the antennas should be selected to provide adequate clearance for the antenna during its movement within the swept volume sphere. The two antennas should be located so that the satellite is always in view of one with a minimum overlap of 10 degrees between sectors and hemispherical coverage from 10 degrees above the horizon to zenith.
n. SPECIAL CONSIDERATIONS — Radiation hazard cam operating switches are located in the antenna pedestal to disable the transmitting system at relative antenna azimuths wherever radiation may be hazardous to ship personnel. The hazardous azimuths are determined at the time of installation, and the cams adjusted to operate the cutout switches when the rotating antennas reach these bearings.
o. REFERENCES — (1) Technical Manual, Tactical Satellite Communications System AN/WSC-1(V), NAVELEX 0967-439-7010 and 0967-467-9010. (2) NAVSEA Drawing 825-4436022.

ANTENNA AS-3018A/WSC-1(V)

a. TYPE/USE — The AS-3018A, though electrically identical to the AS-2410 and AS-3018, is entirely different physically. Rather than rectangular in form, with an array of radiators, it is circular with a single crossed-dipole element recessed within a drum reflector/director. This antenna is part of the OE-82C/WSC-1 antenna group and is mounted on a pedestal that permits rotation through 360° in azimuth and from horizon to zenith in elevation. Normally, two antennas are used to prevent masking of the antenna system by the ship's superstructure. If one antenna's coverage is blocked, the control system automatically switches operation to the other antenna. The antennas transmit and receive simultaneously in the UHF range, provide the necessary azimuth and elevation controls to direct the antenna at the desired satellite, and maintain tracking during ship movement.

b. PHYSICAL CONFIGURATION — [See Figure 2-119].

c. DIMENSIONS — (1) Diameter: 54 inches; (2) Depth: 33½ inches; (3) Weight: 294 pounds.

d. FREQUENCY RANGE — 240 to 318 MHz.

e. INPUT IMPEDANCE/VSWR — 50 ohms, VSWR not greater than 1.5:1 between 240 and 290 MHz and 1.2:1 from 290 to 312 MHz.

f. RF POWER RATING — 400 watts CW, 800 watts PEP.

g. POLARIZATION — Right-hand circular.

h. GAIN — 12 dB minimum.

i. TYPE OF FEED — Unbalanced, coaxial cable.

j. PRIMARY POWER REQUIRED — For antenna drive motors, antenna position, and control: 400 Vac, three-phase, three-wire, 50/60 Hz, supplied from ship's power source or from radio set group, at 2500 volt-amperes maximum, 0.92 power factor.
Auxiliary gyro references: 115 Vac, one-phase, 400 Hz (nominal), 25 volt-amperes (approx.) 1.0 power factor.
Own ship's course information: 115 Vac, one-phase, 400 Hz (nominal), 25 volt-amperes (approx.), 1.0 power factor.

k. OPERATING TEMPERATURE — Minus 28°C to +65°C.

l. INPUT CONNECTOR — ⅞-inch EIA flange.

m. INSTALLATION REQUIREMENTS — The antenna pedestal mounting base has 12 equally spaced ¾-inch holes on a 16½-inch bolt circle for use in the installation of the antenna. The overall diameter of the base plate is 17½ inches. The antenna may be deck mounted or installed on a suitable mounting base. Normally, two antennas are installed with each system to give maximum view area of the satellite within the physical constraints of the ship's structure and fittings. The location of the antennas should be selected to provide adequate clearance for the antenna during its movement within the sweep volume sphere. The two antennas should be located so that the satellite is always in view of one with a minimum overlap of 10 degrees between sectors and hemispherical coverage from 10 degrees above the horizon to zenith.

n. SPECIAL CONSIDERATIONS — Radiation hazard cam operating switches are located in the antenna pedestal to disable the transmitting system at relative antenna azimuths wherever radiation may be hazardous to ship's personnel. The hazardous azimuths are determined at the time of installation, and the cams are adjusted to operate the cutout switches when the rotating antennas reach these bearings.

o. REFERENCES — (1) Technical Manual NAVELEX 0967-LP-614-0040. (2) NAVSEA Drawing RE-D5034197.

ANTENNA AS-2815/SSR-1

a. TYPE/USE — The AS-2815/SSR-1 is a unit of the AN/SSR-1 fleet satellite broadcast receive set. It consists of two loops and two parasitic dipoles per unit. The loops are driven in phase quadrature, and the dipoles are parasitically excited. An AN/SSR-1 system is comprised of four AS-2815/SSR-1 antennas to provide 360° coverage and diversity reception. Figure 2-120 depicits an AS-2815 installed aboard ship.

b. PHYSICAL CONFIGURATION — [See Figure 2-120].

c. DIMENSIONS — (1) Height: 36 inches; (2) Width: 31 inches; (3) Depth: 31 inches; (4) Weight: 13 pounds.

d. FREQUENCY RANGE — 248 to 255 MHz.

e. INPUT IMPEDANCE/VSWR — 50 ohms, VSWR not greater than 1.5:1.

f. RF POWER RATING — Not applicable, receive only.

g. POLARIZATION — Right-hand circular.

h. TYPE OF FEED — Unbalanced, coaxial cable.

i. PRIMARY POWER REQUIRED — None.

j. OPERATING TEMPERATURE — Minus 28°C to +65°C.

k. INPUT CONNECTOR — 44AW Andrews.

l. INSTALLATION REQUIREMENTS — The antenna mounting base has four $^{25}/_{32}$-inch bolt holes on a 5½-inch diameter bolt circle for use in the installation of the antenna; the overall diameter of the base plate is 7⅛ inches. The antenna may be deck mounted or installed on a suitable mounting base. The four antenna locations should be as widely separated as possible with one located in each quadrant of the ship to provide hemispherical receiving capability.

m. SPECIAL CONSIDERATIONS — Each antenna operates in conjunction with an AM-6534/SSR-1 amplifier-converter. Each amplifier-converter must be located so that the coaxial cable from its associated antenna does not exceed ten feet. A stainless steel spacer is required between the aluminum antenna base and any dissimilar metal to which it may be attached. Stainless steel bolts must be used to secure the antenna to the mounting base.

n. REFERENCE — Technical Manual, Satellite Signal Receiving Set, AN/SSR-1, NAVELEX 0967-440-1010.

2-4.2 SHF Satellite Communications Antennas

In addition to the UHF AN/WSC-3 FLTSATCOM system which has been installed virtually fleet wide, a microwave SATCOM system, the AN/WSC-6, is now being employed for naval service. To be used in situations of very high volume communications traffic such as exists on flag command ships, the AN/WSC-6 provides near worldwide, jam-resistant, high capacity data, teletype, and voice communications via the DSCS satellites. Large, four-foot diameter, high gain, parabolic dish antennas are used with the WSC-6 system. Because of their size and weight, and the need to provide hemispherical coverage, they normally can be accommodated only by the larger ships of the fleet.

ANTENNA AS-3399/WSC

a. TYPE/USE — The AS-3399/WSC SATCOM antenna, a large microwave dish, is illustrated by Figure 2-121. It is part of the OE-279/WSC-6(V) antenna group that provides continuous, automatic tracking of the selected satellite while simultaneously transmitting and receiving wideband, high bit rate data, teletype, or voice signals. The antenna unit consists of the reflector, feed structure, waveguide assembly, servo and drive system, vertical reference gyro, and radome. Major antenna components located within the radome are depicited in Figure 2-122.
   The remaining antenna equipment to supply pointing information and power for the antenna drive motors, is installed below deck. The Digital Servo Controller (DSC) commands the antenna to search for signals from the satellite. Once the satellite signals have been acquired by the track receiver, an automatic feature keeps the antenna locked on the satellite. Additionally, servo error signals are generated which correct the antenna position due to ship or satellite motion.
b. PHYSICAL CONFIGURATION — [See Figures 2-121, 2-122, and 2-123].
c. DIMENSIONS — (1) Diameter: 72 inches (radome); (2) Height: 86.5 inches; (3) Weight: 580 lbs.
d. FREQUENCY RANGE — 7.25 to 8.4 GHz.
e. GAIN — 37 dB nominal.
f. VSWR — Not to exceed 1.3:1.
g. RF POWER RATING — 6 kW.
h. POLARIZATION — Transmit: right hand circular; receive: left hand circular.
i. TYPE OF FEED — Waveguide, CPR 137 flange.
j. PRIMARY POWER REQUIRED — 115 Vac, 60 Hz, one-phase, 1100 volt-amperes.
k. OPERATING TEMPERATURE — Minus 28°C to +65°C.
l. INSTALLATION REQUIREMENTS — Since it is essential that full hemispherical coverage be attained, the normal practice is to use a pair of AN/WSC-6 antennas aboard ship, located as clear and high as possible. However, the size and weight, stability requirements, and the waveguide length limitations preclude installation at very great height. Therefore, the antennas will usually be located port and starboard just above the mast base, or on such stable platforms as illustrated by Figure 2-124.
   The antenna is designed to meet MIL-S-901 Grade A shock requirements and to have a resonant frequency above 33 Hz. Moreover, the antenna must acquire and track the satellite under severe environmental condi-

tions (i.e. sea state 6 and 100 knot winds). Therefore, the antenna foundation must be designed to meet the same requirements.

In fact the static flexibility of the supporting structure is considered to be a critical factor in achieving an operational AN/WSC-6 system under all environmental conditions. It is important to note, too, that the antenna base must be at least two feet above its mounting platform to allow personnel entry through a trap door in the radome base plate.

The Low Noise Amplifier (LNA) weighs about 200 pounds and is limited to a maximum waveguide distance from the antenna of 50 feet. It is recommended that, even though the LNA is designed to be installed on the weather deck exposed to the atmosphere, every effort should be made to install this unit internally below deck.

The transmit waveguide is M85/1-067 [(WR-137)(RG-50/U)] and is limited to a maximum length of 150 feet. The mating flange on the antenna transmit port is CPR-137G. The receive waveguide, from antenna to LNA is also M85/1-067, but is limited to a maximum of 50 feet as stated above. The mating flange on the antenna receive port is CPR-137G. The waveguides and LNAs require dry air between 2 and 5 PSIG fed at the top of each waveguide run.

m. SPECIAL CONSIDERATIONS — Because the high gain of the antenna results in an effective radiated power in excess of 40 megawatts, concentrated in an effective beamwidth of approximately 3 degrees, high intensity RF energy between 7.9 and 8.4 GHz could cause severe *EMI* and *RADHAZ* problems. It is estimated that a personnel hazard from continuous or intermittent exposure to main beam radiation extends for at least 1000 feet with a maximum exposure of one second, per 30 second interval. Furthermore, the AN/WSC-6 is expected to exceed *HERO SAFE* limits out to approximately 300 feet in the main beam, thereby creating a hazard to ordnance on nearby ships and aircraft.

n. REFERENCES — (1) Basic Requirements for Installing the AN/WSC-6(XN-1) (V2) — SHF Satellite Communications Set, NAVELEX, Vallejo, CA. (2) Integrated Logistic Support Plan for AN/WSC-6(XN-1) (V) Shipboard Satellite Communications Set, NAVELEX, Vallejo, CA. (3) Service Test Manual, NAVELEX 523-0564148-B3D.

REFERENCES

1. L.S. Howeth, *History of Communications-Electronics in the United States Navy* (Washington, D.C.: U.S. Government Printing Office, 1963), p. 30.
2. *Ibid.*, p. 28.
3. *Ibid.*, p. 51.

4. *Ibid.*, p. 107.
5. *Ibid.*, p. 168.
6. S.S. Robison, *Manual of Radio Telegraphy and Telephony for Naval Radiomen* (Annapolis: The United States Naval Institute, 1927), pp. 351, 388, 509-516.
7. Howeth, *op.cit.*, p. 167.
8. Robison, *op.cit.*, pp. 351, 389, 509.
9. R.L. Linton, Jr., "The Folded Fan as a Broadband Antenna," *Proceedings of the IRE*, November 1951, pp. 1436–1444.
10. "Wire-Rope Communications Antennas," Electronics Engineering Bulletin No. 006 (Washington, D.C.: Naval Ship Engineering Center, 19 March 1979), p. 5.
11. *Shipboard Antenna Systems Installation Details*, NS 0967-177-3020 (Department of the Navy, September 1973), pp. 2–10.
12. W.W. Stockton and G.E. Bartz, *Standardized Wire-Rope Antenna Hardware*, TD 282 (San Diego: Naval Electronics Laboratory Center, 2 October 1973), p. 14.
13. W.M. Chase and C.W. Tirrel, *TRED HF Communications System Analysis*, TR1786 (San Diego: Naval Electronics Laboratory Center, 24 September 1971), p. 7.
14. J.L. Lievens, *Mistuning of AN/URA-38 Antenna Coupler In The Presence of Interfering Signals*, TD-170 (San Diego: Naval Electronics Laboratory Center, 20 March 1972), p. 1.
15. Louis A. Gebhard, *Evolution of Naval Radio-Electronics and Contributions of the Naval Research Laboratory* (Washington, D.C.: Naval Research Laboratory, Report 8300, 1979), pp. 80–82.
16. G.E. Bartz, *Standardized Trussed Monopole Communications Antenna*, TD-376 (San Diego: Naval Electronics Laboratory Center, 1 Dec. 1974), p. 8.
17. Gebhard, *op.cit.*, p. 79.
18. Henry Jasik, *Antenna Engineering Handbook* (New York: McGraw-Hill Book Company, 1961), p. 3-24.
19. *Miniloop High Frequency Antenna Final Project Report* (Newport, Rhode Island: Commander, Destroyer Development Group, Technical Report No. 24-72, 15 July 1972.)
20. C.E. Hobbis et. al., *Evaluation of the MLA-2/B Miniloop Antenna* (Washington, D.C.: Naval Research Laboratory, NRL Report 2494, September 1972.)

21. J.L. Lievens and D.W. DuBrul, *MLA-1 Miniloop Antenna — A Technical Evaluation* (San Diego: Naval Electronics Laboratory Center, NELC Report TD317, 10 May 1974.)
22. *Ibid.*, p. 7.
23. Howeth, *op.cit.*, p. 610.
24. Robison, *op.cit.*, p. 465.
25. Gebhard, *op.cit.*, p. 96.
26. Howeth, *op.cit.*, p. 540.
27. Gebhard, *op.cit.*, p. 101.
28. *Ibid.*, p. 115-136.
29. *Getting Acquainted With FLTSATCOM: Introducing the Navy's Fleet Satellite Communications System* (Washington, D.C.: Naval Electronics Systems Command, April 1975), pp. 1-1, 1-2.

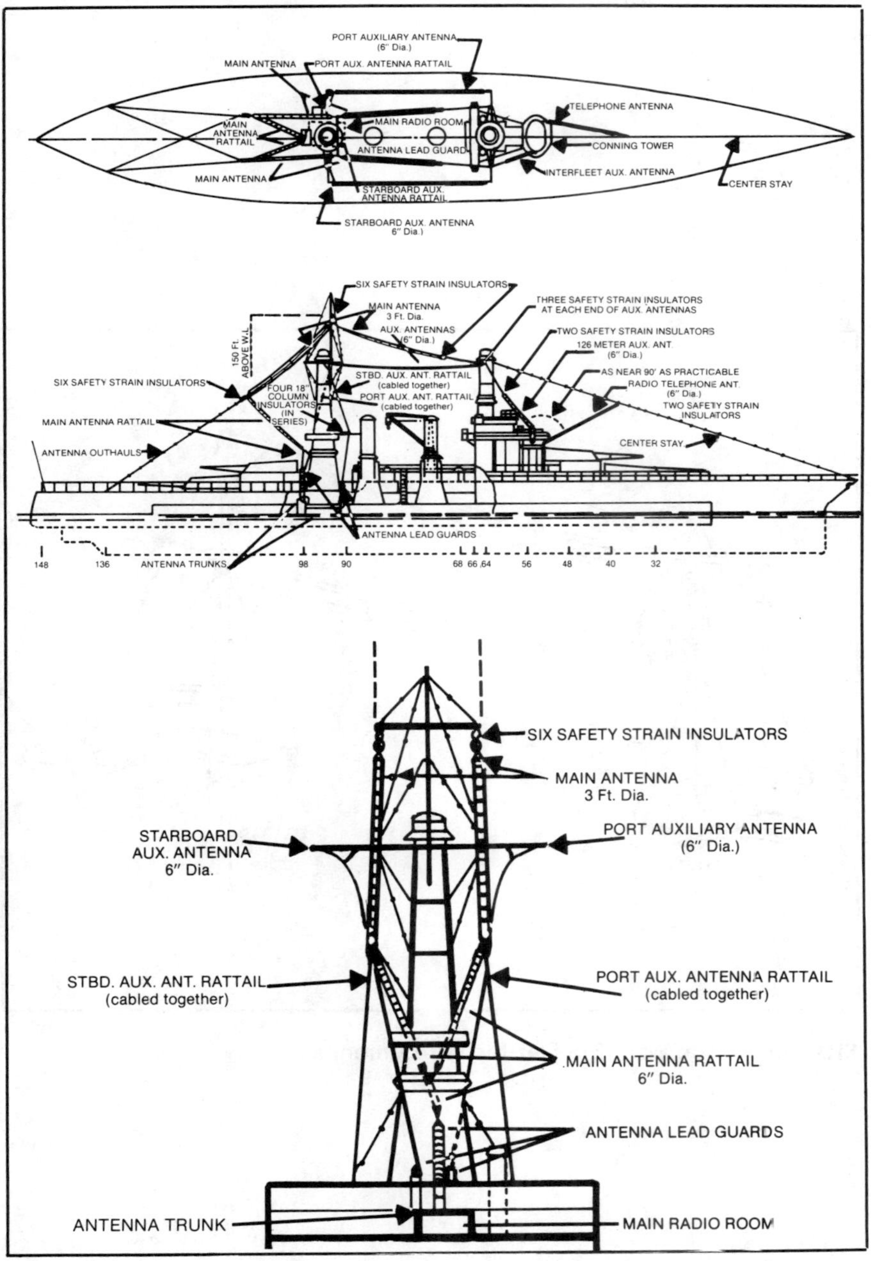

FIG. 2-1 Early Shipboard Long-wire Antennas (circa World War I)

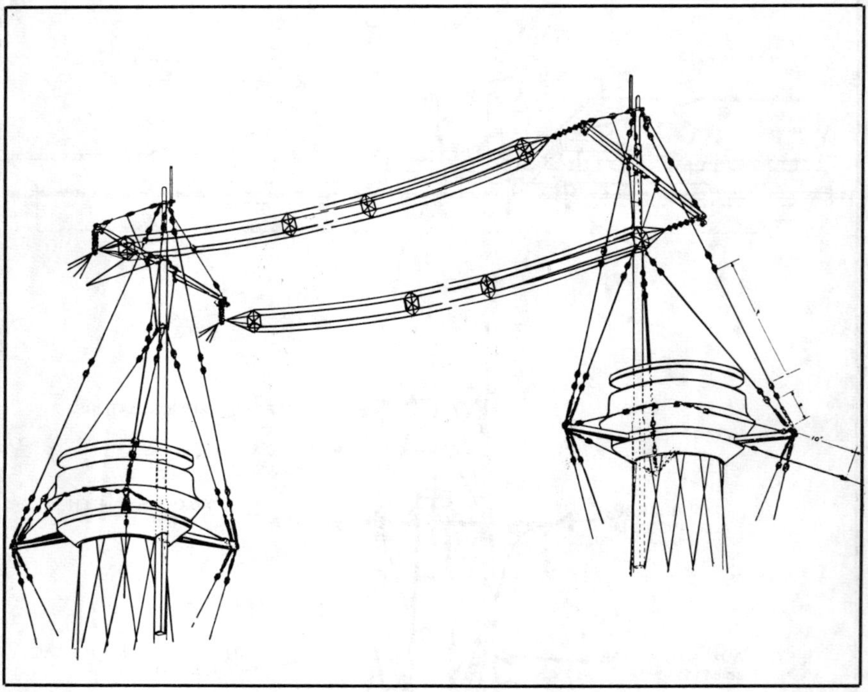

FIG. 2-2 World War I Era Flat-top Wire Antennas

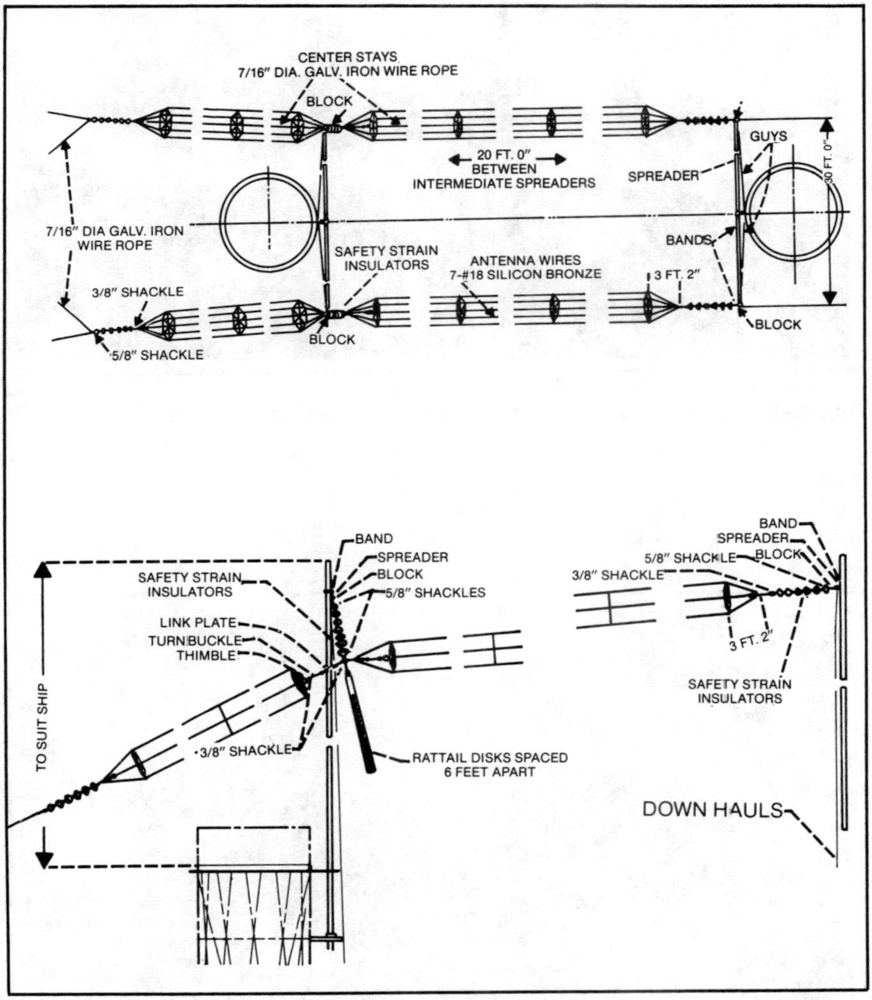

FIG. 2-3 World War I Era Shipboard Wire Antenna Details

FIG. 2-4 Inclined Longwire Antenna — *USS Curtiss* AV-4 World War II

FIG. 2-5 Multi-element Horizontal Long-wire Antennas — Bow To Stern *USS Essex* CVS-9 World War II

FIG. 2-6 LF/MF Flat-Top Antennas

FIG. 2-7 Typical Flat-top Antenna

FIG. 2-8 Typical Flat-top Antenna

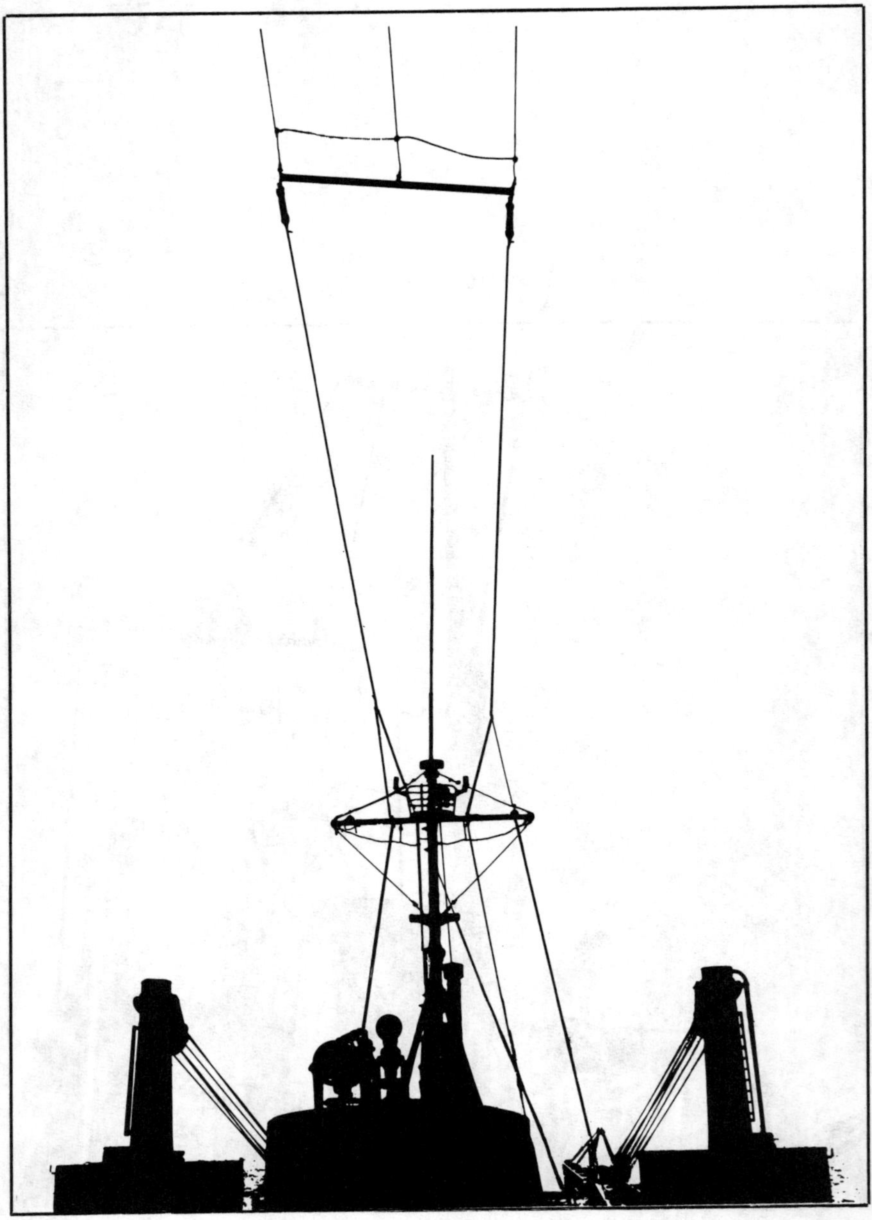

FIG. 2-9 Typical LF/MF Flat-top Antenna

FIG. 2-10 Single LF/MF Long-wire Antenna

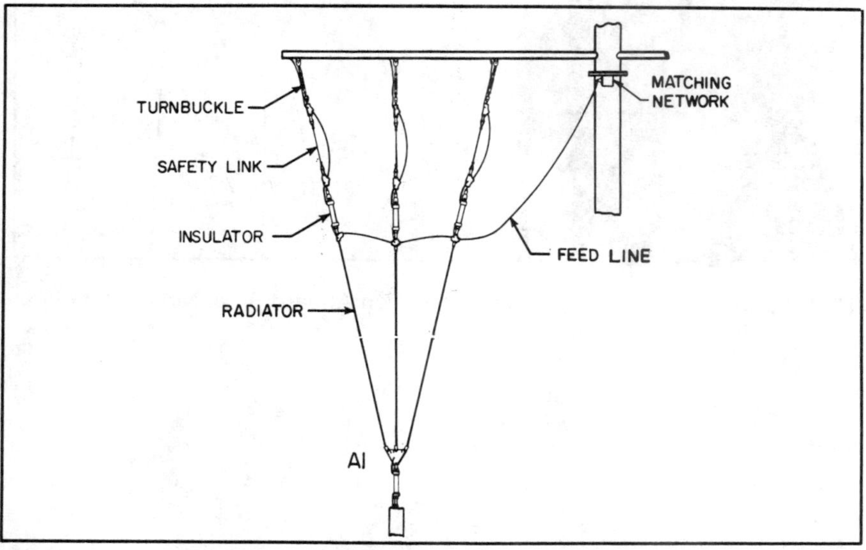

FIG. 2-11 Wire-rope Fan Antenna Components (not to scale)

FIG. 2-12 Typical Wire-rope Fan Installation Aboard Ship. Note Safety Link In Upper Portion.

FIG. 2-13 Wire-rope Antenna Safety Link Details

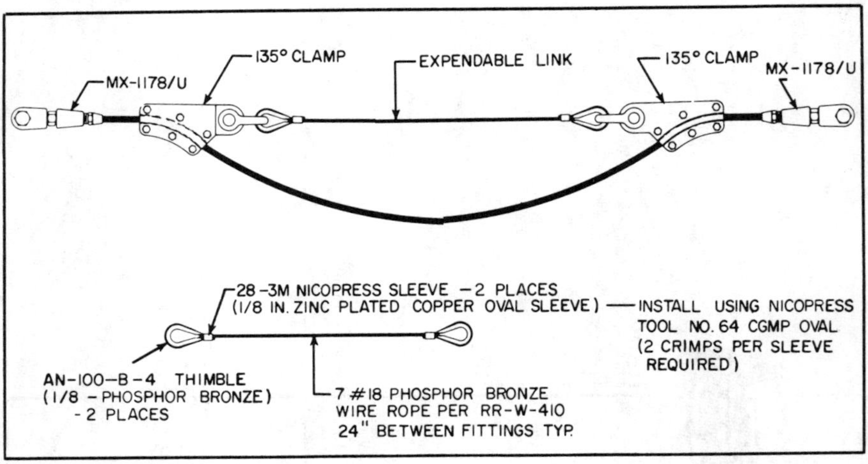
135° CLAMP
EXPENDABLE LINK
135° CLAMP
MX-1178/U
MX-1178/U
28-3M NICOPRESS SLEEVE -2 PLACES
(1/8 IN. ZINC PLATED COPPER OVAL SLEEVE)
INSTALL USING NICOPRESS TOOL NO. 64 CGMP OVAL (2 CRIMPS PER SLEEVE REQUIRED)
AN-100-B-4 THIMBLE (1/8 - PHOSPHOR BRONZE) - 2 PLACES
7 #18 PHOSPHOR BRONZE WIRE ROPE PER RR-W-410
24" BETWEEN FITTINGS TYP.

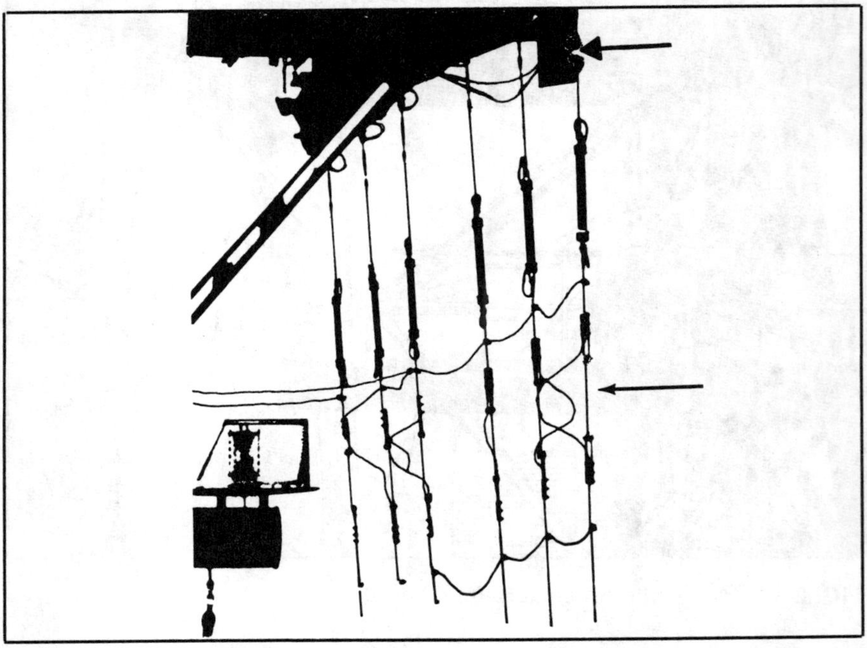

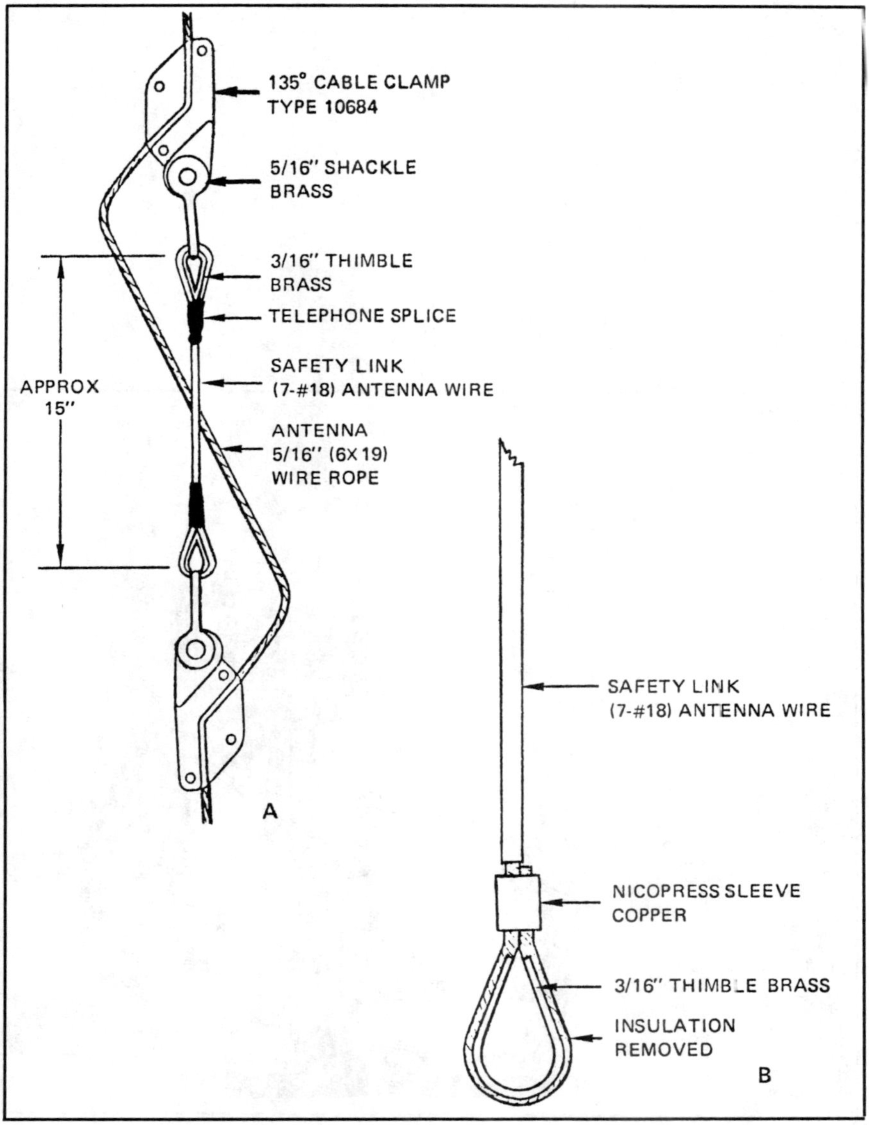

FIG. 2-14 Alternative Wire-rope Safety Link Concept

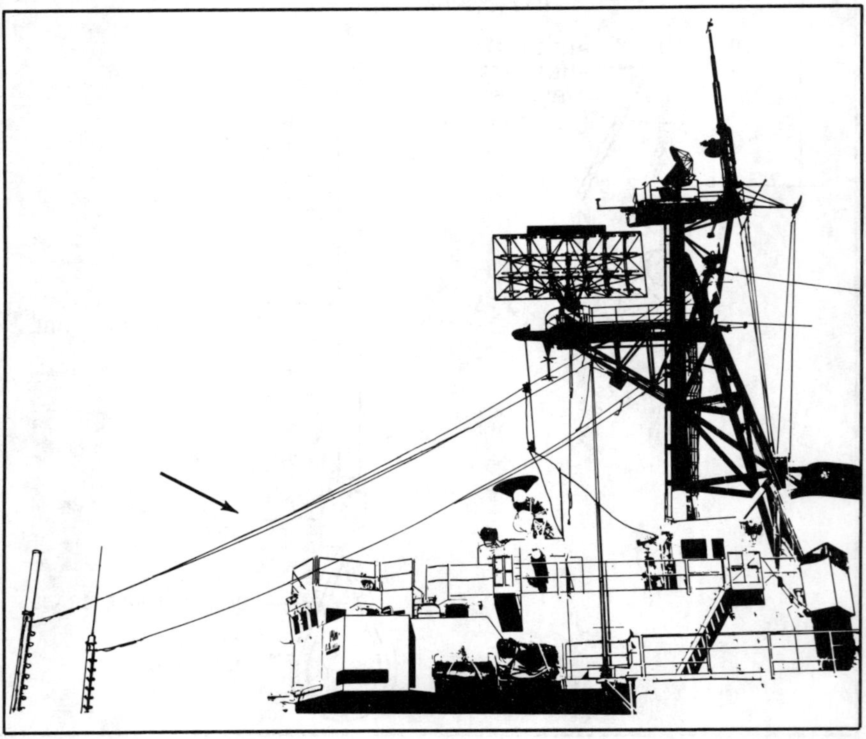

FIG. 2-15 Three-wire Fan

FIG. 2-16 Three-wire Twin Fan

FIG. 2-17 2 to 30 MHz HF Single Fan Antenna

FIG. 2-18 2 to 6 MHz Yoked Twin Fan

FIG. 2-19 Single Section Of 2 to 6 MHz Yoked Twin Fan Antenna

FIG. 2-20 4 to 12 MHz Multi-element 5-Wire HF Single Fan Antenna

FIG. 2-21 2 to 6 MHz Multi-element 15-Wire Yoked Twin Fan (foremast portion in view) *AND* 6 to 18 MHz Multi-element 15-Wire Yoked Twin Fan (center-between masts; both sections in view).

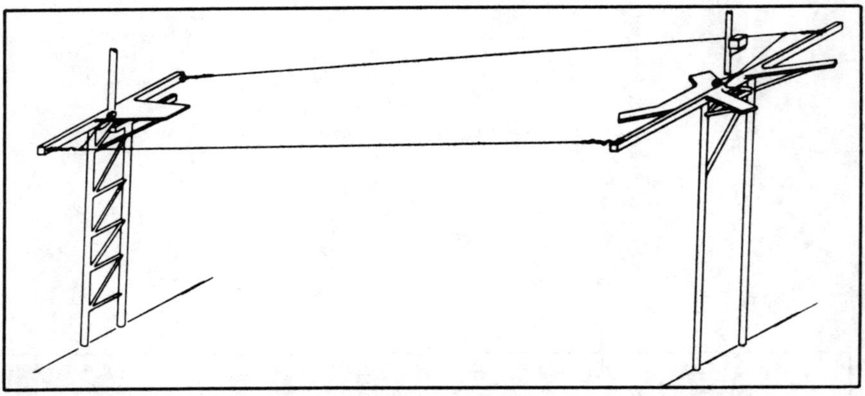

FIG. 2-22 AS-2801/SRT LF/MF Wire-rope Antenna

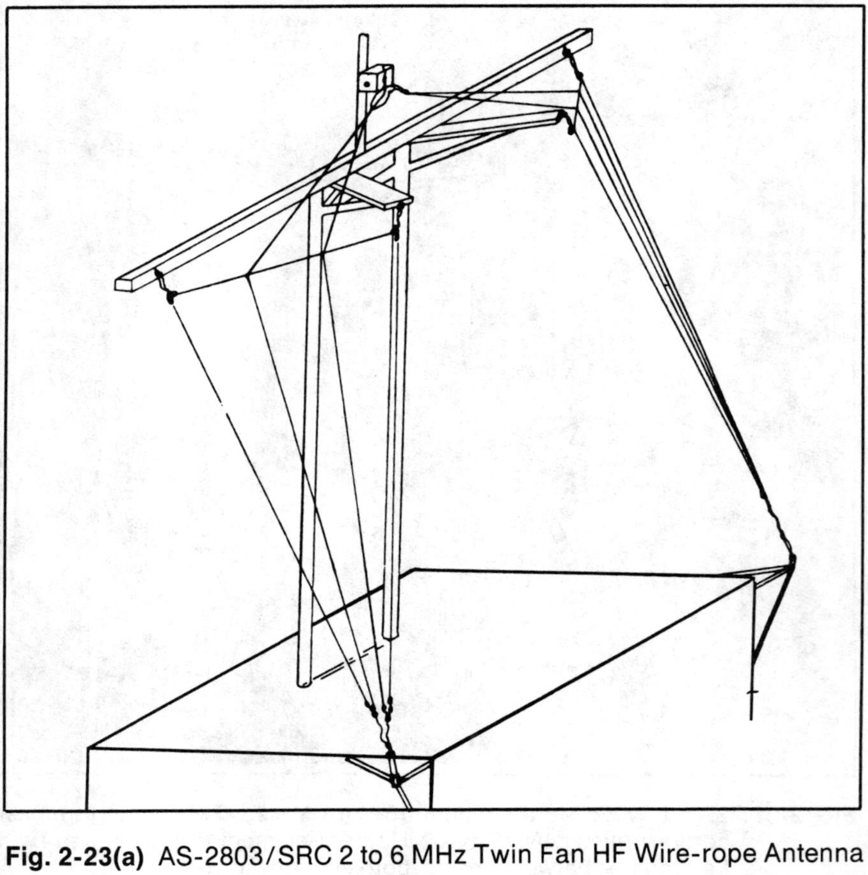

Fig. 2-23(a) AS-2803/SRC 2 to 6 MHz Twin Fan HF Wire-rope Antenna

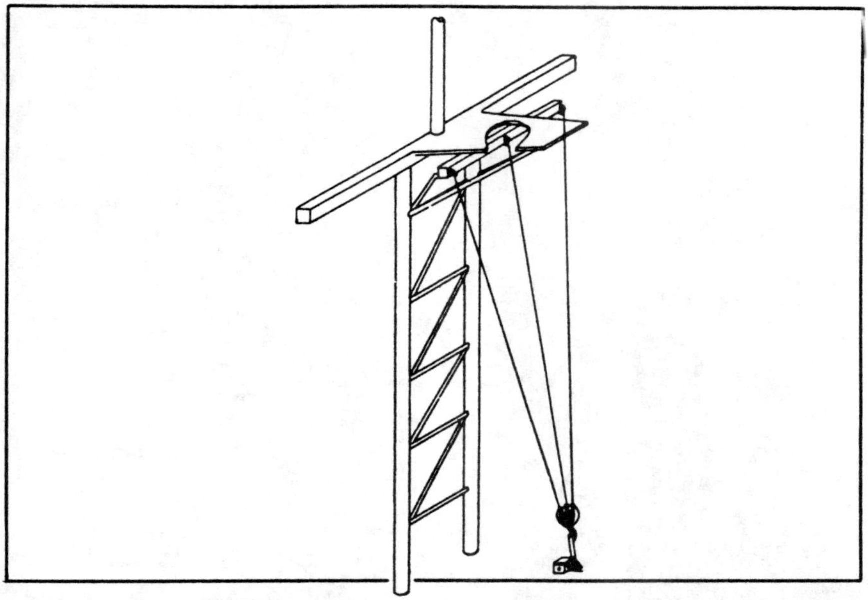

Fig. 2-23(b) AS-2804/SRR Single Fan HF Wire-rope Antenna

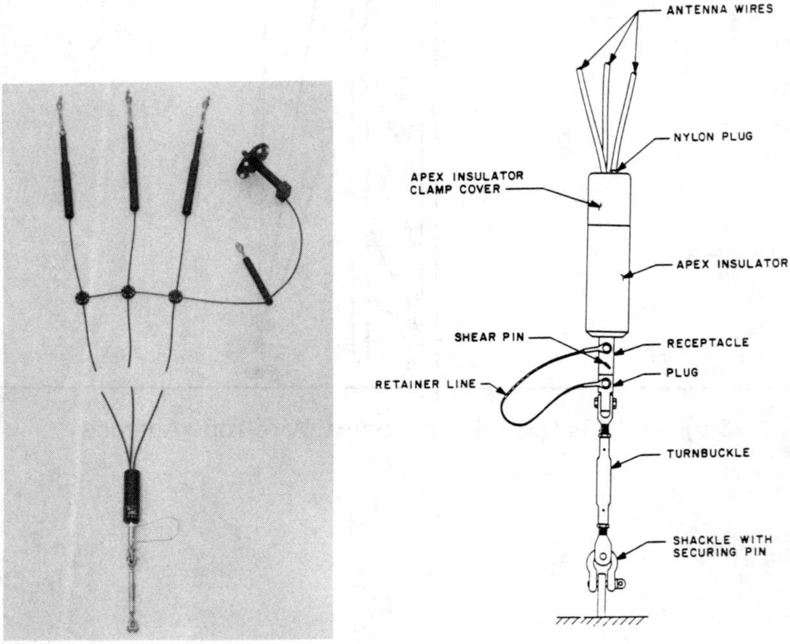

(a) (b)

Fig. 2-24 (a) New Lightweight Fan Antenna Components; (b) LWCA Apex Insulator/Shear Pin Details

Fig. 2-24(c) Lightweight Fan Antenna (LWCA) Shipboard Installation

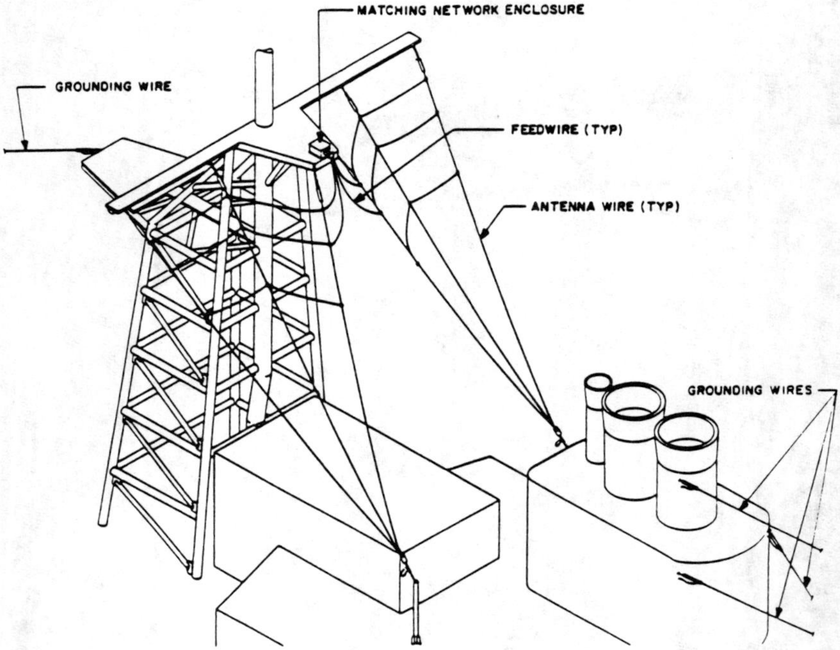

Fig. 2-24(d) AS-3727(V) Lightweight Fan Antenna (LWCA)

FIG. 2-25 35-Foot Whip Monopoles Along Carrier Deck Edge

FIG. 2-26 35-Foot Whip Monopoles On Superstructure

FIG. 2-27 NT-66046 28-Foot Aluminum Whip

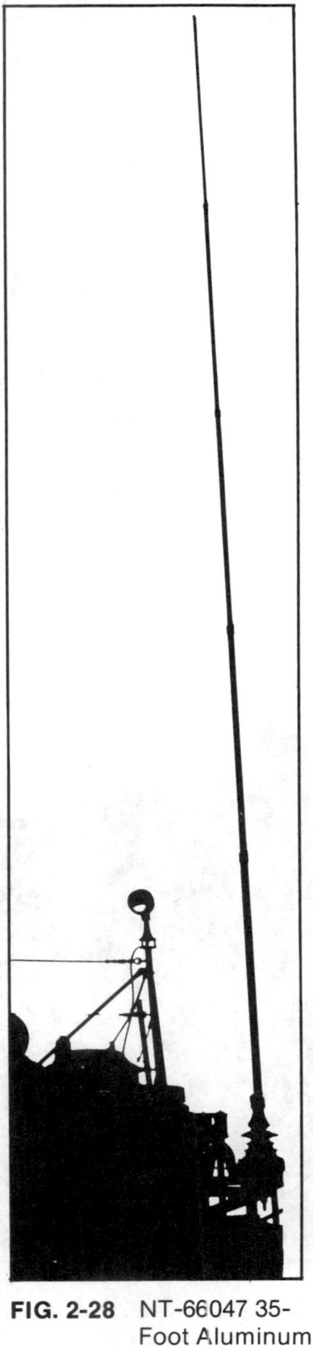

FIG. 2-28 NT-66047 35-Foot Aluminum Whip

FIG. 2-29 AS-2537 35-Foot Fiberglass Whip with Enlarged View Of Base Construction

FIG. 2-30 AS-2537A 35-Foot Fiberglass Whip with Enlarged View Of Base Construction

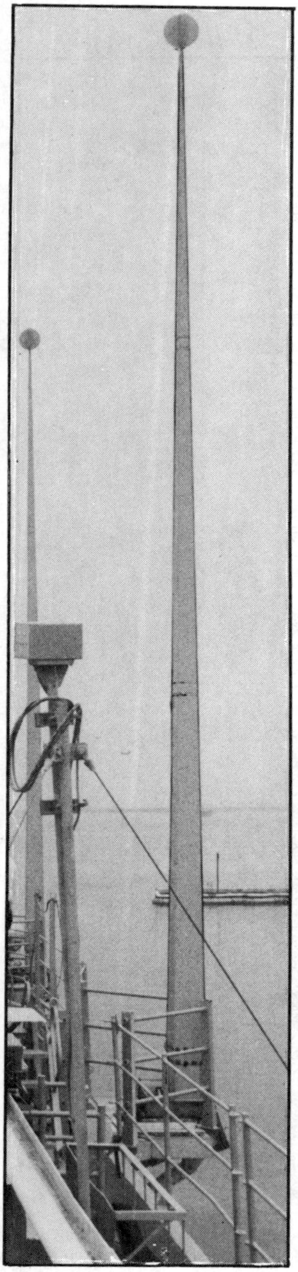

FIG. 2-31 AS-2807 35-Foot Aluminum Whip

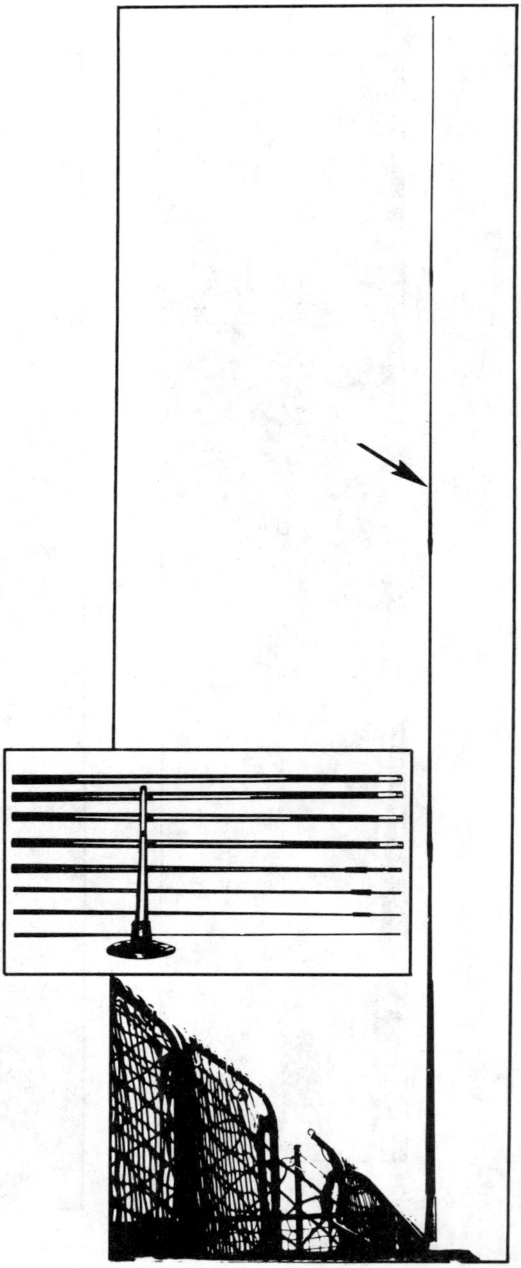

FIG. 2-32 AT-1011 and AT-1047 35-Foot Fiberglass Thin Whips

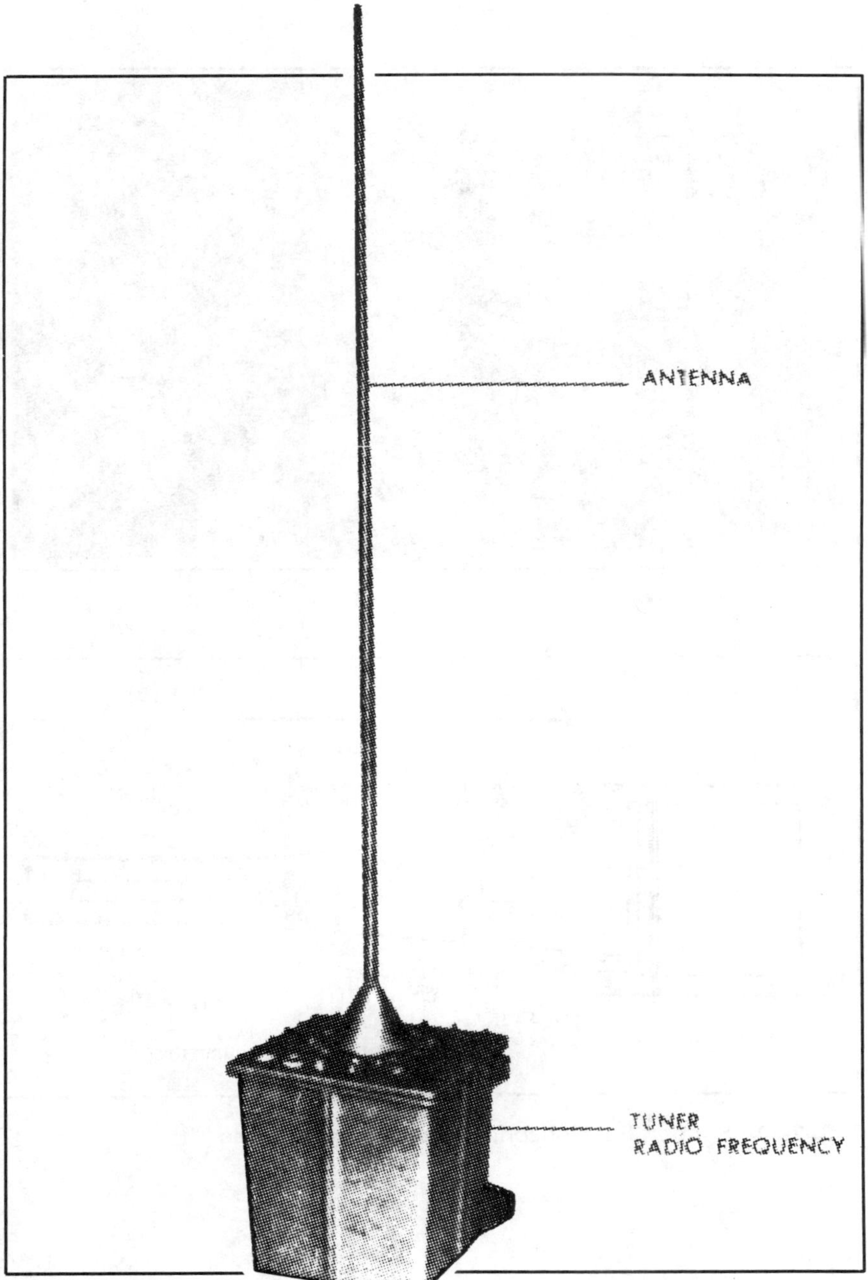

FIG. 2-33 AS-1857 5-Foot Fiberglass Tuner Whip

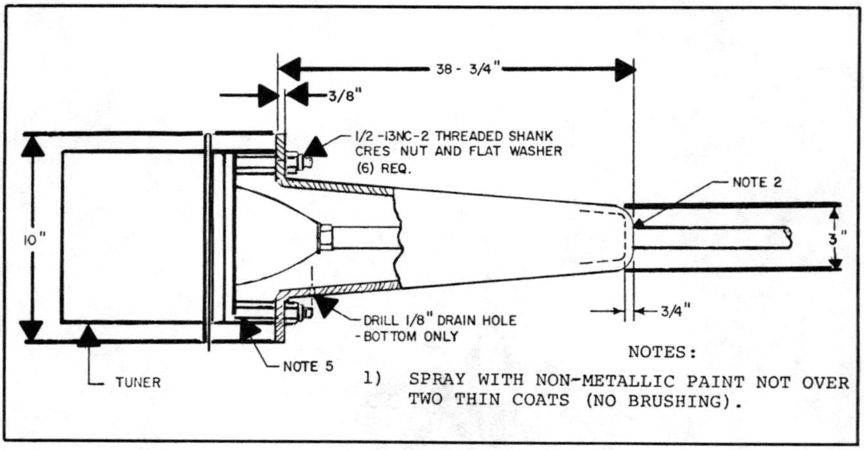

FIG. 2-34 Support For Horizontal Tuner Whip Antennas

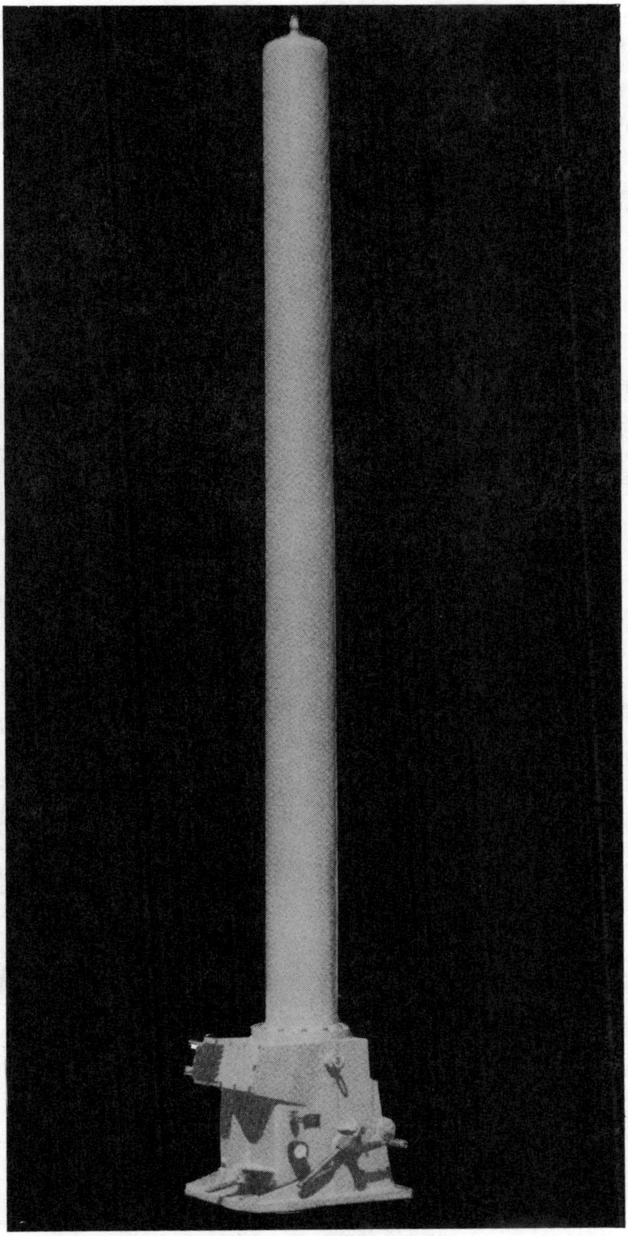

FIG. 2-35 AS-2875 Helical Monopole

FIG. 2-36 AS-2876 Helical Monopoles

FIG. 2-37 35-Foot 3-Wire Open-tip Trussed Whip

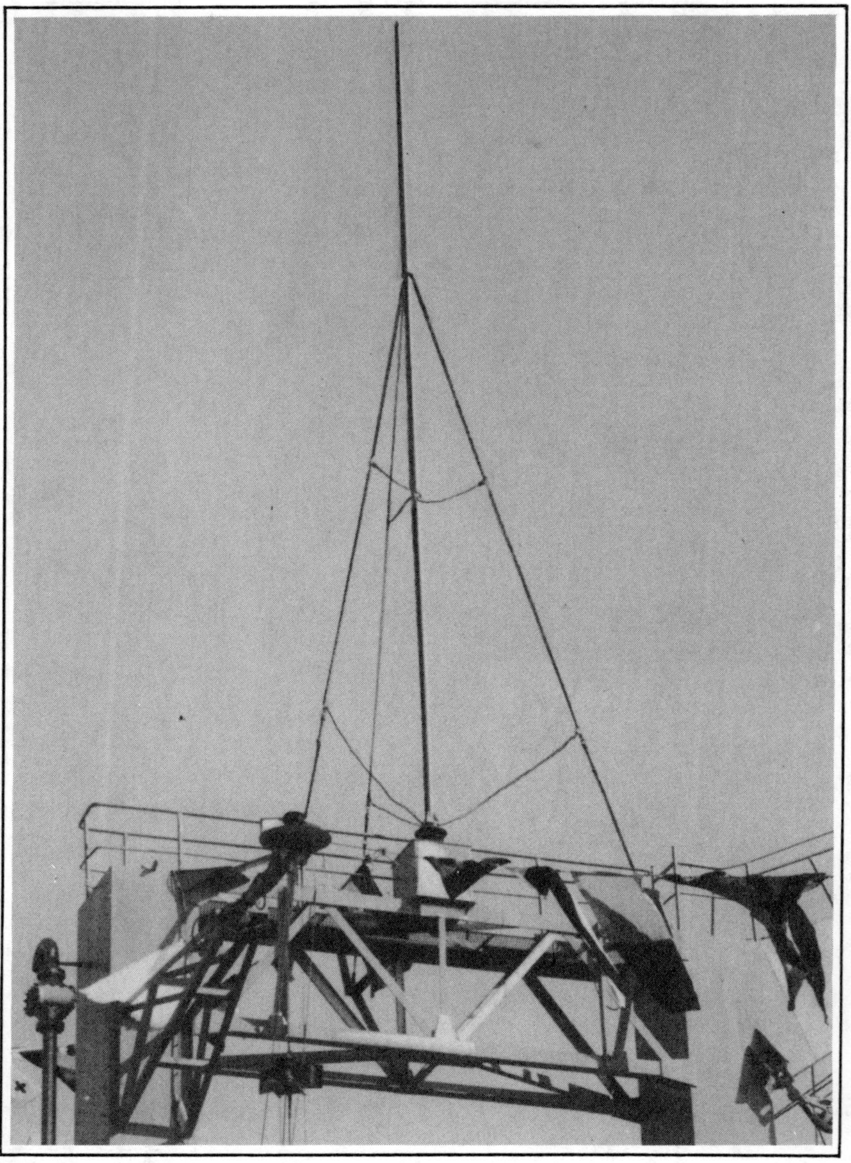

FIG. 2-38 35-Foot 3-Wire Open-tip Trussed Whip

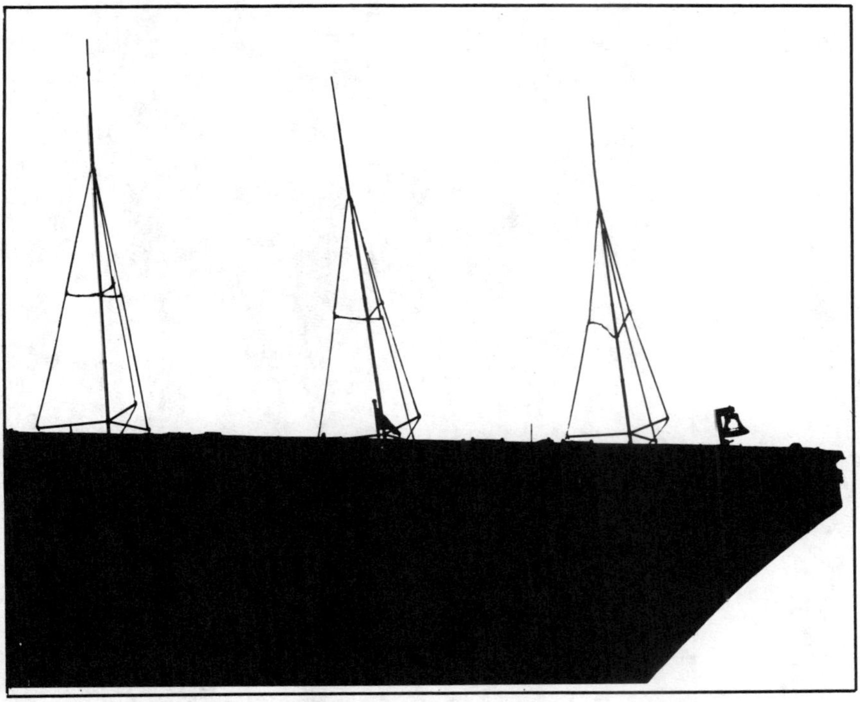

FIG. 2-39 35-Foot 3-Wire Open-tip Trussed Whips

FIG. 2-40 4-Wire Open-tip Trussed Whip

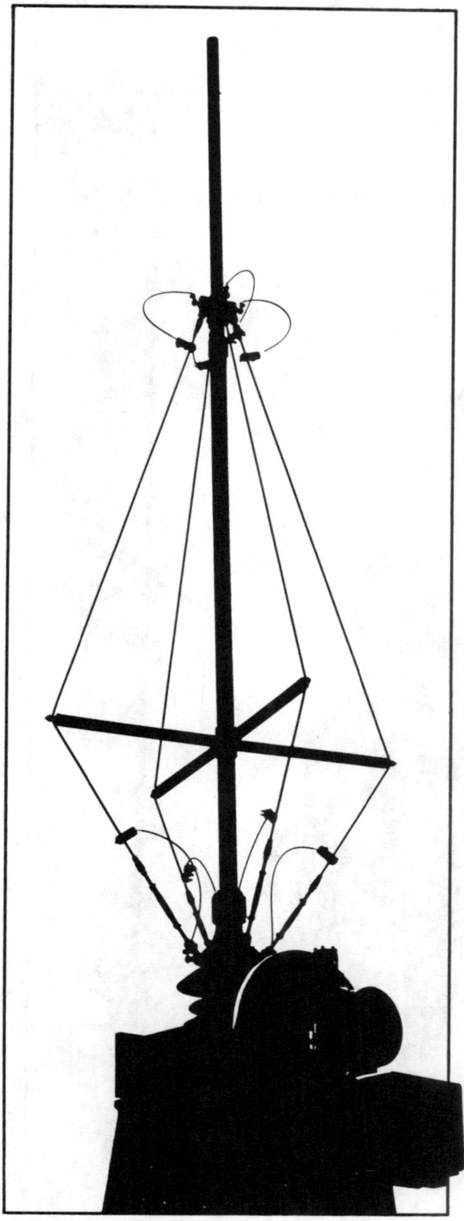

FIG. 2-41 4-Wire Open-tip Trussed Whip

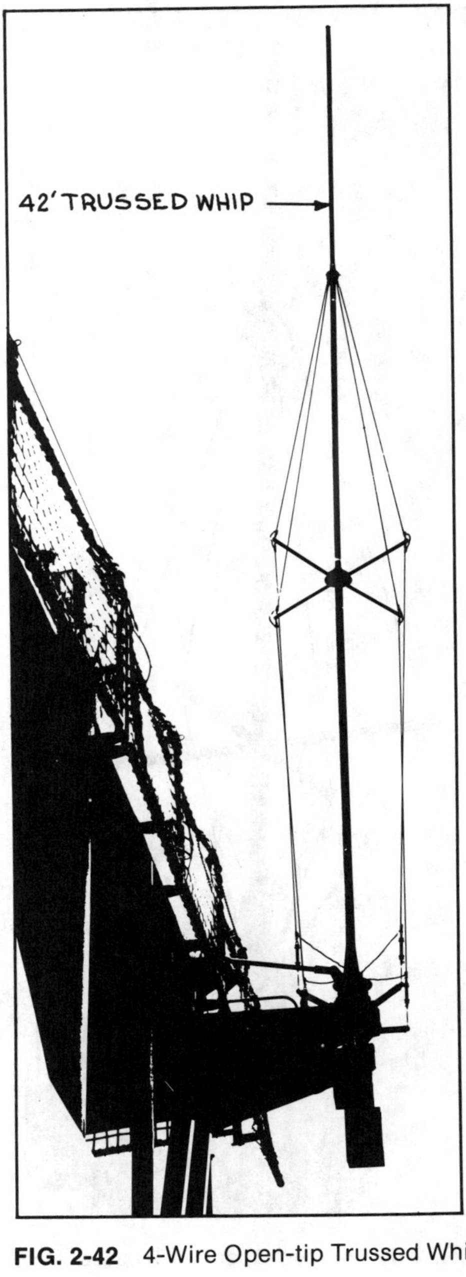

FIG. 2-42 4-Wire Open-tip Trussed Whip

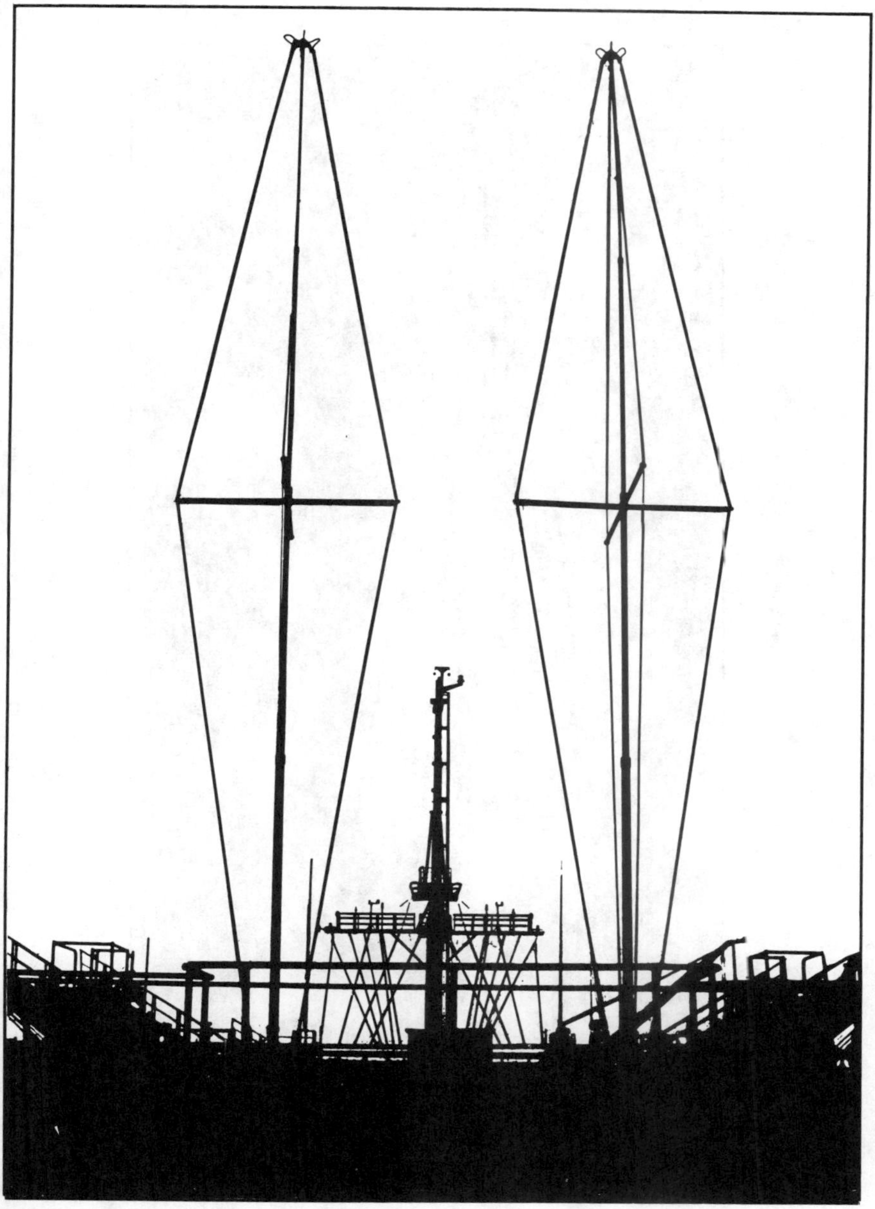

FIG. 2-43 4-Wire Closed-tip Trussed Whips

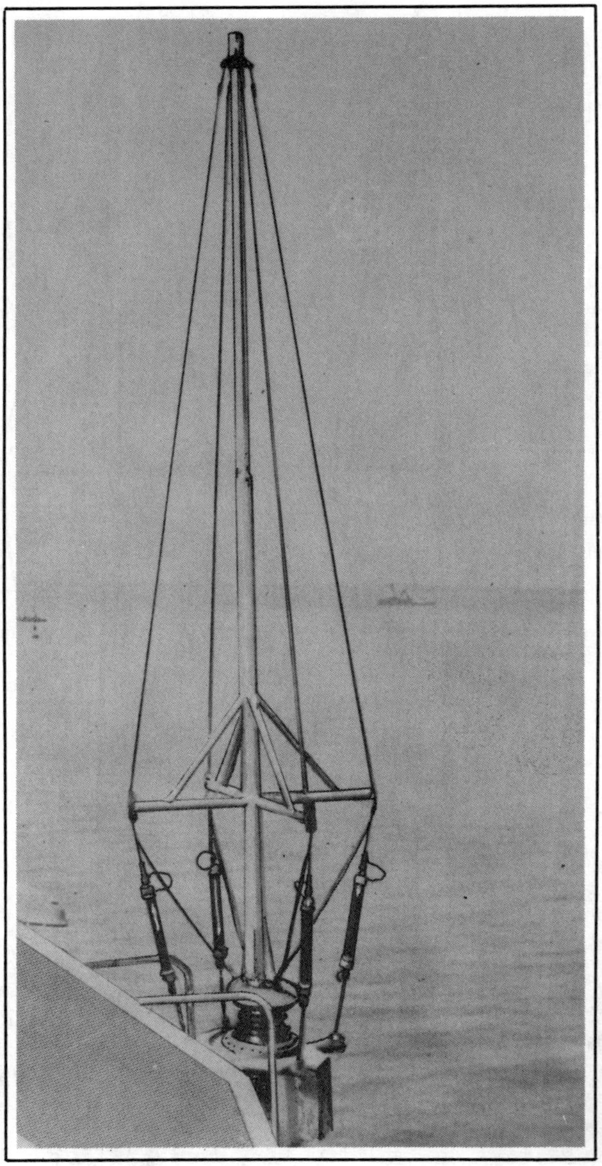

FIG. 2-44 4-Wire Closed-tip Trussed Whip

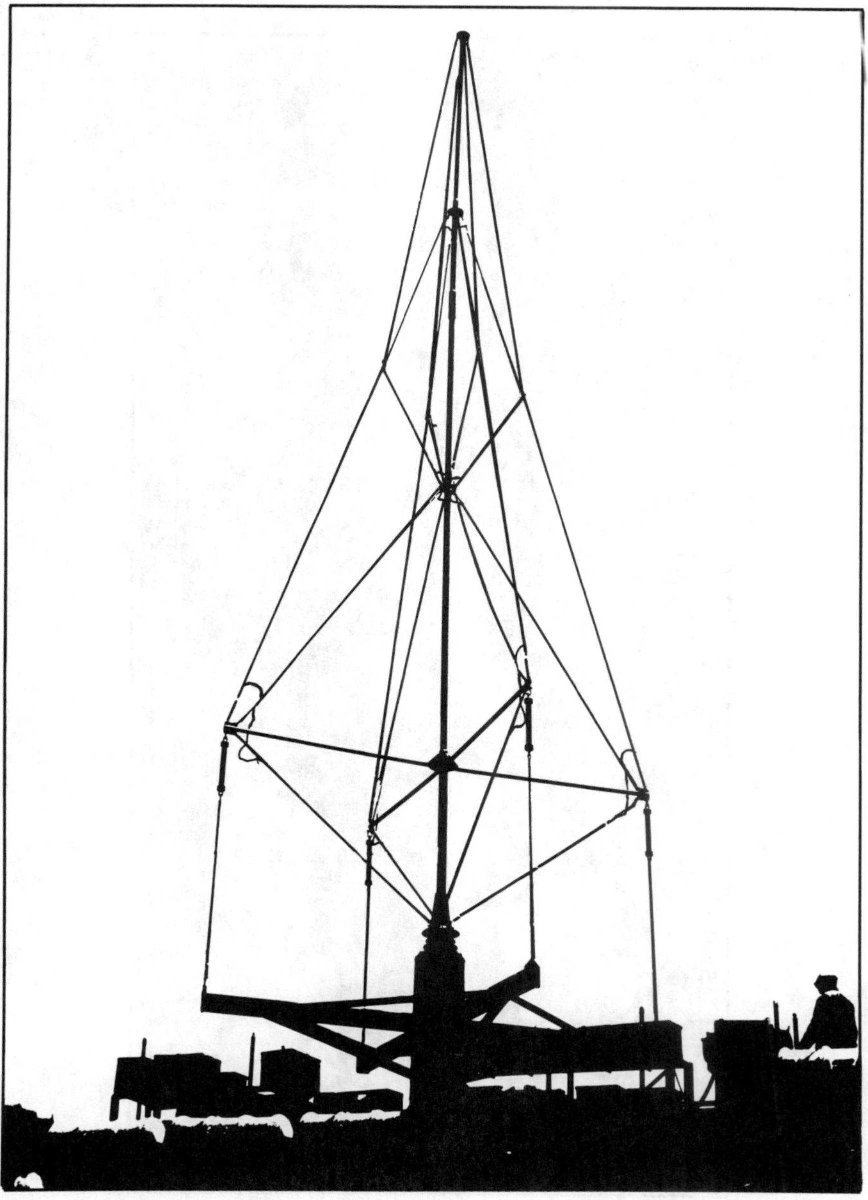

FIG. 2-45 4-Wire Closed-tip Trussed Whip

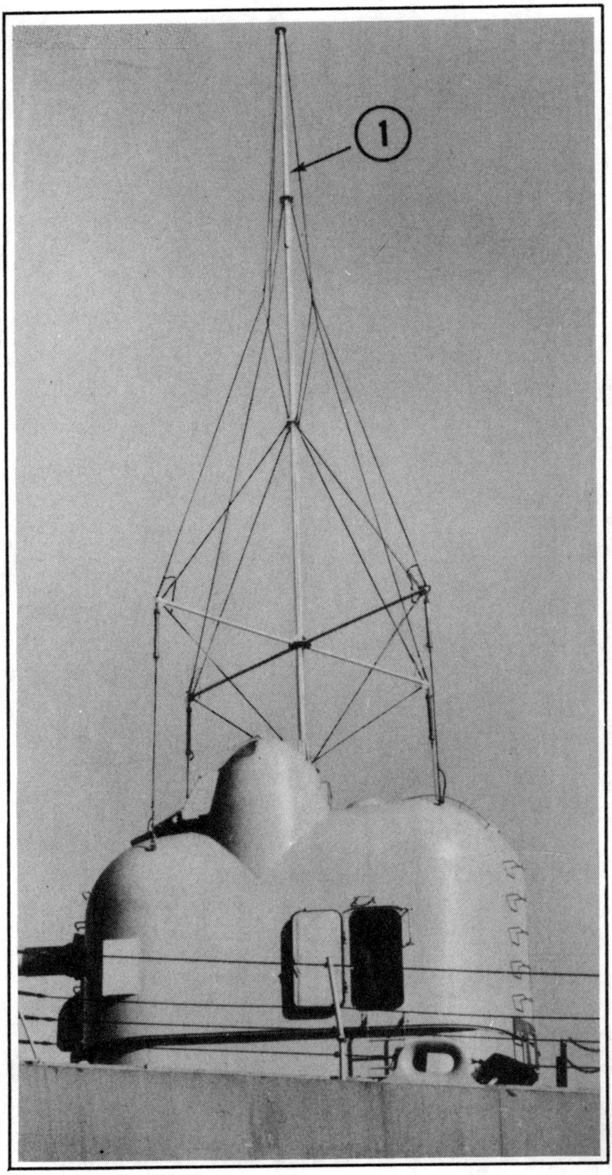

FIG. 2-46 4-Wire Closed-tip Trussed Whip

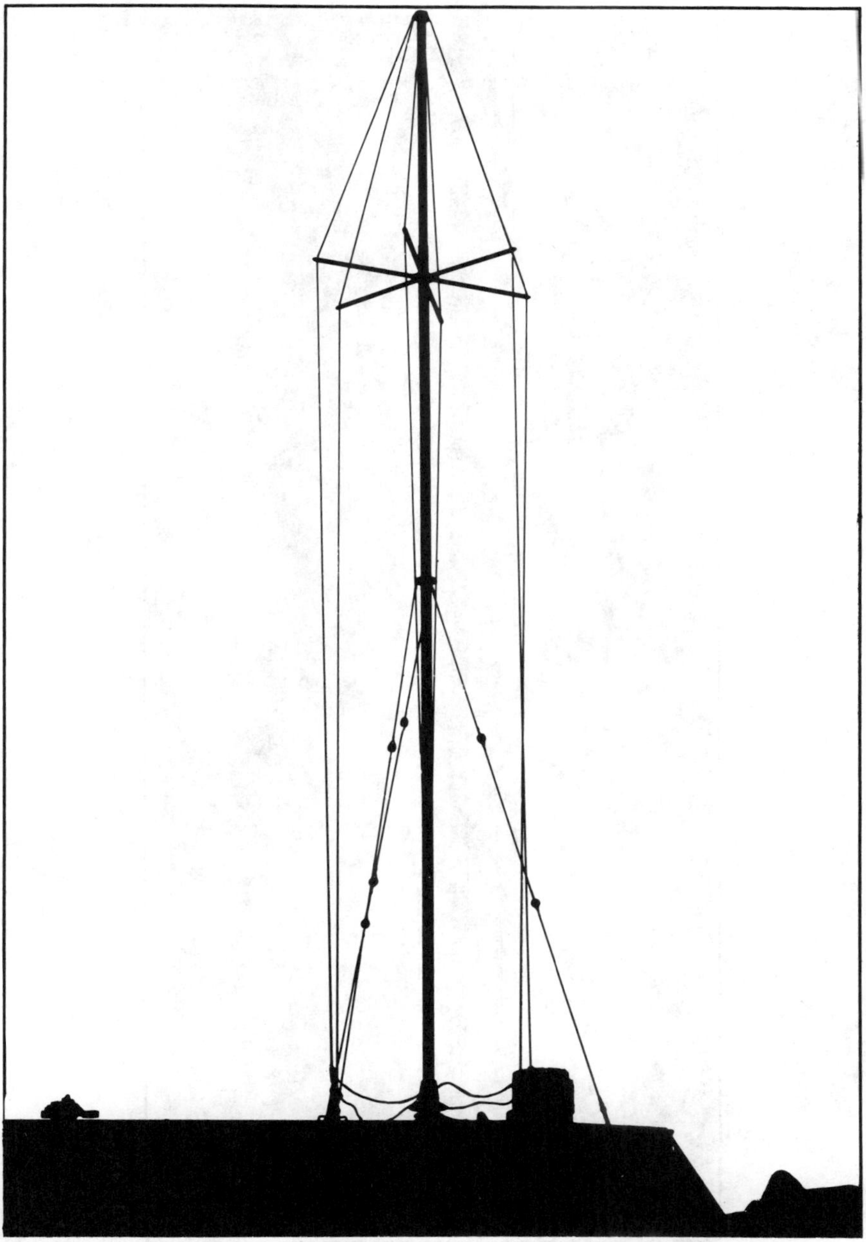

FIG. 2-47 6-Wire Closed-tip Trussed Monopole

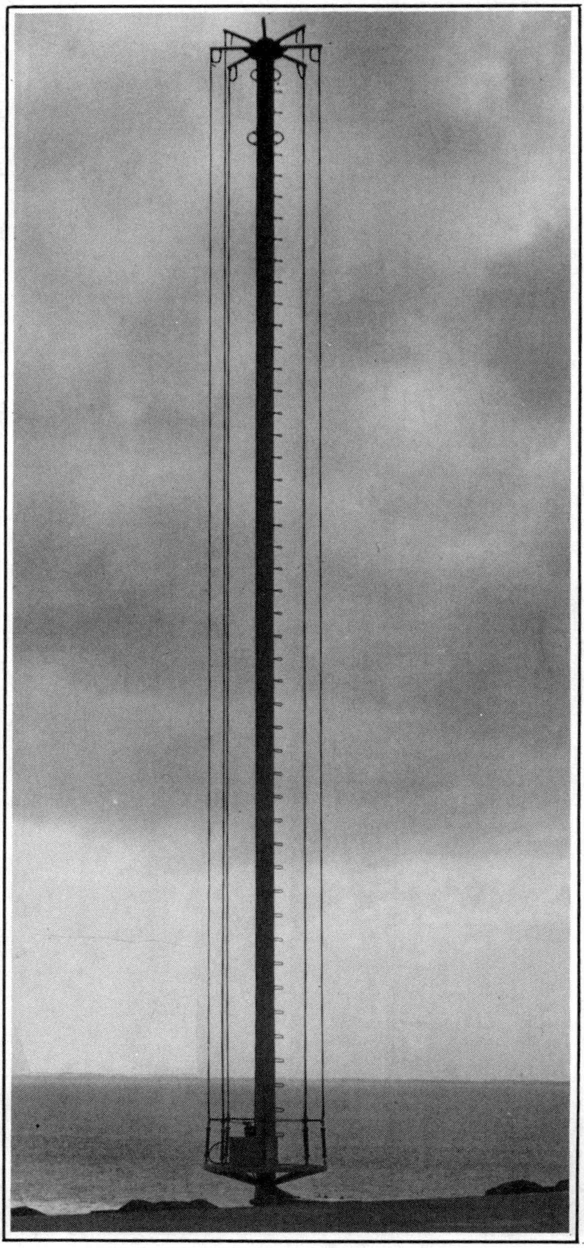

FIG. 2-48 Cylinder Cage Monopole

FIG. 2-49 Multi-element Broadband Monopole

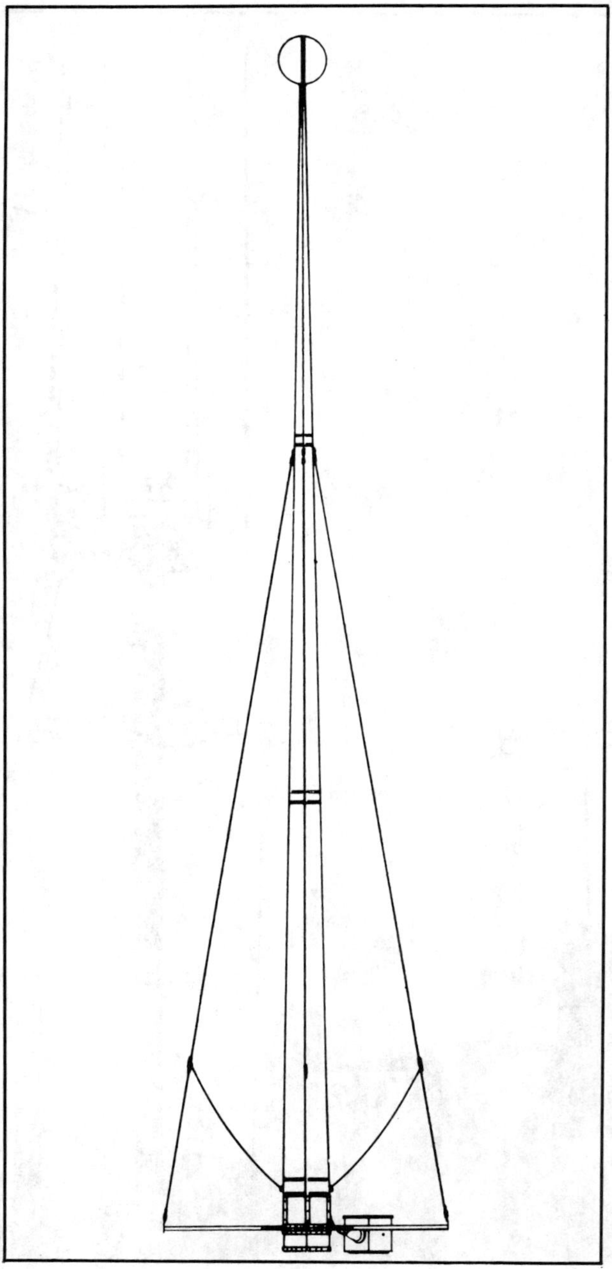

FIG. 2-50 AS-2805 35-Foot Trussed Whip

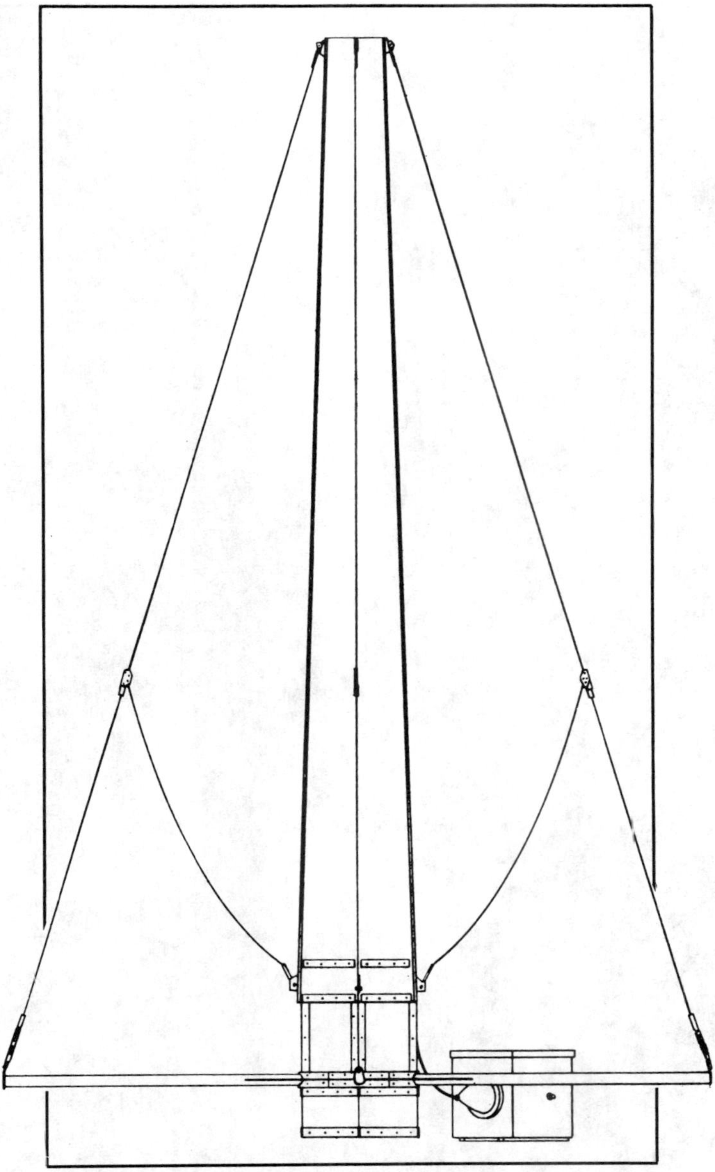

FIG. 2-51 AS-2806 13-Foot Trussed Whip

FIG. 2-52(a) 35-Foot Twin Aluminum (AS-2807) Whip

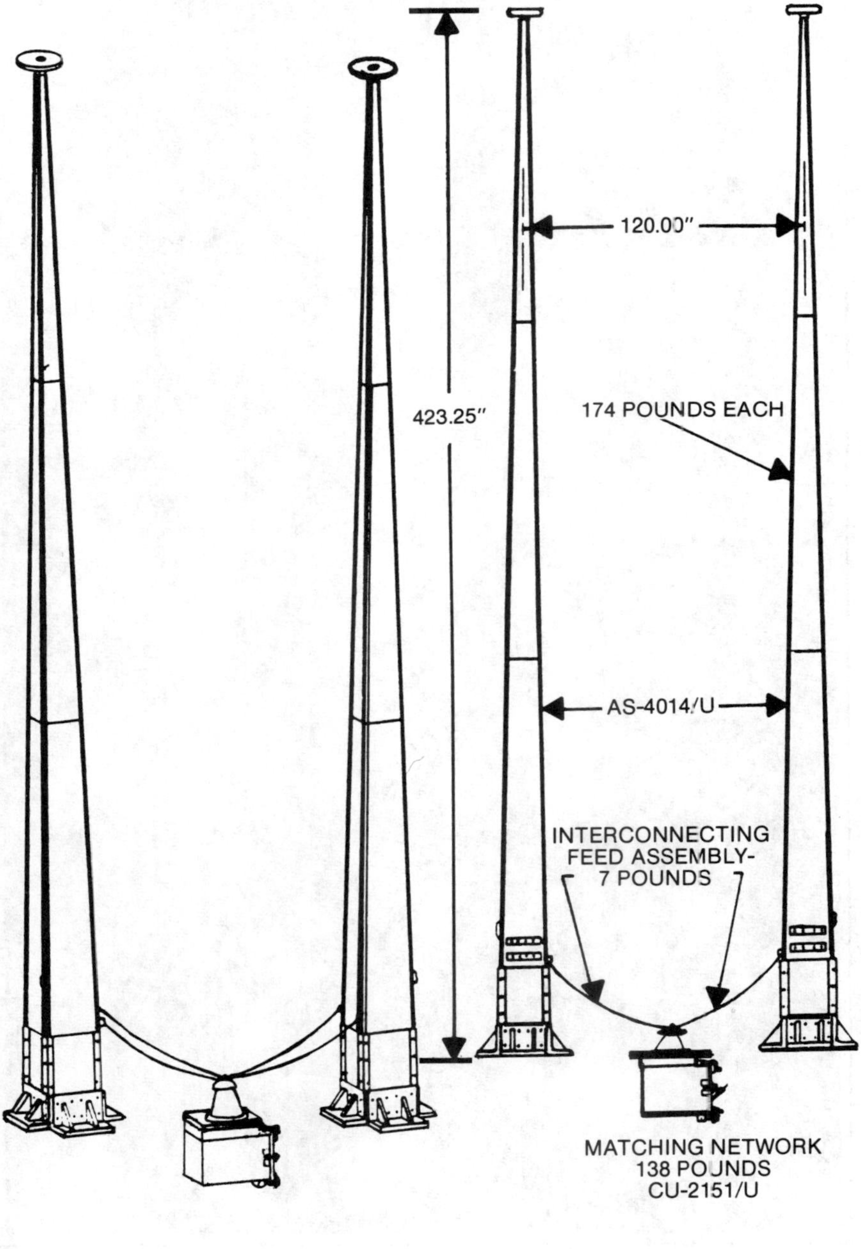

FIG. 2-52(b) OE-214/U 35-Foot Twin Whip Assembly (Figure courtesy of Chu Associates)

FIG. 2-53 35-Foot Twin Fiberglass (AS-2537A) Whip

FIG. 2-54 35-Foot Twin (NT-66047 type) Whip In Permanent Tilt

FIG. 2-55 35-Foot Twin Trussed (NT-66047) Whips

FIG. 2-56(a) AS-4037 Stubby Twin Whip

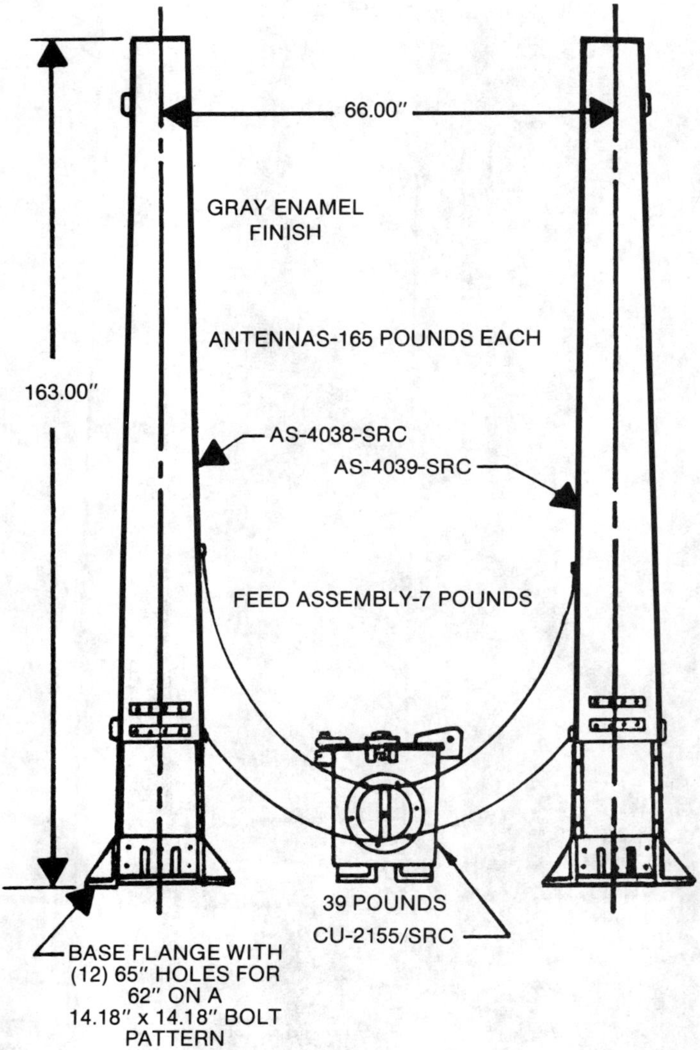

FIG. 2-56(b) AS-4037 Stubby Twin Whip Antenna

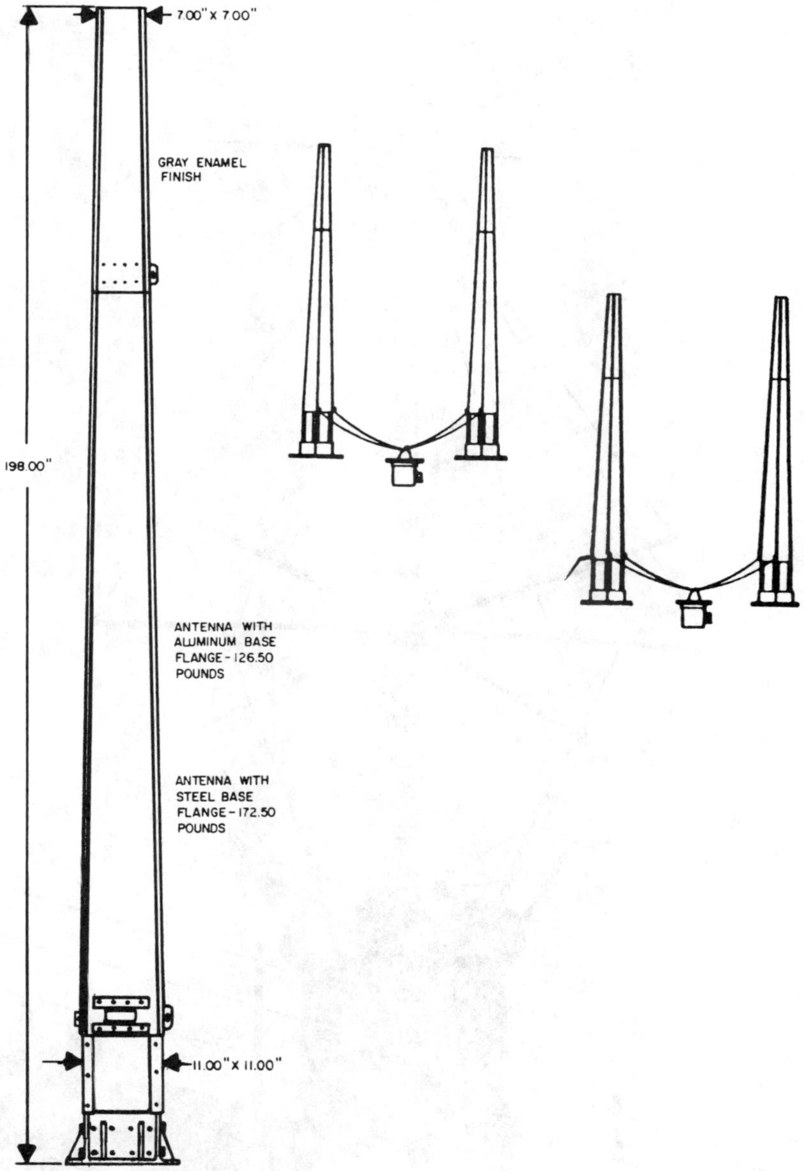

FIG. 2-56(c) OE-213/U 10 to 30 MHz Twin- Whip Dual Array (Photo courtesy of Chu Associates)

FIG. 2-57 HF Discage Antenna

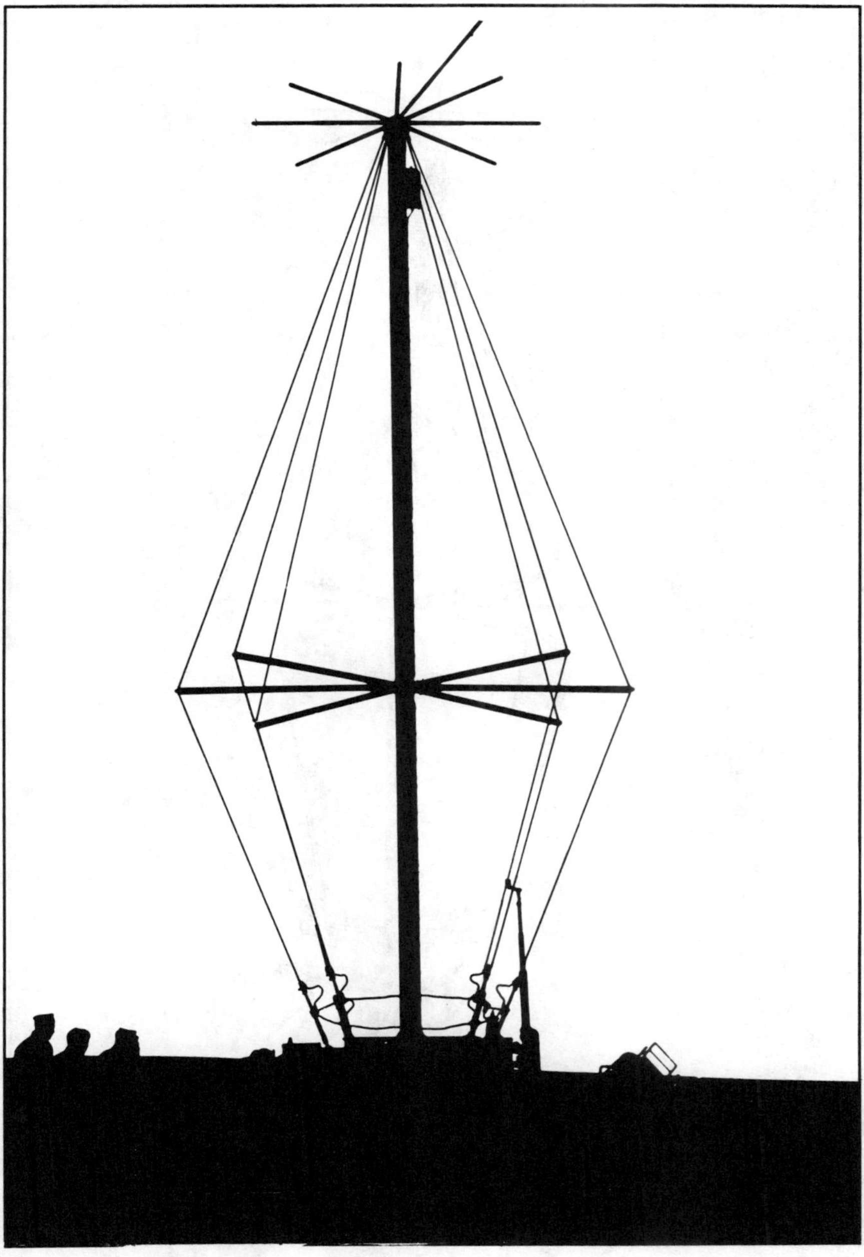

FIG. 2-58 HF Discage Antenna

FIG. 2-59 HF Discage Antenna

FIG. 2-60 HF Discage Antenna

FIG. 2-61 HF Discage Antenna

FIG. 2-62 HF Discage Antenna

FIG. 2-63 HF Discage Antenna

FIG. 2-64 AS-2802 HF Discage Antenna

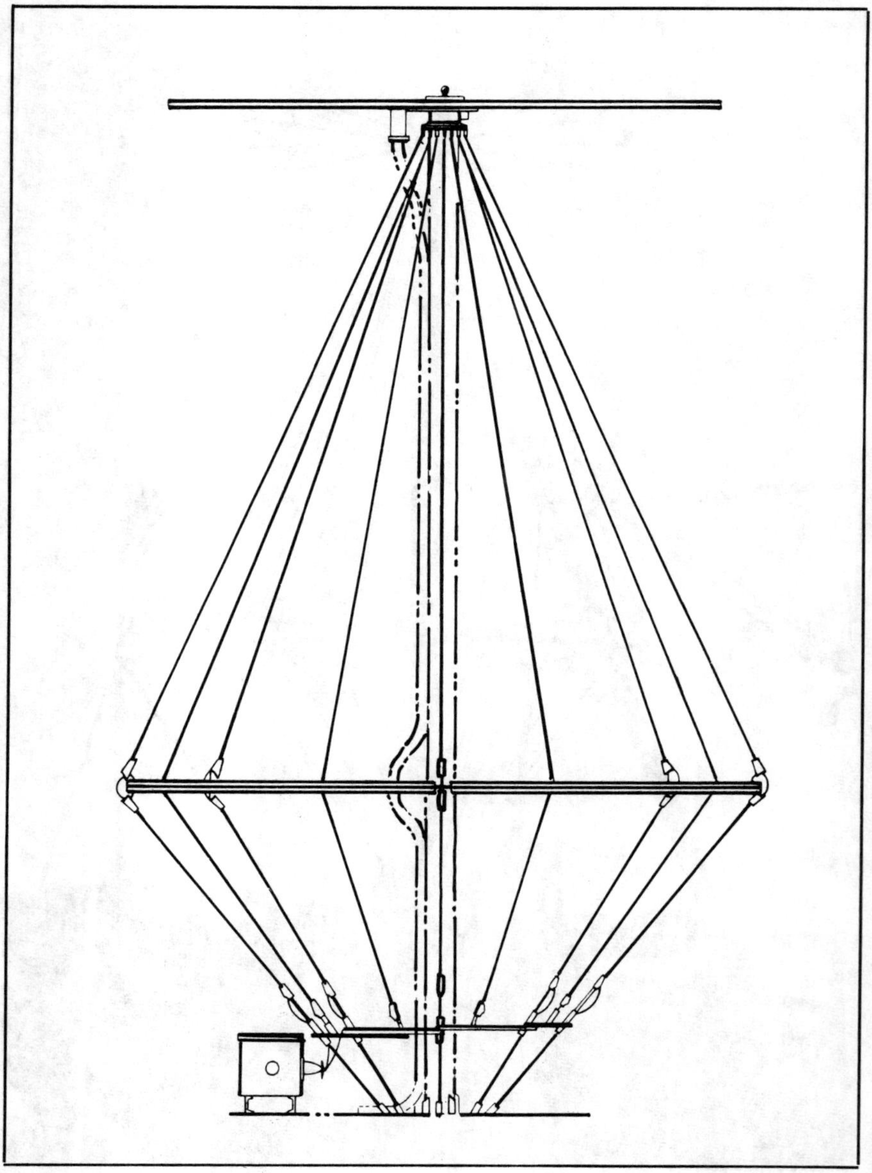

FIG. 2-65 AS-2802 HF Discage Antenna

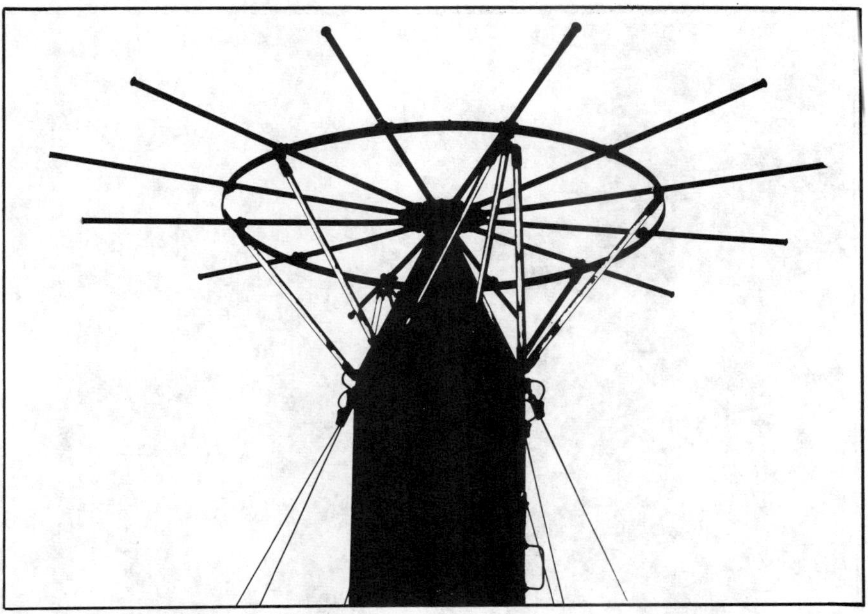

FIG. 2-66 HF Discone Antenna

FIG. 2-67 HF Discone Antenna

FIG. 2-68 HF Discone Antenna

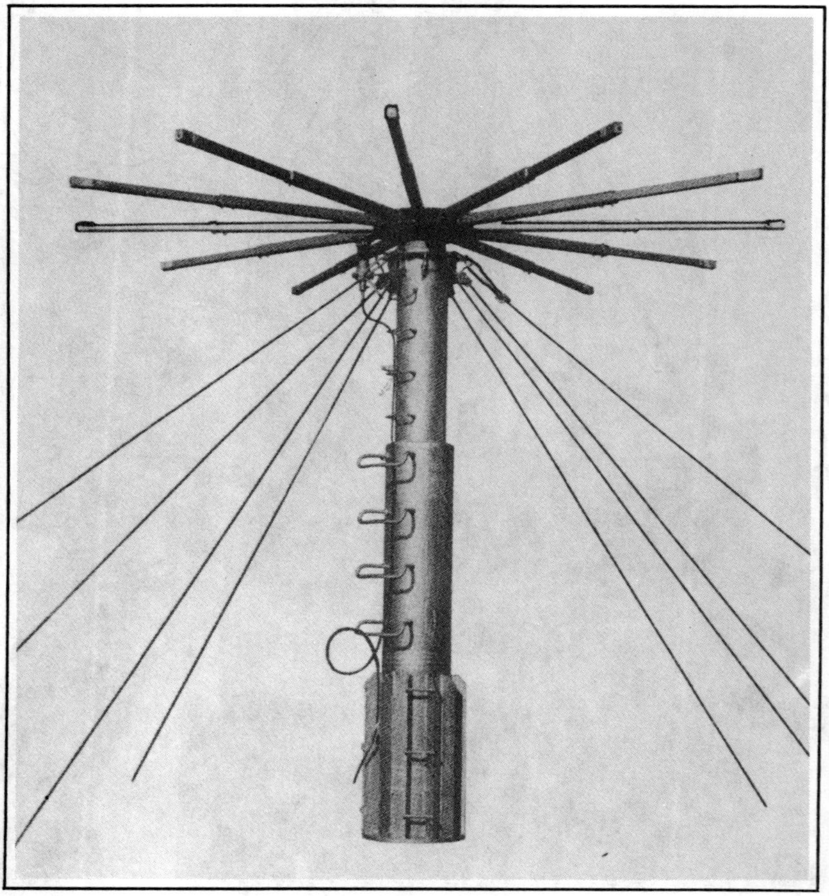

FIG. 2-69 AS-2865/SRC HF Discone Antenna

FIG. 2-70 HF Inverted Cone Antenna

FIG. 2-71 HF Inverted Cone Antenna

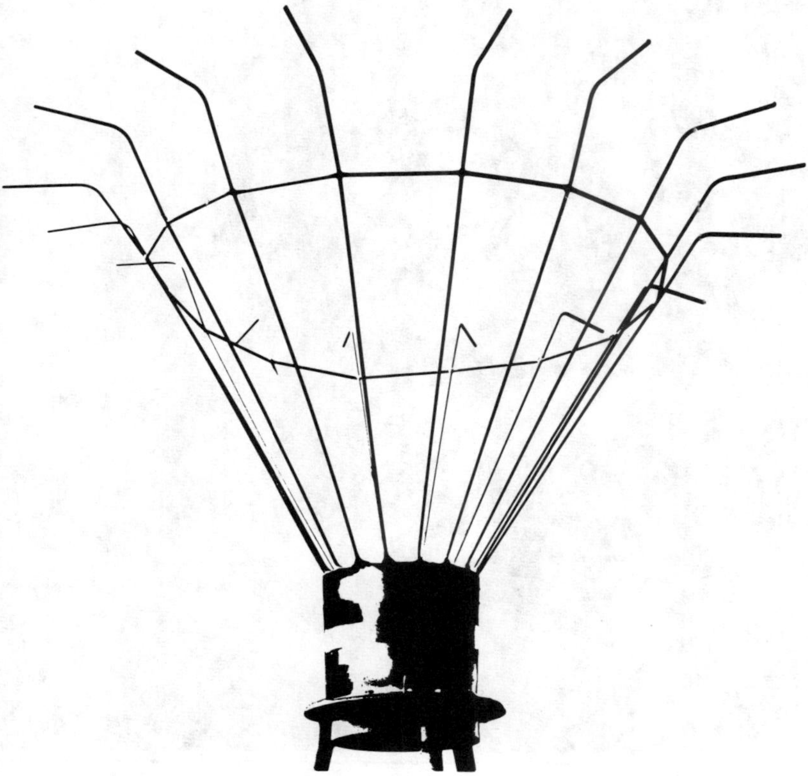

FIG. 2-72 AS-2866 HF Inverted Cone Antenna

Fig. 2-73 AS-2071 Minaject HF Antenna

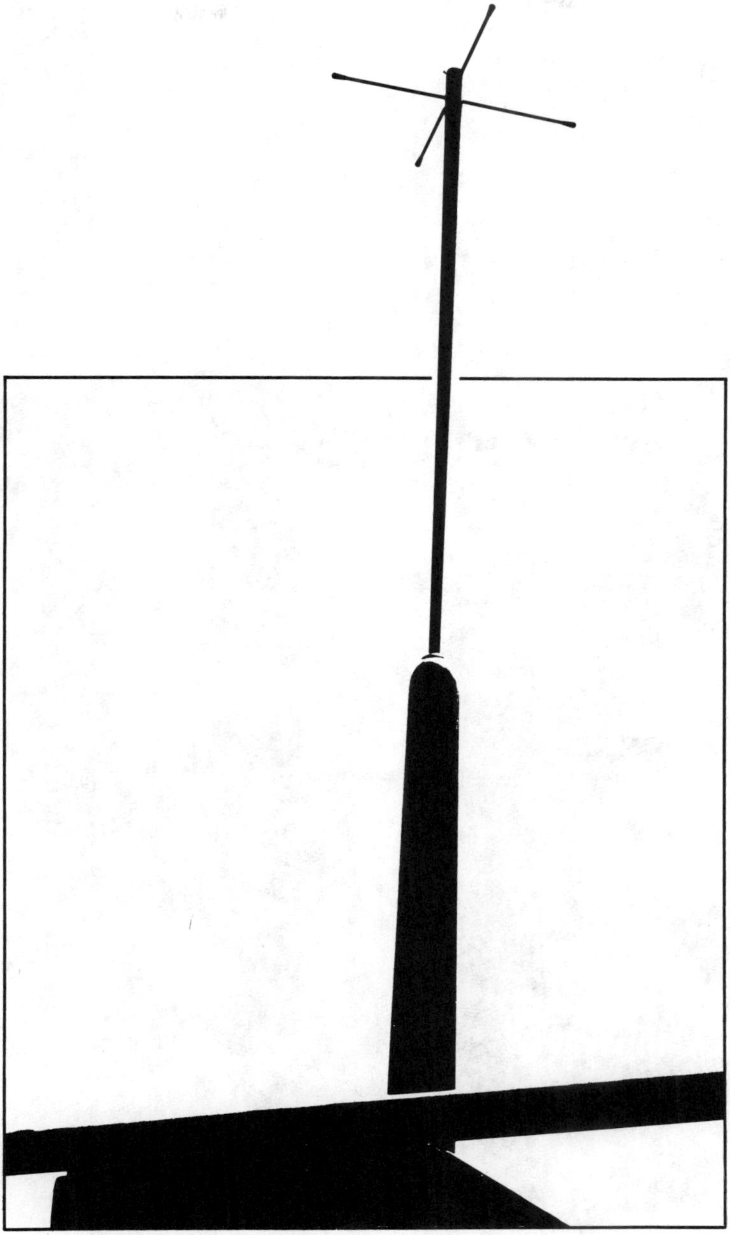

FIG. 2-74 AS-2071 Minaject HF Antenna

FIG. 2-75 AT-924/SRA-17 LF/MF 10-Foot Tuner Whips

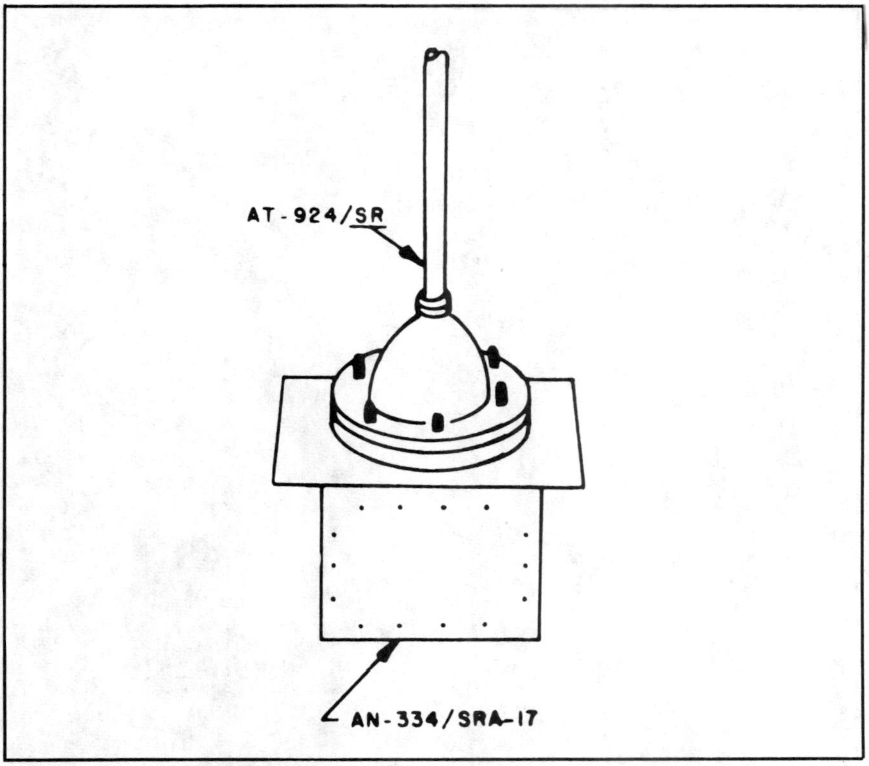

FIG. 2-76 Antenna AT-924A/SR and AT-924B/SR Tuner Whips

FIG. 2-77 Miniloop Antenna

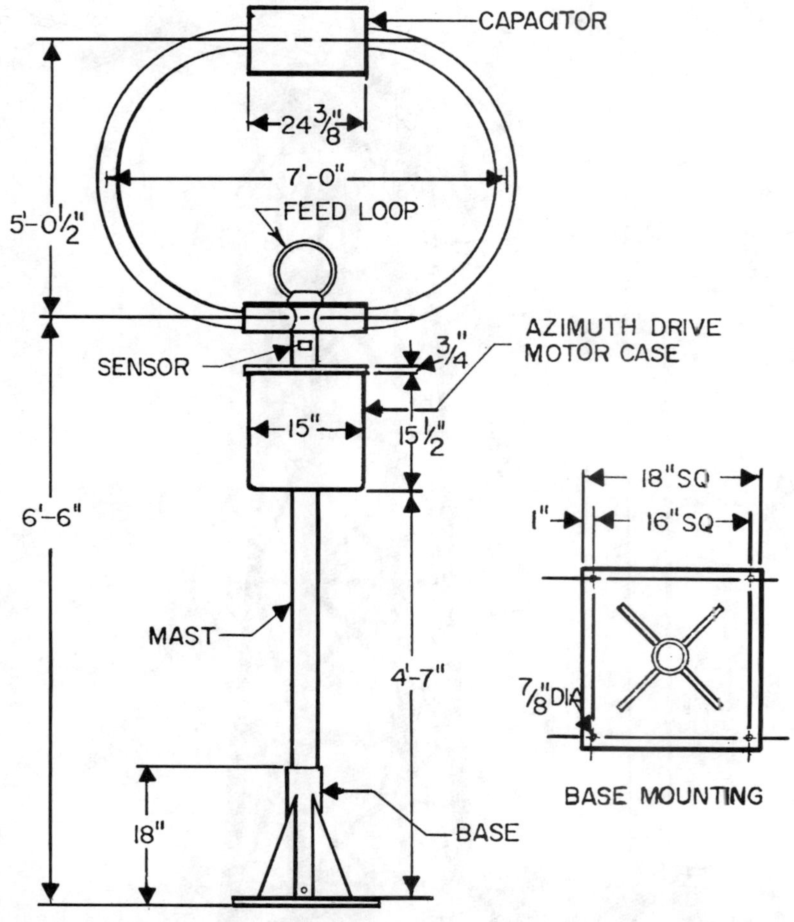

FIG. 2-78 Outline Dimensions Of MLA-1B Miniloop Antenna and Base Mount
(Photo courtesy of Antenna Research Associates)

FIG. 2-79 Shipborne Rotatable Log-Periodic HF Antenna

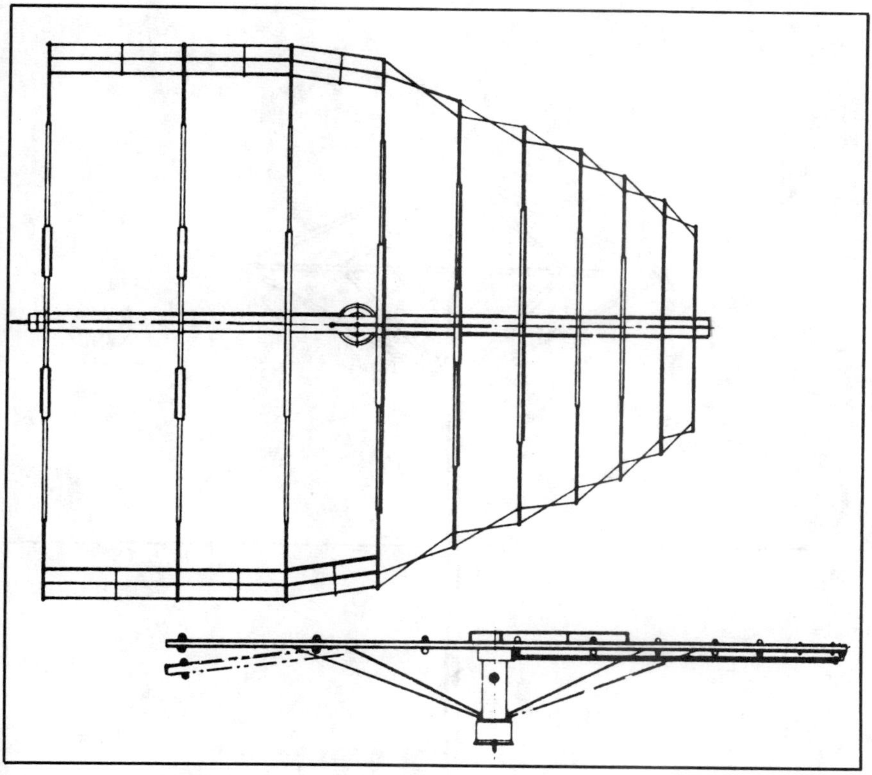

FIG. 2-80 AS-2874/SRC Rotatable Log-Periodic Antenna

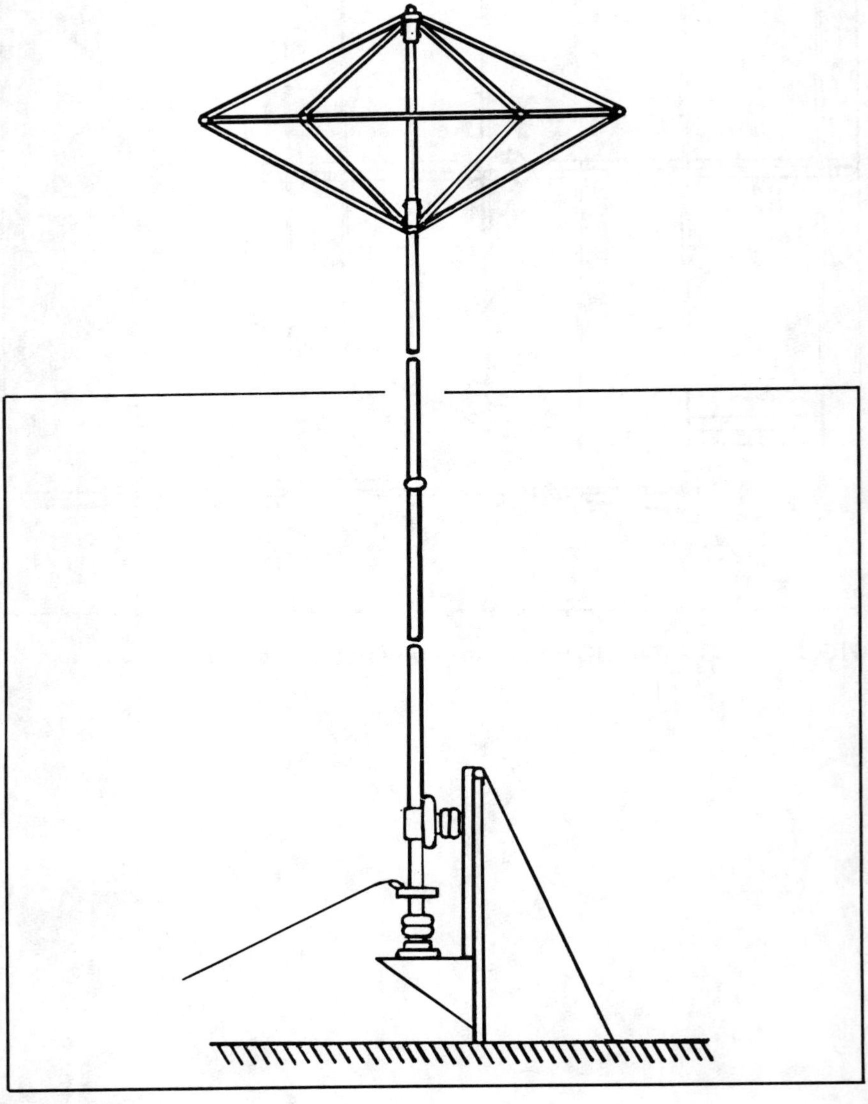

FIG. 2-81 MR106-38A Vertical Marine Antenna
(Photo courtesy of ITT Mackay Marine)

FIG. 2-82 AS-1729 VHF Whip Antennas

FIG. 2-83 AS-1729 VHF Whip Antenna Installed Properly With Ground Plane (Counterpoise) Radials

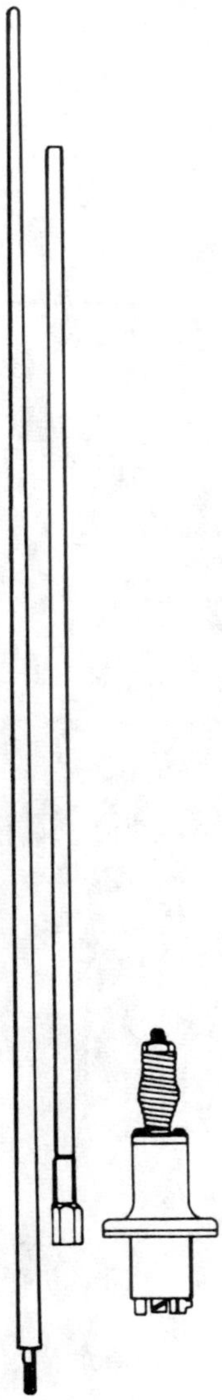

FIG. 2-84 Antenna AS-1729/VRC, Physical Configuration

FIG. 2-85 NT-66095 VHF Vertical Dipole

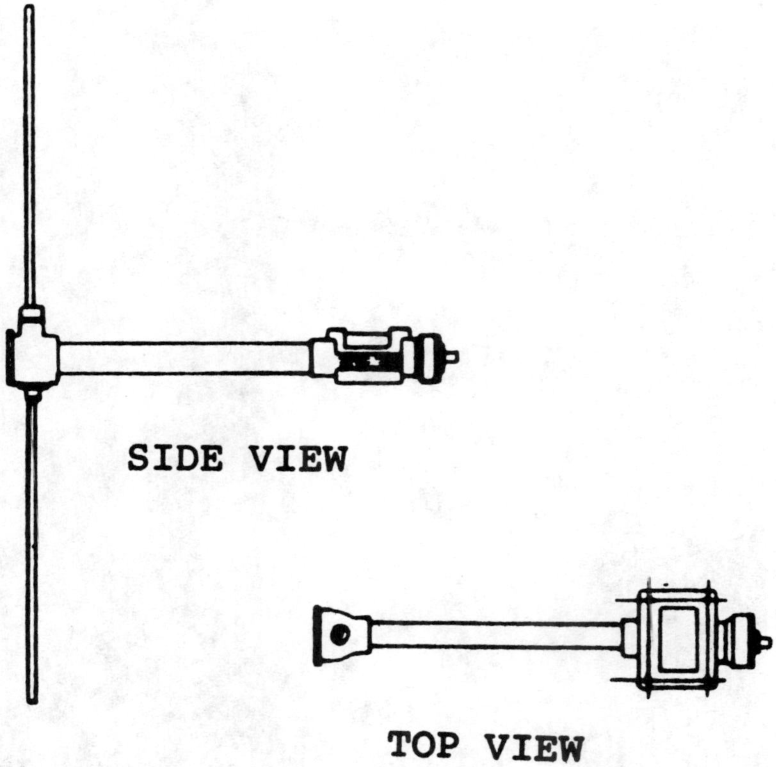

FIG. 2-86 NT-66095 VHF Vertical Dipole Assembly

FIG. 2-87 AS-2231 VHF Biconical Dipole (deck mounted)

FIG. 2-88 AS-2231 VHF Biconical Dipole (yardarm mounted)

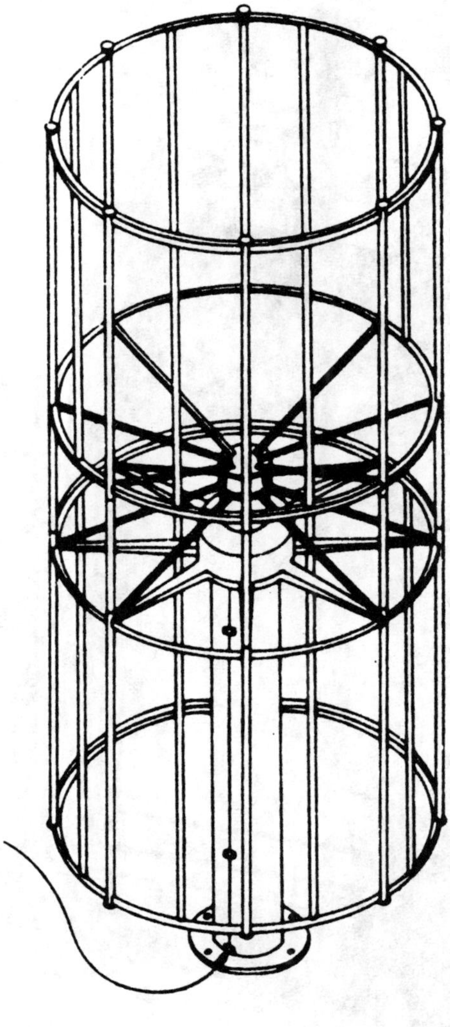

FIG. 2-89 AS-2231 VHF Biconical Dipole Physical Configuration

(a)

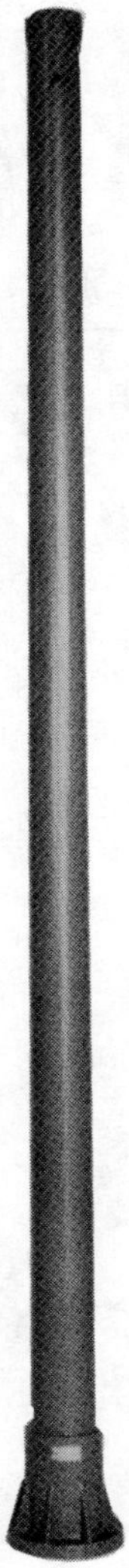

(b)

FIG. 2-90 AS-2808 VHF Vertical Dipole (a)

FIG. 2-91 AS-2809 VHF Vertical Dipole (b)

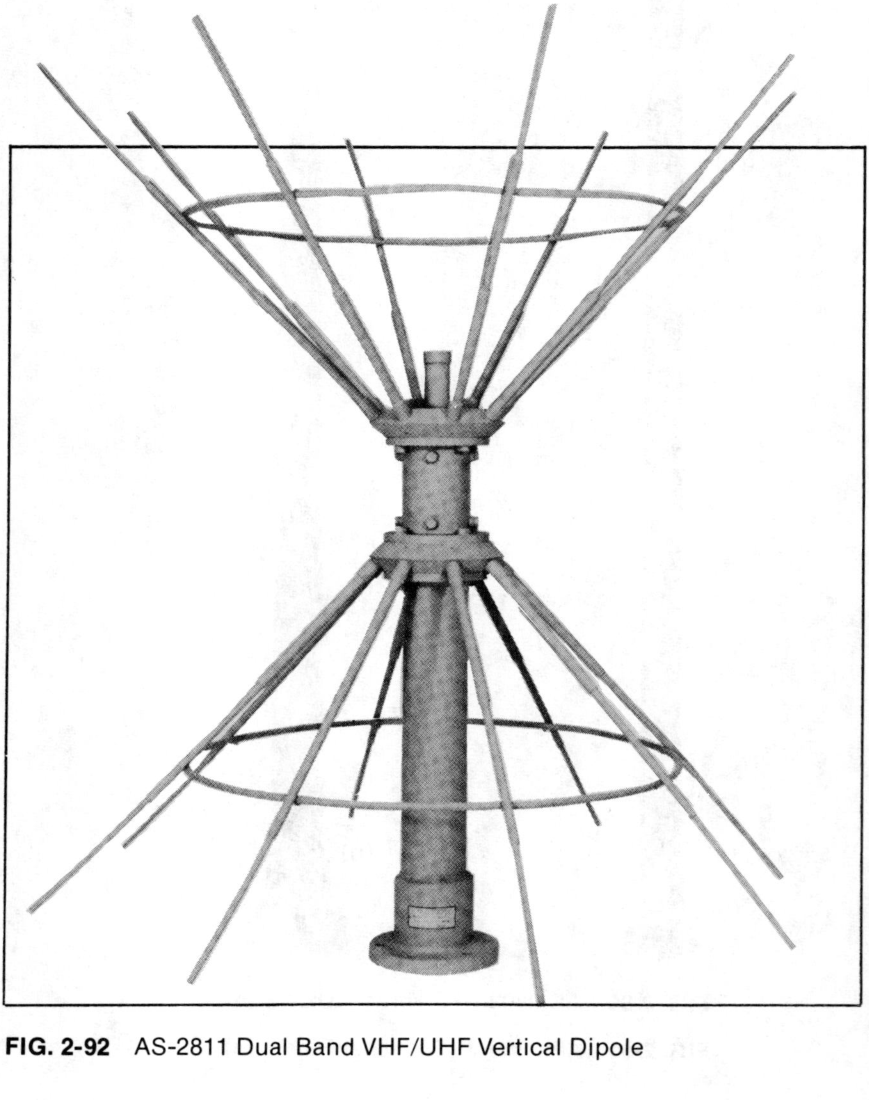

FIG. 2-92 AS-2811 Dual Band VHF/UHF Vertical Dipole

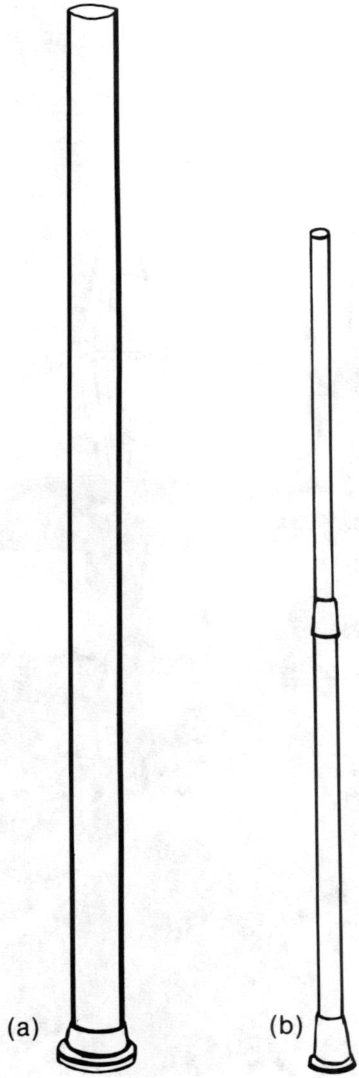

FIG. 2-93 AS-3226(a) and AS-3227(b) VHF Vertical Dipole Antennas

Fig. 2-94(a) AS-2867 VHF/UHF Broadband Discage Antenna with Ground Plane

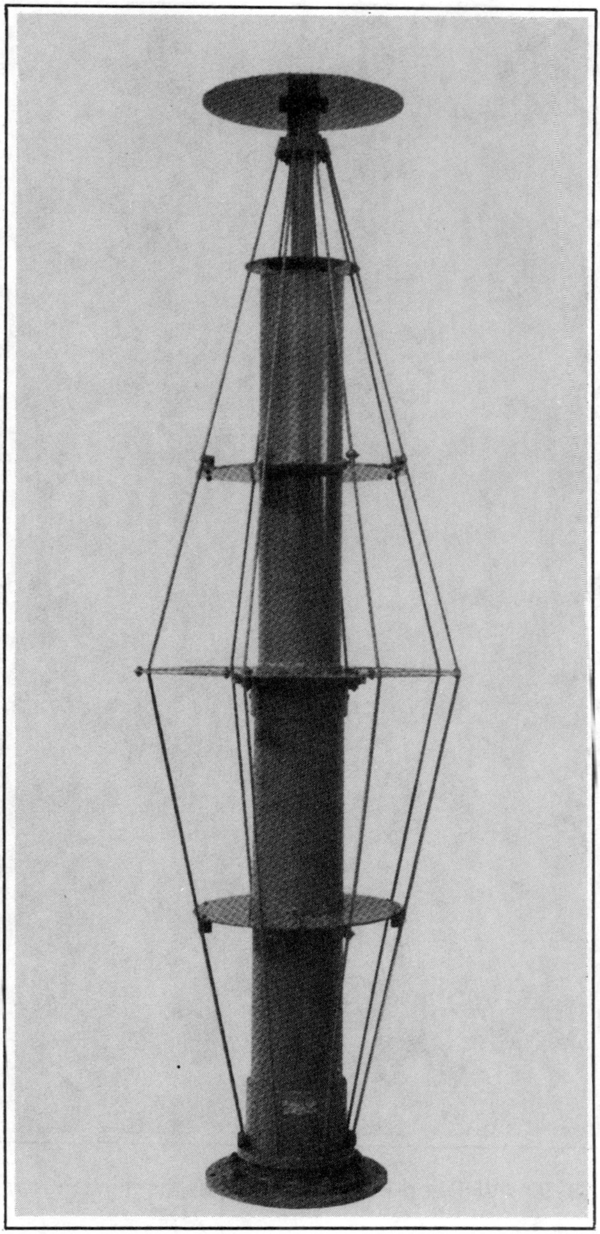

FIG. 2-94(b) AS-2867/SRR VHF/UHF Broadband Discage

FIG. 2-95 AS-2814 VHF Log-Periodic Antennas Installed Port and Starboard

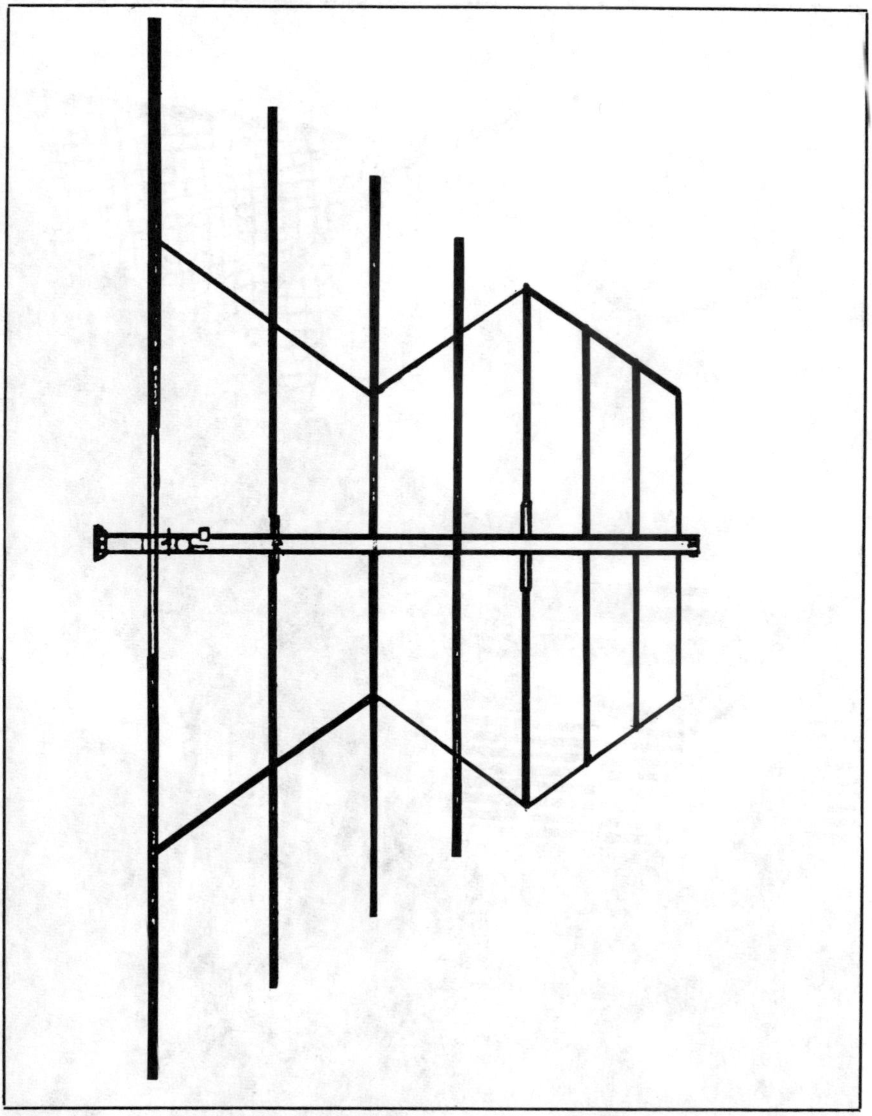

FIG. 2-96 Antenna AS-2814/SRC Physical Configuration

Fig. 2-97 AT-150 UHF Dipole Antennas Mounted on Aircraft Carrier Life Net Edge

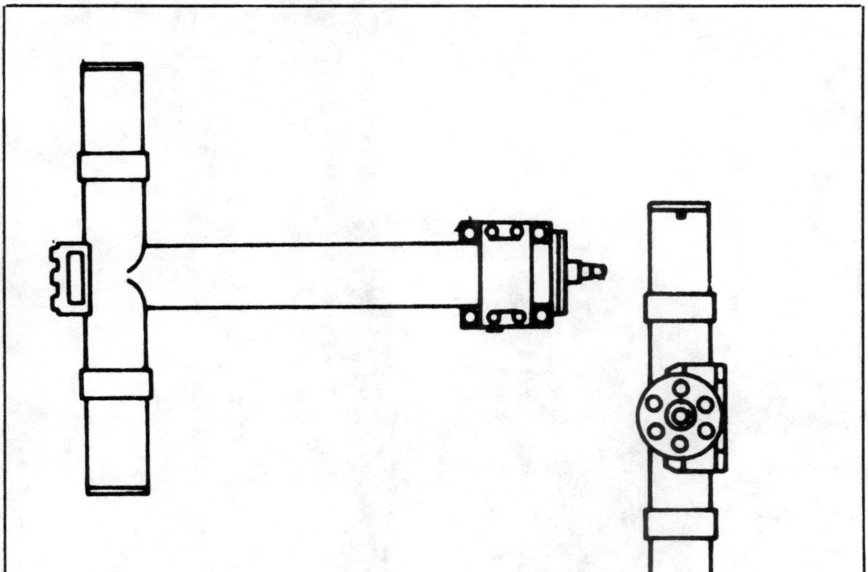

FIG. 2-98 Antenna At-150/SRC Physical Configuration

FIG. 2-99 AS-1735 UHF Wrap-Around Antennas

FIG. 2-100 AS-1735 UHF Antenna Physical Configuration

FIG. 2-101 AS-2812 UHF Biconical Dipole

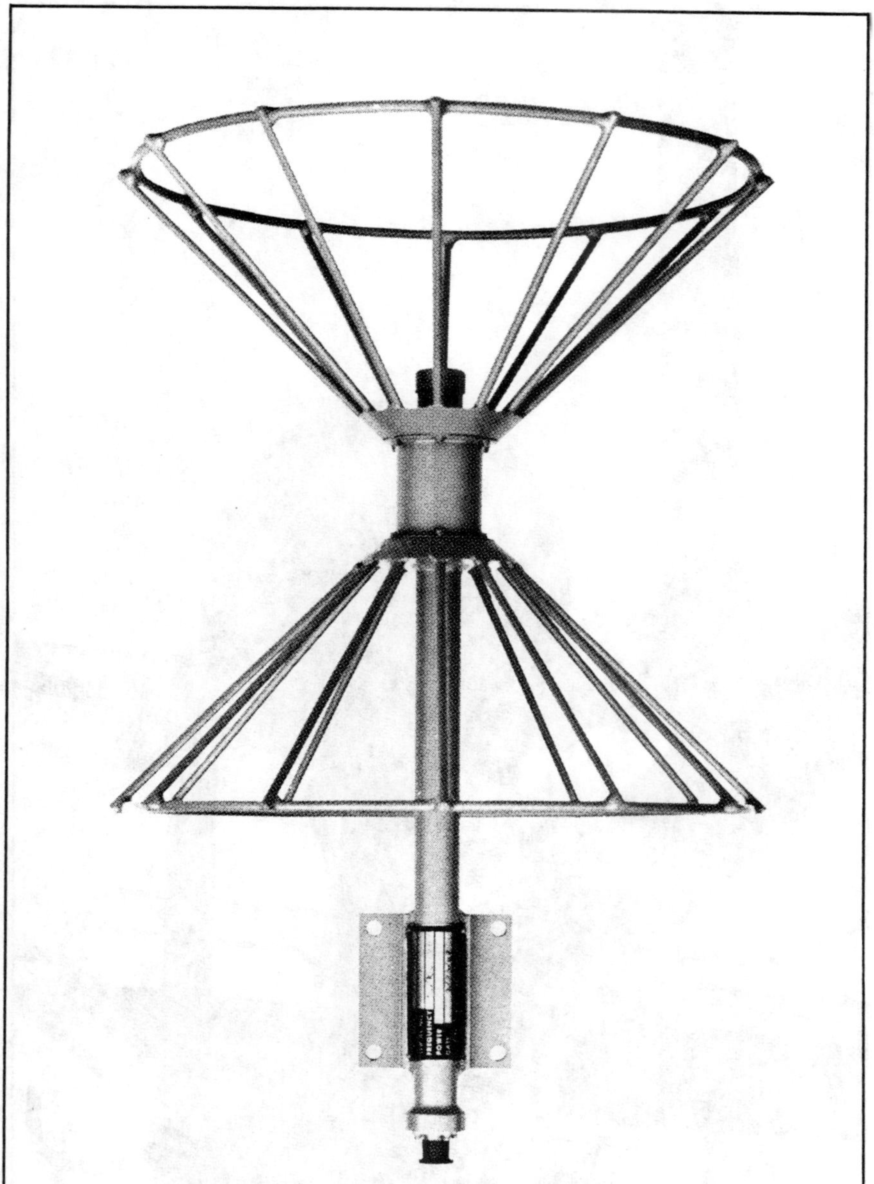

FIG. 2-102 AS-2812 UHF Biconical Dipole Physical Configuration

FIG. 2-103 AS-1018 UHF Vertical Dipole Antenna

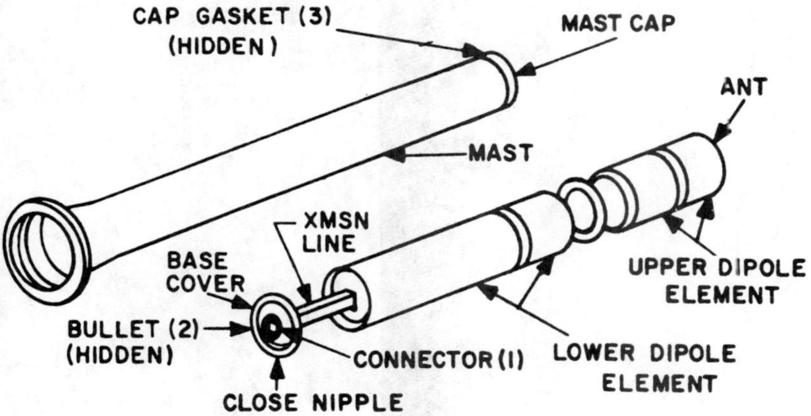

FIG. 2-104 AS-1018/URC Physical Configuration

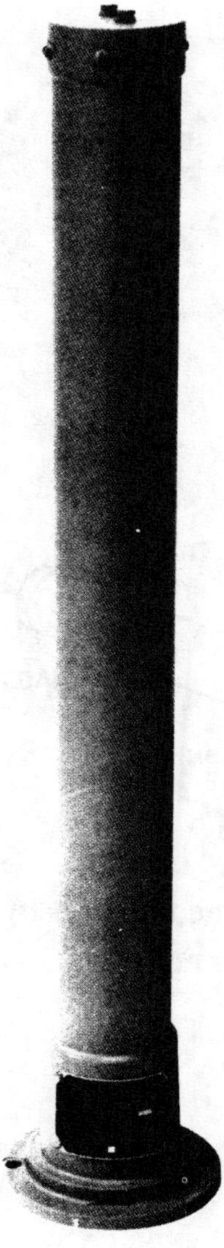

FIG. 2-105 AS-2810 UHF Vertical Dipole

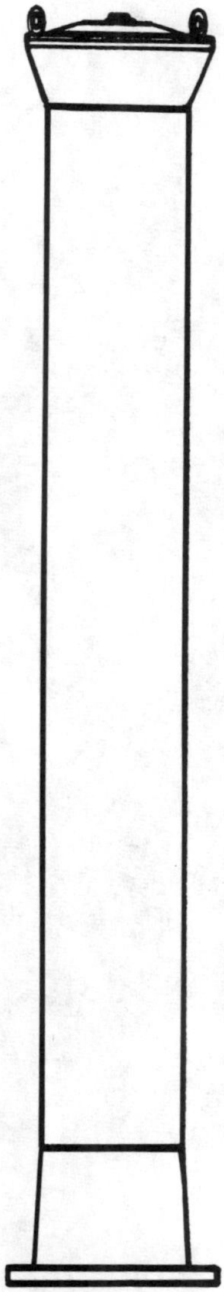

FIG. 2-106 Antenna AS-2877/SRC Physical Configuration

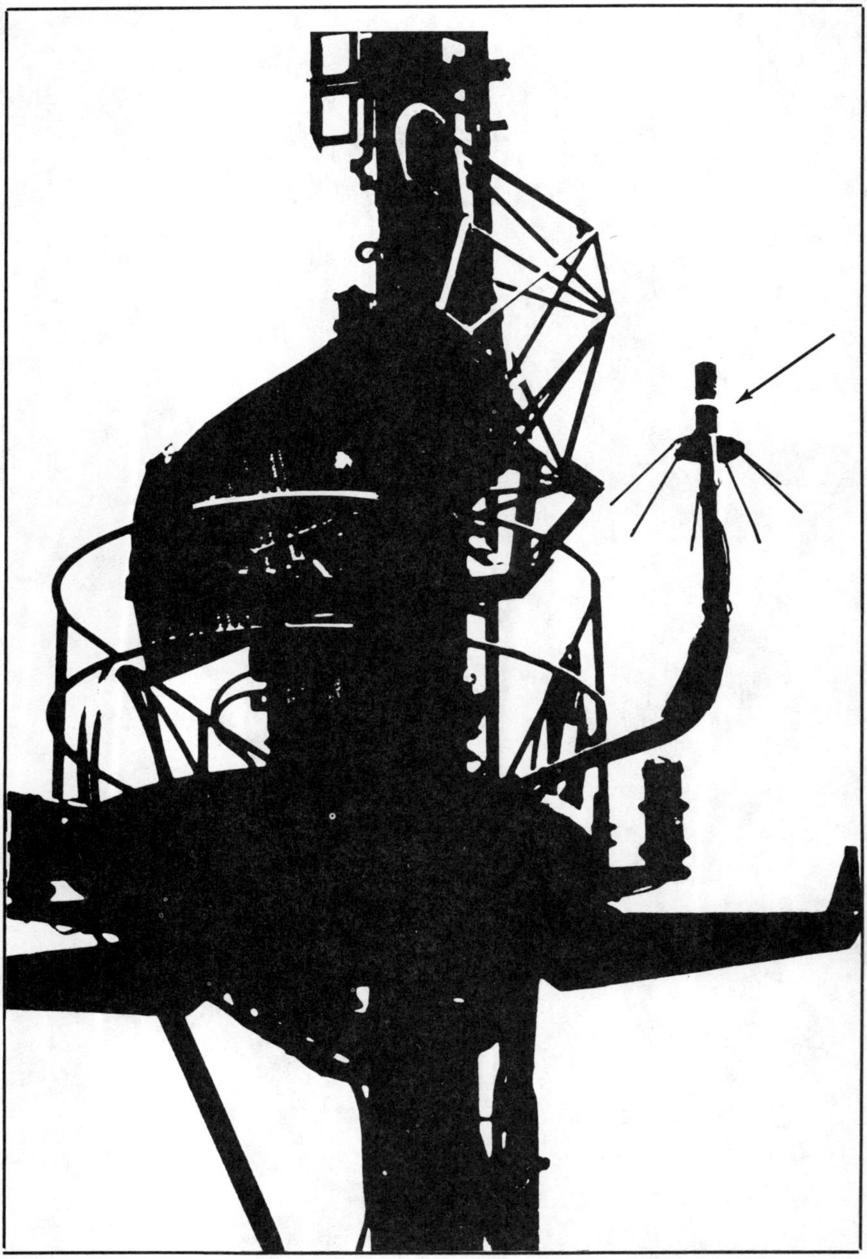

FIG. 2-107 AS-390 Vertical Stub UHF Antenna (mounted on mast)

FIG. 2-108 AS-390 Vertical Stub UHF Antenna (mounted off life net)

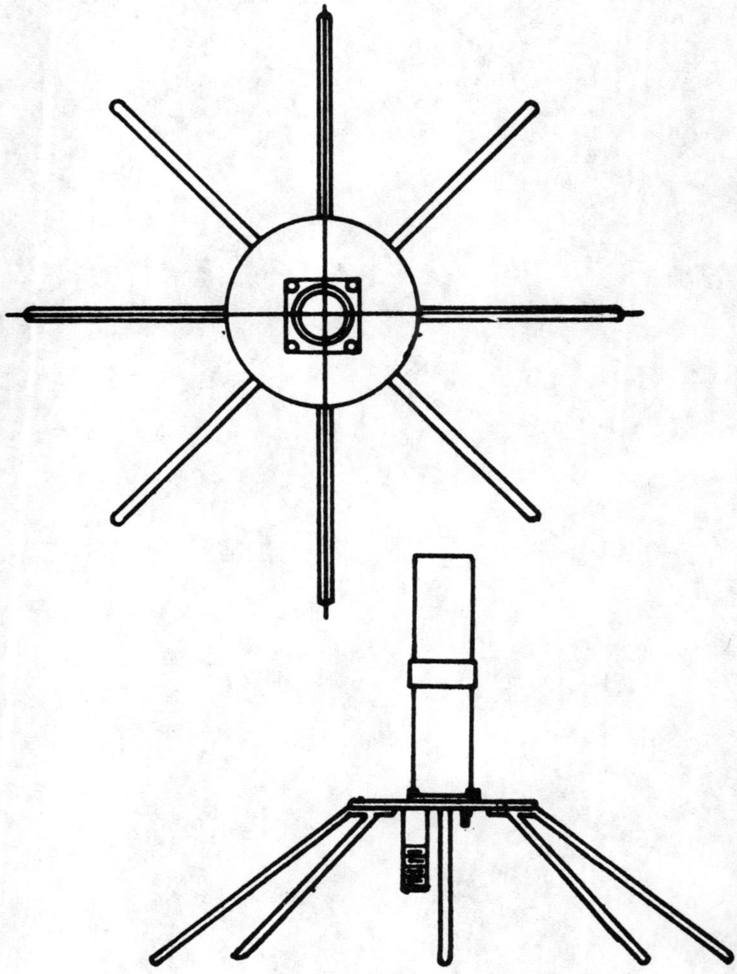

FIG. 2-109 Antenna Assembly AS-390/SRC Physical Configuration

FIG. 2-110 Phasor-90 UHF Dual Antennas AS-1496 and AS-1497

FIG. 2-111 Phasor-90 UHF Dual Antennas

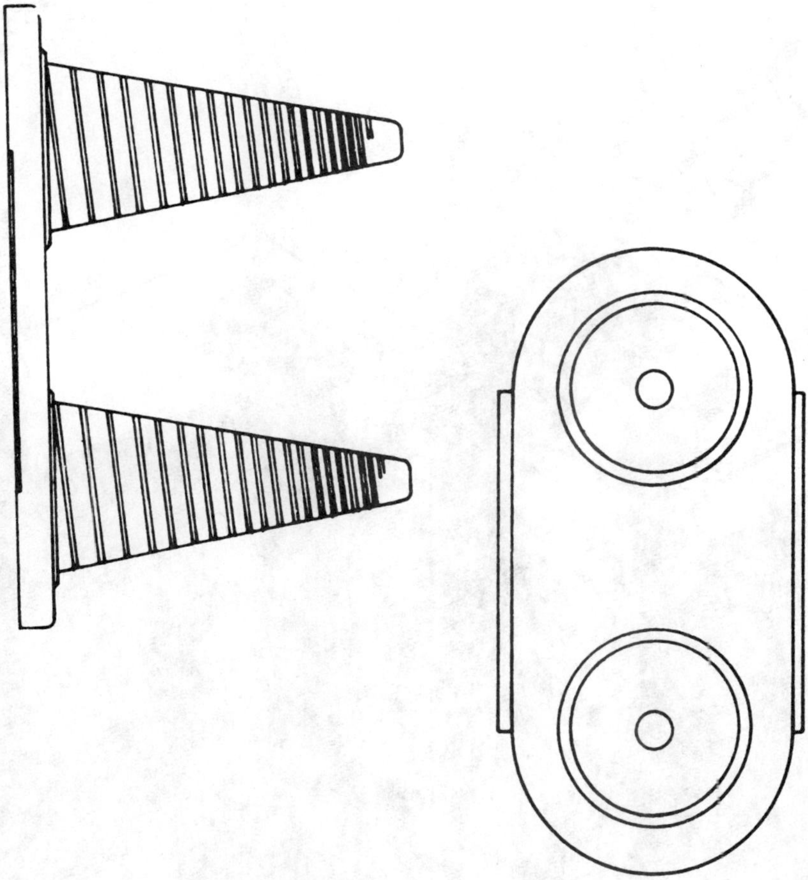

FIG. 2-112 Antennas AS-1496/U and AS-1497/U Physical Configuration

FIG. 2-113 AS-2493 UHF Cross-Log Periodic Sector Antennas

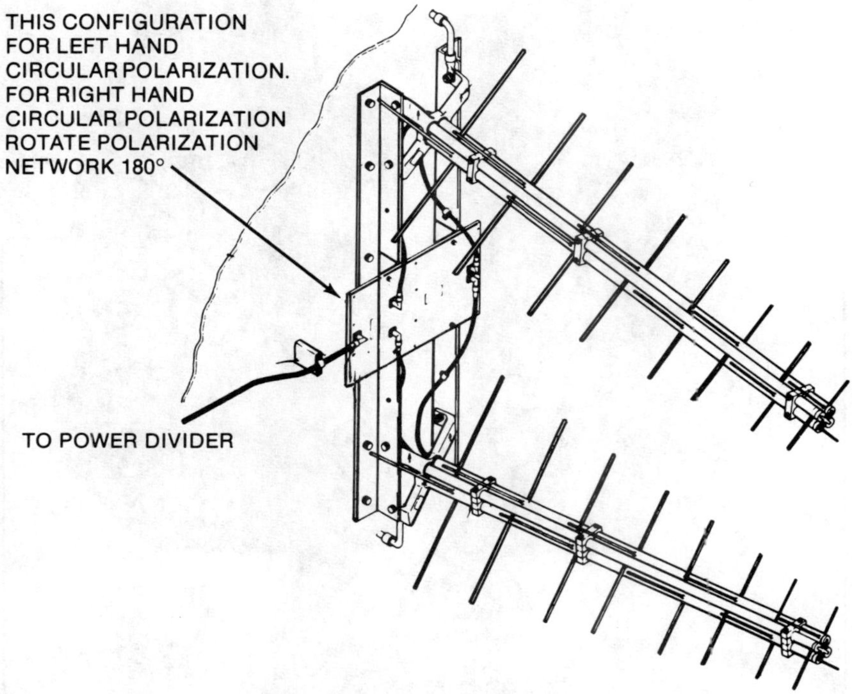

FIG. 2-114 Antenna AS-2493/SRA-62 Physical Configuration

Fig. 2-115 AS-2410/WCS-1(V) UHF SATCOM Antenna

FIG. 2-116 AS-3018/WSC-1(V) Antenna and C-9597/WSC-1(V) Antenna Control

FIG. 2-117 AS-3018A/WSC-1(V) UHF SATCOM Antenna

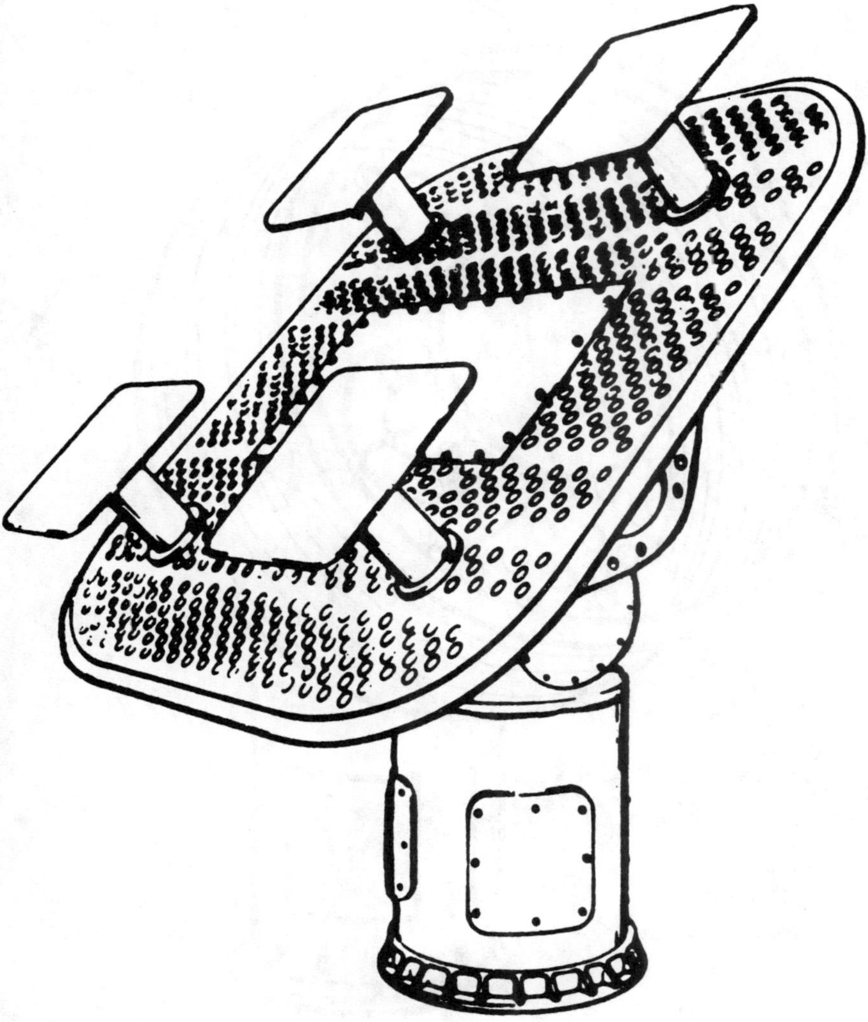

FIG. 2-118 AS-3018/WSC-1(V) UHF SATCOM Antenna

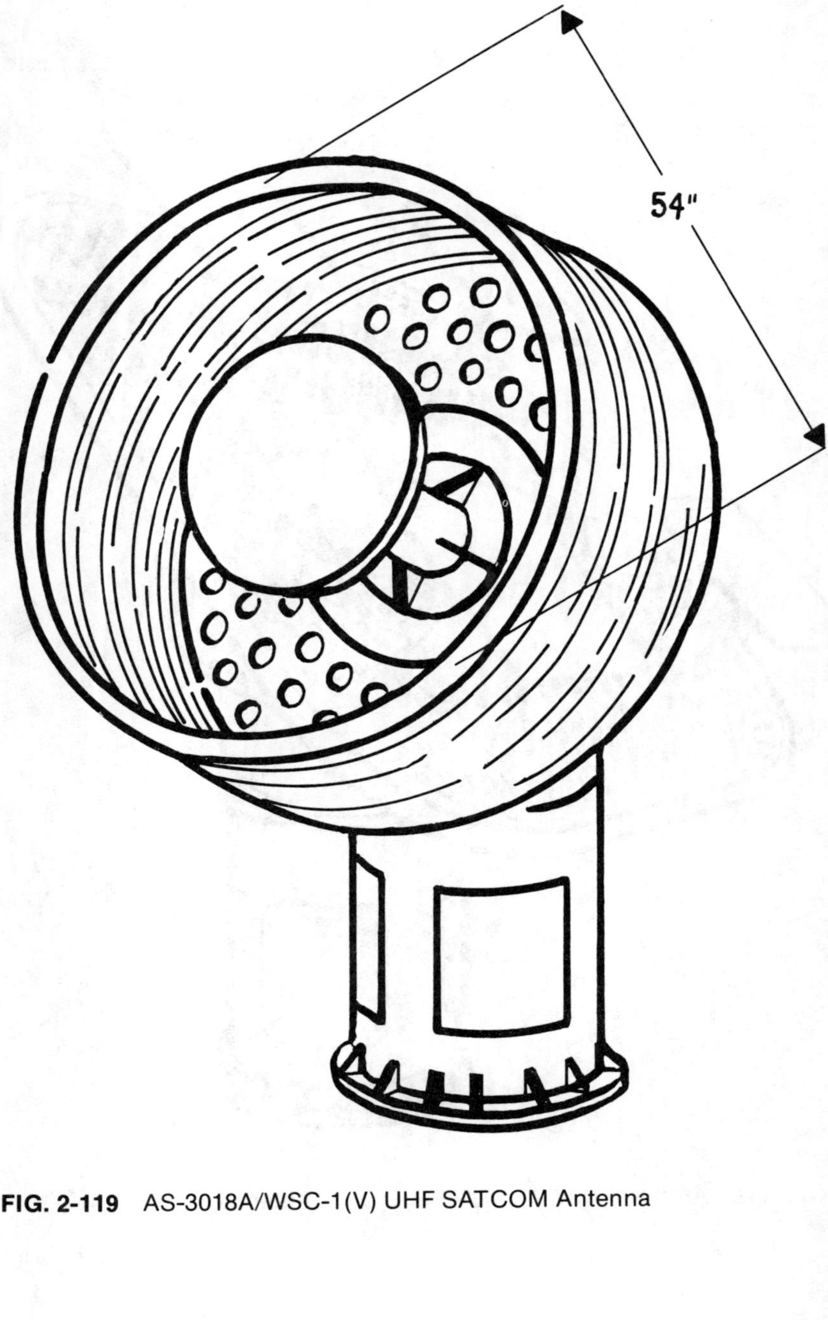

FIG. 2-119 AS-3018A/WSC-1(V) UHF SATCOM Antenna

FIG. 2-120 AS-2815/SSR-1 UHF SATCOM Receive Antenna

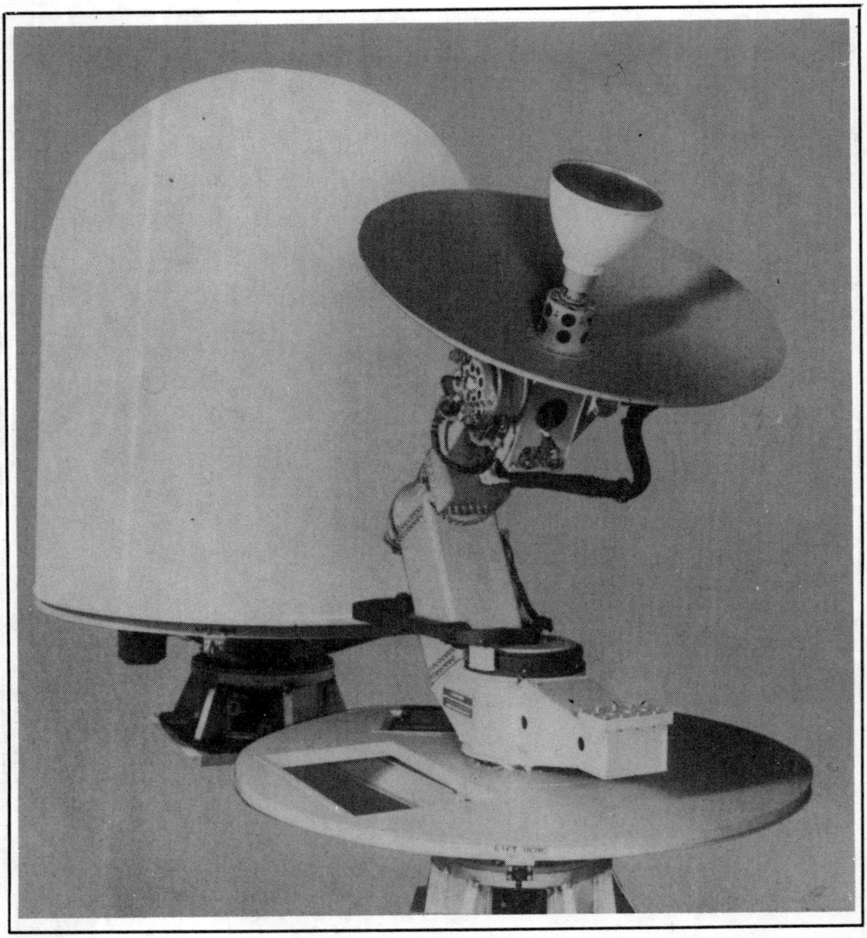

FIG. 2-121 AN/WSC-6 SHF SATCOM Antenna

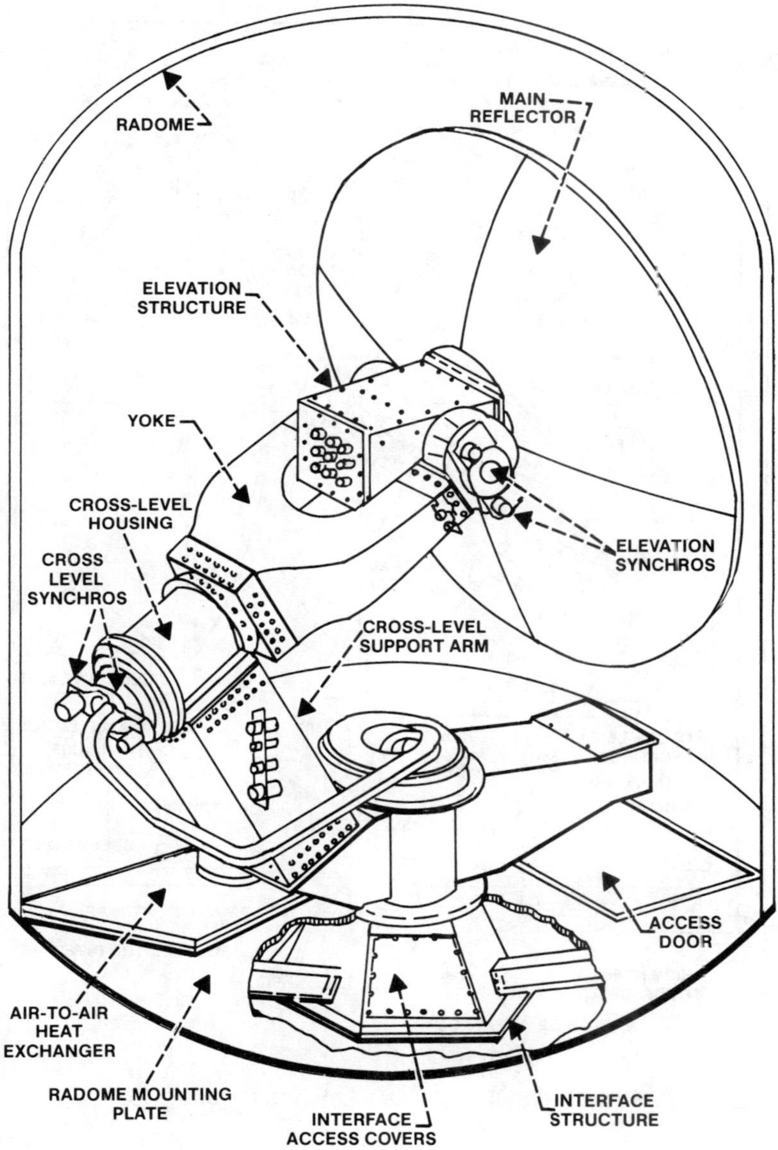

FIG. 2-122 AN/WSC-6 SHF SATCOM Antenna Main Components

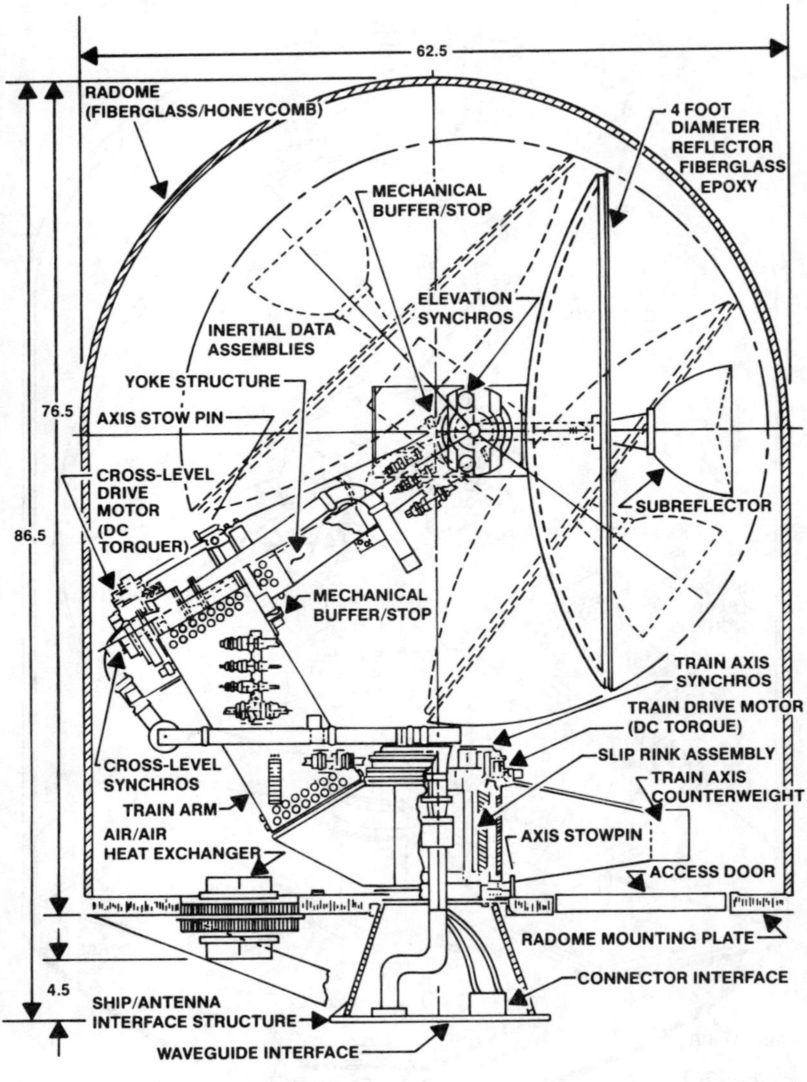

FIG. 2-123 AN/WSC-6 SHF SATCOM Antenna Details

FIG. 2-124 AN/WSC-6 SHF SATCOM Antenna Deck Mounting Platform

Chapter 3

Navigation Antennas

3-0 INTRODUCTION

The early adaptation of radio as a navigational aid was second only to wireless communication at sea. The directive characteristics of propagated electromagnetic waves were quickly recognized and efforts were made immediately to develop equipment which would take advantage of these properties. In fact, naval interest in the so-called "radio compass" began at about the same time as the first shipboard wireless communication installation. In 1906, as a result of testing a radio direction finder placed onboard the *USS Lebanon*, a report to the Secretary of Navy stated:

> The Bureau has recently been making experiments with a wireless telegraph direction finder on board the *Lebanon*. The results thereon indicate that a development of the system will have a far-reaching effect on the safety of vessels at sea, and will possibly play an important part in naval warfare by making it feasible to locate the direction of an enemy's fleet. Indeed the subject is of such importance, in the opinion of the Bureau, that it is deemed advisable, without delay, to thoroughly test out the system on large vessels under sea conditions.[1]

Little more than a decade later, events would prove how dramatically prophetic that report would be.

Nevertheless, at the time, the early direction-finding equipment did not give satisfactory results. The primary reason for this lack of success was that the shipboard antenna used with the system was fixed in azimuth. This required turning the ship to intercept the maximum signal for determining the transmitter bearing, a cumbersome and inaccurate procedure. As a consequence, interest in radio as a navigational aid waned. There was no further attempt to procure direction-finding equipment until 1913, when tests performed on the *USS Wyoming* proved again to be a failure. It was found that, among other things, reradiation from the ship's topside structures introduced large errors in bearing information.

Then, in 1915, Dr. Frederick A. Kolster, of the U.S. Bureau of Standards, discovered that a coil of wire wound on a rotatable, rectangular frame could be so oriented as to determine the direction of received radio waves. When the coil was turned such that it was in a plane parallel to the incoming waves, maximum current was induced. Conversely, when the coil was in a plane normal to the wavefront, minimum signal was received. Thus, if the loop was connected to a receiver which was tuned to a prefixed signal, then rotated through a maximum and minimum indication, the direction to the transmitter could be ascertained. Before the end of 1916, twenty of the Kolster devices were installed on battleships and cruisers. Modifications were soon made to increase their accuracy by using the sharper "null" method of minimum current, rather than the broader maximum signal.

Still, radio direction finding was not greeted with much enthusiasm. Its possibilities were either not realized or ignored by officers of the fleet who were quite reluctant to replace centuries-old, tried-and-proved methods of navigation with "new-fangled" radio devices.[2] Not until America entered World War I was there a sharp turnaround in interest. By that time German submarines were devastating Allied shipping, and a means was being sought desperately to counter the menace. Radio direction finders would provide the solution.

In May 1917, Dr. L.W. Austin, eminent scientist at the Navy Radio Research Laboratory, found that the addition of a single vertical wire to the direction-finder antenna system would eliminate one of the two nulls in the conventional figure-eight pattern, thereby giving a positive direction of the transmitting target.[3] This system, much superior to earlier equipment, was rushed into service by the summer of 1917. Shore equipment was installed around the approaches to the harbor at Brest, France, to determine enemy U-boat positions and to re-route convoys to safer courses. Every available U.S. destroyer was equipped and fresh hunter-killer operations got underway. Never suspecting the Allies' new secret method of detection and tracking, the German subs went right on transmitting RF energy and in turn became, one by one, the destroyed. So successful were the new direction finders that by early 1918, the German Submarine Service was completely demoralized and ineffective.[4]

With the utility of radio navigation now firmly established, the Navy installed a series of radio compass stations along the Atlantic, Gulf, and Pacific coasts using a standard frequency of 375 kHz. By 1923, a total of 46 direction-finder shore stations were in operation around the United States providing navigational position information to over 5000 ships a month out to 100 miles.[5]

The radio direction finder was so specialized for sea operations that, until 1928, the Navy developed and fabricated all its own equipment (a schematic diagram of which is shown in Figure 3-1) and when it did contract for its first commercial equipment in 1928, the delivered sets were so unsatisfactory that the Navy resumed designing and building its own Models DK and DL, which were "superior to any direction-finder apparatus in service" and remained in the fleet through World War II.[6]

As electronics technology accelerated, several new radio navigation techniques began to supplement (and in many cases supplant) the original MF loop direction finder. Systems in heavy use aboard ships at sea today include LORAN, OMEGA, TACAN, and NAVSAT. Soon to come along is the NAVSTAR Global Positioning System. Each of these will be discussed in this chapter. The specialized cases of aircraft landing and control and close-in harbor navigation, although certainly navigational aids, will be addressed separately under the subject of radar.

3-1 VLF/LF/MF NAVIGATION ANTENNAS

3-1.1 OMEGA Long-Range Radio Navigation

A worldwide, all-weather radio navigation system, OMEGA takes advantage of the high-phase stability and low attenuation characteristics of very low frequency (VLF) propagation to provide continuous en route position fixes with an accuracy of one or two nautical miles for ships, submarines, and aircraft. Eight 10 kW transmitting stations situated strategically around the earth ensure global coverage, with each station supplying continuous navigation pulses at ranges up to 8,000 nautical miles. Selecting any four transmit stations, the shipboard set receives and tracks the VLF signals (10.2, 11.05, 11.33, 13.6 kHz) and computes the phase differences to visually display hyperbolic navigation lines of position. Described below are the antennas used aboard ship to intercept OMEGA transmissions:

ANTENNA AS-2283/SRN-12

a. TYPE/USE — The AS-2283/SRN-12 is a unit of the 10.2 kHz basic navigation frequency AN/SRN-12 Omega receiving set. It is a ten-foot VLF fiberglass probe whip, mounted atop an integral base coupler used to match the antenna to the receiver coaxial transmission line.
b. PHYSICAL CONFIGURATION — [See Figure 3-2].
c. DIMENSIONS — (1) Height: 125 inches; (2) Base Diameter: 6 inches; (3) Weight: 9½ pounds.
d. FREQUENCY RANGE — 10.2 kHz.
e. POLARIZATION — Vertical.
f. TYPE OF FEED — Unbalanced, coaxial cable.
g. INPUT CONNECTOR — UG-421 B/U.

h. INSTALLATION REQUIREMENTS — The antenna has a base-mounting flange with four bolt holes to accommodate 3/8-24 inch bolts. Mounting may be either horizontal or vertical on yardarms, mast platforms, or other parts of the superstructure that provide areas relatively free of obstruction and sources of electrical interference.

i. SPECIAL CONSIDERATIONS — Because the AS-2283 is a VLF probe, only a fraction of a wavelength being used to intercept vertically polarized energy, it is imperative that the antenna be placed high on the ship superstructure. The only restriction is that for optimum operation the transmission line coax cable from antenna to receiver not exceed 300 feet.

j. REFERENCE — Technical Manual, Omega Receiving Set, AN/SRN-12, NAVSEA 0967-344-0010.

ANTENNA AS-2271/SRN-14

a. TYPE/USE — The AS-2271/SRN-14 is a unit of the Omega Receiving Set AN/SRN-14. A ten-foot beryllium copper probe encased in fiberglass, the AS-2271 antenna receives 10.2 and 13.6 kHz signals from widely separated phase-locked Omega transmitter stations and applies these through the antenna coupler CU-1810 to the receiver input.

b. PHYSICAL CONFIGURATION — [See Figure 3-3(a)].

c. DIMENSIONS — (1) Height: 128¾ inches; (2) Width: 3 inches, base diameter; (3) Weight: 3 lbs.

d. FREQUENCY RANGE — 10.2 to 13.6 kHz.

e. INPUT IMPEDANCE — The CU-1810/SRN-14 provides impedance matching between the antenna and the balanced 95-ohm transmission line to the receiver.

f. RF POWER RATING — Not applicable, receive only.

g. POLARIZATION — Vertical.

h. TYPE OF FEED — Unbalanced, coaxial cable.

i. INPUT CONNECTOR — Type N.

j. INSTALLATION REQUIREMENTS — The antenna has three equally spaced bolt holes, 25/32 inches in diameter, on a ten-inch bolt circle for use in mounting. The antenna may be located on yardarms, mast platforms, or other parts of the superstructure that provide areas relatively free of obstruction and sources of interference. Since it is a VLF probe used to intercept vertically polarized energy, it is imperative that the antenna be placed high on the ship.

k. SPECIAL CONSIDERATIONS — The length of the coaxial cable between the antenna and coupler must not exceed two feet, and, for optimum operation, the distance between the coupler and receiver indicator should not exceed 300 feet.

l. REFERENCE — Technical Manual, Omega Receiving Set, AN/SRN-14, NAVSEA 0967-869-9010.

ANTENNA AS-2960/SRN-17

a. TYPE/USE — The AS-2960/SRN-17 is a unit of the Omega Receiving Set AN/SRN-17. A ten-foot long, ⅛-inch diameter phosphor bronze rod, supported by a tapered fiberglass overcoating, the AS-2960 antenna receives 10.2, 11.33, and 13.6 kHz sequenced Omega signals. This antenna is identical in every respect to the AS-2283/SRN-12 shown in Figure 3-2.

ANTENNA LTN-211 (Model 6-714AM042S)

a. TYPE/USE — The LTN-211 antenna is a dual-channel, magnetic field (H-field) active receive VLF/LF loop assembly. Originally designed for aircraft use, it is being employed as a specialized, low-interference antenna on ships as part of the LTN-211 digitized OMEGA receiver system. The antenna is designed with an extremely low-noise pre-amplifier, and crossed loops, in a balanced ferrite frame with high shielding to make it virtually immune to precipitation static and nearby electric field (E-field) noise interference effects. Very small, lightweight, and rugged, the antenna is easily accommodated in the ship topside upon completion of a hull skin mapping to determine optimum, low-noise siting. An orthogonal figure-eight (cloverleaf) reception pattern with greater than 40 dB nulls is provided from use of the crossed loop array.
b. PHYSICAL CONFIGURATION — [See Figure 3-3(b)].
c. DIMENSIONS — (1) Height: 1.88 inches: (2) Width: 6.25 inches by 6.25 inches; (3) Weight: 5.5 pounds.
d. FREQUENCY RANGE — 10 kHz to 14 kHz.
e. TYPE OF FEED — Multiwire cable supplied from LTN-211 OMEGA Receiver.
f. INPUT CONNECTOR — Type DBA-33H-14-15 PN multipin connector on antenna to mate with type M83723-86R-14-15N cable connector.
g. INSTALLATION REQUIREMENTS — Four threaded inserts 1/4-28 UNF 3/8-inch deep on antenna accommodate bolt mounting. Normal installation is flat on a metal base plate (i.e., horizontal in ship topside).
h. SPECIAL CONSIDERATIONS — Prior to installation of the antenna, an RF skin mapping measurement is performed to determined the optimum EMI-free location.
i. REFERENCE — SPEARS Associates, Inc., Technical Sheet Model 6-714AM042S Loop Antenna.

3-1.2 LORAN

For forty years, from the beginning of World War II to the present time, the mainstay of Navy long-range radio navigation has been the pulsed signal system known as LORAN. Basically an electronic instrument for measuring very small (microsecond) periods of time that elapse between the arrival of RF pulses from pairs of coastal transmitters, the

LORAN set aboard ship establishes a fix at the intersection of two or more hyperbolic lines of position independent of weather, location or shore transmitting stations, and direction of signal received.

The most widespread technique, in general use for many decades, has been the system known as LORAN-A. Using master and slave stations, centered on the MF frequency of 1850 kHz, coverage is provided out to 1200 miles. LORAN-A, however, is currently being phased out of service, to be replaced by the more accurate LF system known as LORAN-C. In addition to being more precise (accuracy to within 1/4 mile) and greatly simplified in operation, LORAN-C, transmitting at a pulsed carrier frequency of 100 kHz, differs from LORAN-A in that a master and two slave stations are used and a group of pulses, rather than a single pulse, is transmitted. Moreover, LORAN-C extends the range of operation (distance) over which a ship may determine its geographical location. Available now in small, lightweight equipment employing microprocessors to compute and visually display the ship's position,[7] LORAN-C will remain in widespread use by commercial marine and recreational boating for the indefinite future. Nevertheless, it too is being methodically removed by attrition from naval service in favor of the even more accurate OMEGA and Satellite Navigation systems.

A variety of LORAN equipment has been used by the Navy over the years employing short, thin, whip antennas as an electromagnetic probe input to active tuners. Representative of these are the AT-979/U for LORAN-A, and the AS-1982/SRA-42 for LORAN-C, described below. Other tuner whips, such as the AT-924/SRA-17 discussed in Chapter 2, also are used.

ANTENNA AT-979/U

a. TYPE/USE — The AT-979/U antenna is an integral part of the TN-357/GRA-40 RF tuner, which, along with the antenna control unit, comprises the antenna coupler group AN/GRA-40, used primarily with LORAN-A receiving equipment. Mounted on a base insulator atop the RF tuner housing, the AT-979/U is a five-foot, solid, tapered whip.
b. PHYSICAL CONFIGURATION — [See Figure 3-4].
c. DIMENSIONS — (1) Height: 60 inches (whip); (2) Width: 8$\frac{1}{16}$ inches (tuner); (3) Depth: 8¼ inches (tuner); (4) Weight: 35 lbs. (total).
d. FREQUENCY RANGE — 1700 to 2350 kHz.
e. INPUT IMPEDANCE — 50 ohms nominal (tuner input).
f. RF POWER RATING — Receive only.
g. POLARIZATION — Vertical.
h. TYPE OF FEED — Unbalanced, coaxial cable.
i. PRIMARY POWER REQUIRED — Control unit supplies all power requirements for tuner.

j. OPERATING TEMPERATURE — Minus 28°C to +65°C.
k. INPUT CONNECTOR — Terminal board. Watertight fitting for coaxial and control cable.
l. INSTALLATION RQUIREMENTS —The tuner may be mounted in a horizontal or vertical position to a suitable mounting plate by four 5/16-inch bolts on five-inch square centers. Due to the small size, the tuner and antenna are often installed on yardarms, mast platforms, stacks, and other locations not suitable for larger antennas. Care should be exercised in selecting locations that will not cause interference with signal halyards or the ship's rigging.
m. SPECIAL CONSIDERATIONS — All installations should include LORAN bandpass filter type BP-110-D.
n. REFERENCES — (1) Technical Manual, Antenna Coupler Group AN/GRA-40, NAVSEA 0967-112-2040. (2) NAVSEA Drawing RE-D2692131.

ANTENNA AS-1982/SRA-42

a. TYPE/USE — The AS-1982 is a six-foot tubular whip antenna integral with RF tuner TN-443/SRA-42. Together with an antenna control unit, the antenna and tuner comprise the AN/SRA-42 antenna group which provides rapid remote tuning to receive either LORAN-A between 1700 and 2350 kHz, or LORAN-C at 100 kHz. Its most extensive use in recent years has been with the AN/UPN-15 LORAN-C equipment.
b. PHYSICAL CONFIGURATION — [See Figure 3-5].
c. DIMENSIONS — (1) Height: 70 inches overall: (2) Width: 11⅞ inches (tuner); (3) Depth: 10 5/16 inches (tuner); (4) Weight: 16 lbs. (total).
d. FREQUENCY RANGE — 100 kHz (LORAN-C) and 1700 - 2350 kHz (LORAN-A).
e. INPUT IMPEDANCE — 50 ohms nominal (tuner input).
f. RF POWER RATING — Receive only.
g. POLARIZATION — Vertical.
h. TYPE OF FEED — Unbalanced, coaxial cable.
i. PRIMARY POWER REQUIRED — 115 vac, 60 Hz to control unit.
j. OPERATING TEMPERATURE — Minus 28°C to +65°C.
k. INPUT CONNECTOR — UG 943A/U (tuner).
l. INSTALLATION REQUIREMENTS — The whip antenna is mounted on a ceramic bowl insulator atop the RF tuner. Due to the small size, the tuner and antenna unit are most often installed high on mast platforms or yardarms, either in a horizontal or vertical position, using the mounting plate supplied. Care should be exercised in selecting locations that will not cause interference with signal halyards or ship's rigging.
m. REFERENCE — Technical Manual, Antenna Group AN/SRA-42, NAVSEA 0967-274-9010.

LF/MF Radio Direction Finders

As the name implies, a radio direction finder (DF) is used to determine the direction from which a radio signal arrives. For purposes of navigation germane to our present topic, the transmitted signal is information deliberately sent out for use as a navigational aid. Likewise, it is intended to be received by a direction finder being carried at sea on a charted course. There are cases (very important and very much used) where direction finders are employed to intercept and track unintentionally transmitted energy, i.e., to determine the location and source of unfriendly signals. Antennas used for such purposes (the so-called Combat DF), will not be discussed here but in Chapter 5 concerning Electronic Warfare.

Although the old, low-frequency "radio compass" system described at the beginning of this chapter has all but disappeared from U.S. Navy ships, it is still used extensively by commercial shipping and recreational boating. Now called the Radiobeacon System, the first electronic aid to navigation is operated and maintained by the U.S. Coast Guard. Currently, there are about 200 radiobeacons installed on lighthouses, lightships, large buoys, and other coastal stations around the Atlantic, Pacific, and Gulf coasts and the Great Lakes to provide all-weather position-fixing and homing capability to vessels equipped with direction finders. Radiobeacons, emitting nondirectional AM signals in the 285-325 kHz frequency band, are short-range aids to navigation, useful from 10 to 175 nautical miles. The principal value of this system is in its simplicity of operation and relatively low-cost equipment.

Shipboard direction-finding equipment, commonly known as automatic direction finders (ADFs), consist of a directional antenna, receiver, and direction-indicating device, e.g., instantaneous visual display on a scope, or digital readout of azimuthal bearings. The most practical DF antenna at these low frequencies is the crossed loop which is used in conjunction with a vertical sensing whip. A typical example is the Mackay Marine 4005A loop shown atop a ship mast in Figure 3-6, and described below:

ANTENNA 4005A LF/MF CROSSED LOOP

a. TYPE/USE — The 4005A is a fixed, crossed loop antenna used in association with a whip-sensing antenna and goniometer as part of the Mackay Marine Model 4005A automatic radio direction finder. This system provides instantaneous, accurate bearings of distant emitters as an aid to marine navigation, safety, and distress signal search.

b. PHYSICAL CONFIGURATION — [See Figure 3-7].

c. DIMENSIONS — (1) Height: 44.25 inches; (2) Width: 31 inches; (3) Depth: 31 inches; (4) Weight: 31 lbs.

d. FREQUENCY RANGE — 190-550 kHz and 2197 kHz.
e. INPUT IMPEDANCE — 95 ohms.
f. POLARIZATION — Vertical.
g. TYPE OF FEED — RG-111/U (armored coax cable).
h. INPUT CONNECTOR — Terminal board.
i. INSTALLATION REQUIREMENTS — The antenna should be located in as clear an area as possible, but no more than 100 feet from the receiver.
j. REFERENCE — ITT Mackay Marine Products Catalog.

3-2 VHF/UHF NAVIGATION ANTENNAS

3-2.1 VHF/UHF Radio Direction Finders

Portions of the VHF/UHF frequency bands are used quite extensively for radio direction finding at sea. For example, the maritime industry employs RDF for recovery of floating instrument packages and submersible vehicles, locating buoys, off-shore drilling platforms, and fishing sites, and tracking marine life. Similarly, law enforcement agents use RDF to track and locate craft suspected of illegal operations and to determine sources of radio frequency interference (RFI). Military operations use RDF for tracking test targets and finding air-dropped packages.

However, the best-known and most critical use of RDF is that of search and rescue (SAR) missions. Downed aircraft, ships in distress, and survivors adrift at sea rely heavily on emergency locating radio and position-indicating beacons to guide and home-in search craft to the rescue spot. Here again, at these frequencies (e.g., 121.5 MHz, 123.1 MHz, and 243.0 MHz), the crossed loop antenna is often employed. More frequently, though, because of its capabilities for greater accuracy, the Adcock fixed four-dipole array is now used. An example of this type antenna used by the U.S. Coast Guard is the commercial OAR Model 320 shown in Figure 3-8.

The antenna most heavily relied upon by the Navy over the years for SAR operations and air traffic control is also an Adcock array, rotatable two-dipole UHF version, known as the AS-514A/URD-4, shown in Figure 3-9.

ANTENNA AS-514A/URD-4

a. TYPE/USE — The AS-514A antenna is a rotating Adcock array of two vertical dipoles housed within a watertight plastic radome. Also enclosed just beneath the antenna array are the antenna drive motor, two-phase alternator and tone generator. The complete AS-514A assembly is an integral unit of UHF direction finder set AN/URD-4A used to provide a means of rapidly determining the direction of fixed or mobile sources of radio signal transmission in the 225 to 400 MHz frequency range. Directional information out to 20 miles surface-to-surface, and 90 to 125 miles for aircraft-to-surface, is visually displayed on a scope with a bearing accuracy of plus or minus five degrees.

b. PHYSICAL CONFIGURATION — [See Figure 3-10].
c. DIMENSIONS — (1) Height: 73$\frac{1}{16}$ inches; (2) Diameter: 18¾ inches max; (3) Weight: 72¾ lbs.
d. FREQUENCY RANGE — 225.0 to 399.9 MHz.
e. INPUT IMPEDANCE — 50 ohms.
f. RF POWER RATING — Not applicable, receive only.
g. POLARIZATION — Vertical, but designed to furnish satisfactory azimuth information where received signals are within 45 degrees from vertical.
h. TYPE OF FEED — Unbalanced, coaxial cable.
i. PRIMARY POWER REQUIRED — Control voltages supplied from OA-485A/URD-4 receiver group.
j. OPERATING TEMPERATURE — Minus 54°C to +65°C.
k. INPUT CONNECTOR — (1) RF: UG-271/U; (2) Control: 10-42224-7P.
l. INSTALLATION REQUIREMENTS — The antenna should be mounted high enough to clear obstructions and reflecting surfaces in its immediate vicinity so as to minimize signal reflections and unequal signal pickup. The base of the antenna assembly has eight ⅜-inch bolt holes on a 7¾-inch bolt circle for mounting the antenna to the ship's structure. The maximum cable length between the antenna and receiver is limited to 150 feet. The antenna must be oriented in the direction of the ship's bow in accordance with the arrow stamped on the antenna housing.
m. SPECIAL CONSIDERATIONS — The base of the antenna assembly is fitted with a drain plug and vent hole to allow pressure equalization and drainage of accumulated condensation caused by atmospheric temperature variations. The antenna should be pressurized with dry air to 1 psig by replacing the drain plug with a ⅛-inch AN hydraulic fitting and connecting to an air pressure line.
n. REFERENCES — (1) Technical Manual AN/URD-4, Direction Finder Sets, NAVSEA 0967-108-6010. (2) NAVSEA Drawings, RE-C2695362 and RE-D2692836.

ANTENNA AS-3386/URD-10(V)

a. TYPE/USE — The AS-3386 is a lightweight, fixed-dipole, omnidirectional antenna used with the AN/URD-10(V) dual-channel direction finder system currently undergoing naval evaluation. Consisting of four balanced dipoles, equally spaced around a vertical centerline and connected as two pairs on a crossed baseline, the antenna array is electronically steered. It is intended to be a direct replacement for the AS-514A/URD-4.
b. PHYSICAL CONFIGURATION — [See Figure 3-11].
c. DIMENSIONS — (1) Height: 30 inches; (2) Diameter: 21 inches; (3) Weight: 45 pounds.

d. FREQUENCY RANGE — 225 to 400 MHz.
e. INPUT IMPEDANCE — 50 ohms.
f. POLARIZATION — Vertical.
g. TYPE OF FEED — Unbalanced coaxial cable.
h. PRIMARY POWER REQUIRED — 28 Vdc, 17 watts.
i. OPERATING TEMPERATURE — Minus 28° to +95°C.
j. INPUT CONNECTOR — UG-271/U.
k. INSTALLATION REQUIREMENTS — [Same as AS-514A/URD-4].
l. REFERENCE — NAVELEX Technical Manual AN/URD-10(V).

3-2.2 VHF/UHF Navigation Satellite Systems

The Navy Navigation Satellite System, commonly known as TRANSIT, or simply NAVSAT, is an electronic aid to navigation that provides position fixes with high precision (nominal accuracy from 0.25 to 0.5 miles) anytime, anywhere, irrespective of weather conditions. This system, in operation since the middle 1960s, consists of five near-earth satellites in circular polar orbits at an altitude of approximately 600 nautical miles, four tracking and four injection stations, a computer control center, and assorted shipboard satellite navigation equipment such as the AN/SRN-9, AN/WRN-5, and AN/SRN-19. Circling the earth every 107 minutes, these satellites continually transmit orbital information every two minutes on highly stable, phase-modulated, UHF carrier frequencies (150 and 400 MHz, depending on the system) generated by a common frequency standard. The shipboard sets then obtain navigation fixes from a process based on the shift of frequency (doppler) occurring due to the changing relative distance between transmitter (satellite) and receiver (ship). As such, NAVSAT has added a degree of precision, safety, and reliability in sea navigation heretofore unknown.

Nevertheless, plans are now underway for developing and implementing a new generation Department of Defense satellite navigation system of yet higher accuracy and efficiency. Scheduled to be operational in the mid-1980s, this project, known as the NAVSTAR Global Positioning System (GPS), is an anti-jam, precise-ranging system enabling instant, simultaneous determination of position and velocity even under severe dynamic conditions. Made possible by combining satellite navigation and spread spectrum techniques, the GPS is comprised of three major subsystems: the space segment, the control segment, and the user segment. The space segment will have 18 satellites, six in each of three orbit rings, at a nominal altitude of 11,000 feet. Circling the globe every 12 hours, each satellite will continuously transmit a composite signal of time and position data on two UHF (L-band) frequencies, 1227 MHz and 1575 MHz. Properly equipped users will select and automatically lock onto the four satellites most favorably located to provide instantaneous four-dimensional position fixing (longitude, latitude, altitude, and time) as well as velocity. So accurate is this system, the user will know his position to within 30 feet or so, his speed within fractions of a mile per hour, and time within a millionth of a second.

Because the GPS program is still under development, the exact type of shipboard antenna to be used has not been determined as of this writing. Therefore, only the NAVSAT antennas currently in use will be described below.

ANTENNA AS-2227/SRN-9

a. TYPE/USE — The AS-2227 antenna is part of the OE-49/SRN-9 group of the AN/SRN-9 radio navigation set used to determine ship's position from information transmitted by navigational satellites (NAVSAT). In reality the AS-2227 consists of two separate antennas, a center monopole which operates on the high frequency and a conical spiral which operates on the low frequency. Essentially uniform coverage is obtained over the upper hemisphere. An integral base ground-plane serves to reduce the lower hemisphere coverage and to isolate the antenna from the ship's structure. Because of the low power level of received satellite signals, and the generally long cable runs, preamplifier-filter units enclosed within an aluminum cover are mounted at the antenna base as part of the OE-49 group. The antenna assembly is constructed of aluminum tubing while the tuning skirt, groundplane stubs, and base groundplane are made of spun aluminum. The top cap is fiberglass. All aluminum parts are chemically treated with a conductive, protective film before painting.
b. PHYSICAL CONFIGURATION — [See Figure 3-12].
c. DIMENSIONS — (1) Height: 51½ inches; (2) Diameter: 40 inches; (3) Weight: 50 pounds.
d. FREQUENCY RANGE — 150 MHz (conical spiral), 400 MHz (center monopole).
e. INPUT IMPEDANCE/VSWR — 50 ohms, VSWR not greater than 2:1.
f. POLARIZATION — Left hand circular (conical spiral), vertical (monopole).
g. TYPE OF FEED — Unbalanced, coaxial cable.
h. PRIMARY POWER REQUIRED — 15 Vdc for preamplifiers. This voltage is carried over the signal coaxial cable.
i. OPERATING TEMPERATURE — Minus 28°C to +65°C.
j. INPUT CONNECTOR — Type N.
k. INSTALLATION REQUIREMENTS — The antenna group is designed for mast or yardarm mounting. Propagation in the frequency ranges covered by this antenna is primarily line-of-sight. Therefore, the antenna should not be shadowed by any portion of the ship's structure. The

installing activity must supply a pad or foundation in the shape of a cross drilled to match the bolt holes in the stiffeners on the bottom of the antenna groundplane. The foundation should elevate the antenna sufficiently to allow cable access. Coaxial cables are required for receive and test signals.

l. SPECIAL CONSIDERATIONS — Since signals reflect from metallic objects, the antenna should not be located near large rotating structures. Although the input passband of the preamplifiers is narrow and protected from overloads, care should be exercised in not locating the group in close proximity to higher power, medium-frequency radars.

m. REFERENCES — (1) Technical Manual for Radio Navigation Set AN/SRN-9, NAVSEA 0967-306-3010. (2) NAVSEA Drawings RE-B2696405 and RE-B-2697513.

ANTENNA CA-3086/WRN-5(V)

a. TYPE/USE — The CA-3086, a rugged, lightweight, fiberglass enclosed dual-band monopole antenna, is a part of the OE-233/WRN-5(V) antenna group employed with the AN/WRN-5 radio navigation set which is designed to receive and track signals transmitted by satellites of the Navy Navigation Satellite System. A broad-bandwidth antenna, the CA-3086 is capable of receiving both the 150 and 400 MHz NAVSAT frequencies which are then separated by the input diplexer of the preamplifier to be properly channeled and amplified. Vertical polarization provides optimum hemispherical coverage with minimum overhead null.

b. PHYSICAL CONFIGURATION — [See Figure 3-13].

c. DIMENSIONS — (1) Height: 40½ inches; (2) Diameter: 3³⁄₁₆ inches (monopole), 38 inches (ground radials); (3) Weight: 8 pounds.

d. FREQUENCY RANGE — 150 and 400 MHz.

e. INPUT IMPEDANCE /VSWR — 50 ohms/1.75:1.

f. POLARIZATION — Vertical.

g. TYPE OF FEED — Unbalanced coaxial cable.

h. PRIMARY POWER REQUIRED — 15 Vdc for preamplifiers.

i. OPERATING TEMPERATURE — Minus 28°C to +65°C.

j. INPUT CONNECTOR — UG1006 B/U.

k. INSTALLATION REQUIREMENTS — For optimum satellite signal reception the antenna should have a clear view of the horizon and be mounted above large surfaces, as high as possible. It should be installed no closer than 12 feet from vertical obstructions and 3 feet from large horizontal metallic surfaces. Any shadowing of the antenna by an obstruction in its plane should not subtend an arc greater than 5 degrees. Considering all the above criteria, caution nevertheless, must be exercised, to ensure that the cable length not exceed 200 feet.

l. SPECIAL CONSIDERATIONS — Mounting is accomplished by use of ⅜-inch bolts through four holes of the 11¾-inch diameter steel baseplate. The antenna has been designed for use under severe environmental conditions at sea, with a maximum wind loading of 100 knots, 100%

humidity, and 4.5 pounds per square foot ice loading. When installing, care must be taken to avoid damage to the two tuning stub rods located halfway up the radome enclosure.

m. REFERENCES — (1) Technical Manual for Radio Navigation Set AN/WRN-5(V), NAVELEX 0967-LP-453-9011. (2) CHU Associates, Inc., Catalog — Model CA-3086 Antenna.

ANTENNA AS-3330/SRN-19(V)

a. TYPE/USE — Antenna AS-3330 is a linear, single-frequency, cylindrical, fiberglass enclosed monopole, part of the OE-284/SRN-19(V) antenna group used with the AN/SRN-19(V) automatic shipboard navigation set to receive 400 MHz satellite signals omnidirectionally in the horizontal plane. An RF amplifier, also part of the OE-284 group (integral to the antenna base mounting), is provided for initial amplification.
b. PHYSICAL CONFIGURATION — [See Figure 3-14].
c. DIMENSIONS — (1) Height: 16½ inches (27 inches overall with RF amplifer); (2) Diameter: 3 inches (monopole), 9¾ inches (RF amp), 24 inches (ground radials); (3) Weight: total 4 pounds.
d. FREQUENCY RANGE — 400 MHz.
e. INPUT IMPEDANCE/VSWR — 50 ohms/1.5:1.
f. POLARIZATION — Vertical.
g. TYPE OF FEED — Unbalanced coaxial cable.
h. PRIMARY POWER REQUIRED — +12 Vdc to RF amp via RF coax.
i. OPERATING TEMPERATURE — Minus 28° to +65°C.
j. INPUT CONNECTOR — UG-300/U (type TNC).
k. INSTALLATION REQUIREMENTS — Same as the CA-3086 (i.e., OE-233/WRN-5) described above, including 200 feet maximum cable run.
l. REFERENCE — Technical Manual for Radio Navigation Set AN/SRN-19(V)2, NAVELEX 0967-LP-634-9010.

3-2.3 Tactical Air Navigation (TACAN)

We turn now from where the ship, needing radio navigation to remain on course, *receives* continuous position information, to where the ship becomes instead the *transmitting* aid, sending out data vital to the navigation of enroute aircraft. One of the essential shipborne means for accomplishing this is the radio beacon system, known as TACAN.

In the 1920s, with the introduction of the aircraft carrier into the fleet, it became immediately necessary to provide airplanes with navigation equipment to direct them to and from their carrier base at sea, or to air stations ashore.[10] The earliest methods involved rather primitive radio direction finders using cumbersome and awkward loop antennas aboard the planes. Gradually, however, a greatly improved technique known as the aircraft homing system was developed, in which the carrier used a rotary beam antenna to send out coded 246 MHz signals, identifying twelve equal sectors as the antenna rotated through 360° in azimuth. When these audible radio transmissions were received by the aircraft, the particular course sector being occupied was made obvious by recognizing the strongest signal.[11] Furthermore, only a small, simple antenna was required on the aircraft. This homing system proved to be highly successful during World War II, notably in the Pacific theater. It stayed in naval service for many years until it was gradually replaced in the 1950s by TACAN.

It was in fact the Navy's requirement for a homing-beacon service giving more precise bearing information to carrier-based aircraft that gave impetus to development of the TACAN system.[12]

The first of a long series of TACAN equipment implemented throughout the fleet was the AN/URN-3 radio beacon set. All subsequent shipboard TACAN systems have incorporated the same basic principles of operation as this first set; namely that of supplying a sequence of pulse transmissions which contain, in order of priority, bearing relative to the radio beacon referenced to magnetic north, identification of the radio beacon, and distance from the beacon. Processing this data allows the plane to automatically determine its exact geographic location relative to the ship.

Distance information is in reality initiated from the aircraft by interrogations, sent out from its transmitter, having its own peculiar repetition pattern. Reception of these inquiries by the ship's radio beacon results in transmitted replies of the same pattern. The aircraft then receives these replies along with several other possible replies initiated by other planes, and picks out its own pattern by comparing the pulse character with that from its own transmitter. Distance is computed by measuring the total time elapsed between the original interrogative transmission and receipt of the radio beacon (TACAN) reply.

Bearing information is obtained from the radio beacon by electromechanical modulation of the antenna radiation pattern and the transmitting of reference group pulse bursts, spaced at specific intervals. By making phase comparisons of the reconstructed antenna-modulated envelope with time occurrence of the reference bursts, the aircraft is able to determine its bearing. The beacon periodically transmits its identifying call in Morse code, thus enabling the aircraft to be sure with which beacon it is in contact.

Shipboard TACAN sets have a 252-channel capacity and, therefore, are able to receive on any of 126 frequencies in the L-band range of 1025 through 1150 MHz, and transmit on any of 126 frequencies in the range of 962 through 1024 MHz and 1151 through 1213 MHz. Simultaneous operation of up to 100 aircraft within a 300 mile radius is possible. Since modulation of the radiation pattern is an essential factor in making up the composite navigation signal, the TACAN antenna is a key part of the radio beacon system. Several shipboard antenna types will be discussed below:

ANTENNA AS-677/URN-3

a. TYPE/USE — The AS-677 antenna is part of antenna group OA-553/URN-3, used with the shipboard radio beacon set AN/URN-3. Designed for low-band TACAN operation (transmission between 962 and 1024 HMz, and reception between 1025 and 1087 MHz), its primary radiator is a stationary stack of seven UHF biconical dipoles mounted within a supporting fiberglass cylinder which is four inches in diameter and four feet long. This fixed dipole array is at the center of two concentric rotating, fiberglass cylinders. The inner cylinder is about five inches in diameter and contains a few vertical wires imbedded in the fiberglass, so closely spaced as to appear as a single parasitic element rotating to produce a 15 Hz modulation of the radiation pattern. The outer cylinder, about 41 inches in diameter, has nine parasitic elements spaced 40° apart enabling it to produce a 135 Hz (i.e. 9th harmonic of 15 Hz) modulation. These two concentric cylinders are fastened rigidly together to rotate as a single unit at 900 rpm, thus corresponding to a rotational speed of 15 cycles per second. The entire assembly is stabilized for ship pitch and roll, and, as seen in Figure 3-15, protected from the environment by a fiberglass radome.

b. PHYSICAL CONFIGURATION — [See Figure 3-16].

c. DIMENSIONS — (1) Height: 100 inches; (2) Diameter: 76¾ inches; (3) Weight: 975 pounds. Dimensions include radome, antenna, and pedestal.

d. INPUT IMPEDANCE/VSWR — 50 ohms, VSWR not greater than 2:1 over each frequency range.

e. RF POWER RATING — 5 kW peak at 2.5% duty cycle.

f. POLARIZATION — Vertical.

g. TYPE OF FEED — Unbalanced, coaxial cable.

h. PRIMARY POWER REQUIRED — 115 Vac, 60 Hz, 10 amp, one-phase for convenience outlet. All other power and control voltages are supplied from the associated control amplifier group.

i. OPERATING TEMPERATURE — Minus 28°C to +65°C.

j. INPUT CONNECTOR — Type LC.

k. INSTALLATION REQUIREMENTS — The antenna should be placed as high as possible on the ship, preferably at the uppermost part of the topside, and must be above objects in the immediate vicinity. Objects within 15 feet or less should not be above the lower edge of the base if optimum performance is to be obtained. The installing activity must supply a base plate on which to mount the pedestal. The "N" and "S" marks on the pedestal must be lined up paralled to the ship's centerline to an accuracy of 0.1 degree with the "N" mark facing forward. This system requires roll and pitch inputs from the ship for antenna stabilization.

l. SPECIAL CONSIDERATIONS — The cable lengths between the antenna and its control unit must not exceed 150 feet.

m. REFERENCES — (1) Technical Manual, Radio Set AN/URN-3, NAVSEA-0967-052-6010. (2) NAVSEA Drawing RE-F2683997.

NOTE 1: AS-777, also a low-band antenna, is identical to the AS-677 except for the type of drive motor and tachometer assembly.
NOTE 2: AS-678, part of antenna group OA-554, is identical to the AS-677 except that the outer concentric cylinder of parasitic reflector wires is smaller in diameter (35 inches). This is because the AS-678 is a high-band TACAN antenna, transmitting between 1151 and 1213 MHz and receiving between 1088 and 1150 MHz.

ANTENNA AS-889/SRN-6 and AS-890/SRN-6

a. TYPE/USE — The AS-889 (low band) and the AS-890 (high band) antennas, part of the OA-1545 and 1546 antenna group respectively, were used early with the AN/SRN-6 TACAN system. Except for the slight difference in radome appearance, as seen in Figure 3-17, the antennas are virtually identical in description and function to the AS-677 and AS-678 antennas described above.

ANTENNA AS-1075/SRN-6

a. TYPE/USE — The AS-1075 antenna is part of antenna group OA-1801/SRN-6 used with the shipboard TACAN system AN/SRN-6. It is considerably smaller (less than half the size and more than one third lighter) than its predecessor AS-677/URN-3 and AS-889/SRN-6 antennas described above. The primary radiator of the AS-1075 is an array of two dipoles stacked vertically along the centerline of the antenna. As with the previous TACAN antennas, there are two concentric fiberglass cylinders rotating as a unit at 900 rpm about the central array. The inner cylinder contains a single parasitic director, while the outer cylinder has nine elements acting as individual reflectors to modulate the radiation pattern. The antenna components, spin motor, rotating assembly, and bearing reference servo assembly are mounted as a unit on a single base in the pedestal. This base is maintained in a horizontal plane by the roll servo system. Roll correction is limited to 25° by mechanical stops on the pedestal base. The entire assembly is enclosed within a fiberglass radome, and when installed topside (preferably at the top of the mast) appears as in Figure 3-18.

b. PHYSICAL CONFIGURATION — [See Figure 3-19].

c. DIMENSIONS — (1) Height: 39 inches; (2) Diameter: 36 inches; (3) Weight: 275 pounds. Dimensions include antenna and pedestal.

d. FREQUENCY RANGE — Transmit: 962 to 1024 MHz. Receive: 1025 to 1150 MHz.
e. INPUT IMPEDANCE/VSWR — 50 ohms, VSWR not greater than 2:1 over frequency range.
f. RF POWER RATING — 20 kW peak at 2% duty cycle.
g. POLARIZATION — Vertical.
h. TYPE OF FEED - Unbalanced, coaxial cable.
i. PRIMARY POWER REQUIRED — 115 vac, 60 Hz, one-phase for convenience outlet. All other power and control voltages are supplied from OA-2619/SRN-6 amplifier group.
j. OPERATING TEMPERATURE — Minus 28°C to +65°C.
k. INPUT CONNECTOR — Type C.
l. INSTALLATION REQUIREMENTS — The antenna assembly (antenna, pedestal, and radome) should be placed as high as possible and must be above obstructions in the immediate vicinity. Objects within 15 feet or less should not be higher than the lower edge of the pedestal for optimum TACAN performance. The installing activity must supply a base plate on which to mount the pedestal. The "N" and "S" marks on the pedestal must be lined up parallel to the ship's centerline to an accuracy of 0.1 degree, with the "N" mark facing forward. This system requires an input from the ship's stable element to provide roll information for antenna stabilization.
m. SPECIAL CONSIDERATIONS — The cable lengths between the antenna and amplifier group must not exceed 200 feet.
n. REFERENCES — (1) Technical Manual, Antenna Group OA-1801/SRN-6, NAVSEA 0967-886-7010. (2) NAVSEA Drawing RE-D2692148.

ANTENNA AS-2822/SRN-15

a. TYPE/USE — The AS-2822 is a compact, lightweight TACAN antenna used with the AN/SRN-15 Beacon-Transponder. A low-power, simplified TACAN intended for ship-to-helicopter navigation as part of the LAMPS program, the AN/SRN-15 provides coverage out to 100 nautical miles. Since the antenna consists only of a central radiator and single 15 Hz parasitic element (i.e., no 135 Hz parasitic reflector element) the accuracy of azimuthal information is on the order of 3 degrees rather than the nominal 1 degree available from standard TACAN. The antenna assembly is roll stabilized by a pendulum-type mechanism.
b. PHYSICAL CONFIGURATION — [See Figure 3-20].
c. DIMENSIONS — (1) Height: 37 inches; (2) Width: 31¼ inches; (3) Depth: 18 inches; (4) Weight: 118 pounds.
d. FREQUENCY RANGE — Transmit: 962 to 1024 MHz. Receive: 1025 to 1087 MHz.
e. INPUT IMPEDANCE/VSWR — 50 ohms, transmit VSWR not greater than 1.5 to 1 and receive VSWR not greater than 2.8:1 over frequency range.
f. RF POWER RATING — 5 kW peak.

g. POLARIZATION — Vertical.
h. TYPE OF FEED — Unbalanced, coaxial cable.
i. PRIMARY POWER REQUIRED — All power and control voltages are supplied from C-9146/SRN-15 control-transponder set.
j. OPERATING TEMPERATURE — Minus 28°C to +65°C.
k. INPUT CONNECTOR — Type C.
l. INSTALLATION REQUIREMENTS — The antenna assembly should be located at the top of the mast. Objects within 15 feet or less should not be higher than the lower edge of the pedestal for optimum performance. The base of the pedestal has eight equally spaced $^9/_{16}$-inch holes on a 10-inch diameter bolt circle for mounting the antenna to the ship's structure. The reference marks on the antenna base must be aligned with the ship's centerline.
m. SPECIAL CONSIDERATIONS — This antenna is stabilized by pendulum-dashpot action and does not require roll information from the ship.
n. REFERENCE — Technical Manual, Beacon-Transponder Set AN/SRN-15, NAVELEX 0967-434-3010.

ANTENNA AS-4036/SRN-15A

a. TYPE/USE — The AS-4036 is a mechanically rotating, direct drive antenna that provides all-band (252-channel) capability with precise navigation accuracy. This is in contrast to the AS-2822 described above which was low-band only, and afforded limited accuracy. The broadband capacity of the AS-4036 is achieved through subdividing the TACAN band into low, mid, and high segments, and implementing through the use of three quick-change frequency band kits. These kits are color coded — red for low band, yellow for mid band, and blue for high band.
The antenna assembly consists of a 15 Hz parasitic element rotating cylinder having a set of two removable 15 Hz parastic elements, a 135 Hz parasitic element rotating drum assembly, upper and lower radome sections, spin motor drive assembly, three monitor antennas, and three frequency band change kits. Each kit contains one central array and one set of nine removable 135 Hz parasitic elements. One set of two removable 15 Hz parasitic elements is used for all three bands. Access covers are provided in the top of the upper radome and in the antenna-to-mast adaptor spool to allow for a frequency kit change to be made without removing the upper radome. The monitor antennas, provided to sample the RF radiation, are located in the lower radome.
The AS-4036 is part of the AN/SRN-15A Beacon-Transponder set, a lightweight, small TACAN system designed specifically for use aboard ship as an aid to helicopter navigation.
b. PHYSICAL CONFIGURATION — [See Figure 3-21].
c. DIMENSIONS — (1) Height: 36 inches; (2) Diameter: 41 inches; (3) Weight: 125 pounds.

d. FREQUENCY RANGE — (1) Low Band (Red): Transmit 962-1028 MHz; Receive 1025-1091 MHz. (2) Mid Band (Yellow): Transmit 1029-1109 MHz; Receive 1025-1046, and 1092-1150 MHz. (3) High Band (Blue): Transmit 1110-1213 MHz; Receive 1047-1150 MHz.
e. INPUT IMPEDANCE/VSWR — 50 ohms; VSWR not greater than 1.5:1 (transmit) and 2.5:1 (receive).
f. RF POWER RATING — 5.0 kW peak power continuous.
g. POLARIZATION — Vertical.
h. TYPE OF FEED — Unbalanced, coaxial cable.
i. PRIMARY POWER REQUIRED — 208 ± 10 Vac, 60 Hz, 15 amp, three-phase.
j. OPERATING TEMPERATURE — Minus 28° to +90°C.
k. INPUT CONNECTOR — Type N.
l. INSTALLATION REQUIREMENTS — The antenna should be placed atop the mainmast, i.e., the highest item on the ship, with the antenna centerline parallel to the ship's centerline within plus or minus one-half degree. The side of the antenna which faces forward, toward the bow, is identified with the marking "BOW." Two sets of 11/16 inch diameter holes are provided in the base of the antenna for mounting. Only eight of the holes are used (choice of either 14-inch diameter or 11½-inch diameter), employing eight 5/8-11 UNC mounting bolts.
m. REFERENCE — Technical Manual, Beacon-Transponder Set AN/SRN-15A, NAVELEX 0967-434-3014.

ANTENNA AS-1747/URN-20(V)

a. TYPE/USE — The AS-1747 is part of the OA-7203/URN-20(V) antenna group used with the AN/URN-20(V) solid-state shipboard TACAN system. A central array, consisting of two vertically stacked, bi-cylindrical dipoles, forms the primary RF radiator. Two concentric glass fiber-reinforced plastic cylinders are placed around the central array on a rotating aluminum disk. The inner cylinder has a diameter of approximately six inches. Attached to this cylinder are two lengths of wire which act as parasitic elements to intercept some of the energy radiated from the central array and re-radiate it so that it reinforces the signal in one direction and attenuates it in the opposite direction. The pattern is then reshaped from a circle to a cardioid. To obtain the desired amount of modulation, two parasitic elements are actually used. However, their effect is additive and they act as though a single parasitic director were located midway between them.

The outer cylinder has an approximate diameter of 30 inches. Nine groups of thin metallic parasitic elements are attached to the cylinder and act as reflectors, producing a cogwheel effect on the overall radiated pattern. This final radiated pattern is rotated by means of mechanically spinning the two cylinders at 15 revolutions per second. The central array is stationary (electrical energy is connected through the hollow shaft of the spin motor assembly). No direct electrical connection is required to the cylinders. To an aircraft, the radiated signal appears to be amplitude modulated at 15 Hz and 135 Hz rates, thus producing the desired effect for bearing information.

The antenna array spin motor, rotating assembly, and bearing control servo are all mounted on a single base in the pedestal. This base is maintained in a horizontal plane by the roll servo with mechanical stops limiting roll correction to ± 28 degrees. The entire antenna assembly and pedestal are enclosed by a glass fiber, reinforced plastic, protective radome.

b. PHYSICAL CONFIGURATION — [See Figure 3-22].

c. DIMENSIONS — (1) Height: 54½ inches; (2) Diameter: 65 inches; (3) Weight: 480 pounds. These are overall dimensions for the antenna group.

d. FREQUENCY RANGE — Transmit: 962 to 1024 MHz; Receive: 1025 to 1150 MHz.

e. INPUT IMPEDANCE/VSWR — 50 ohms; VSWR not greater than 2:1 over frequency range.

f. RF POWER RATING — 20 kW peak at 2% duty cycle.

g. POLARIZATION — Vertical.

h. TYPE OF FEED — Unbalanced, coaxial cable.

i. PRIMARY POWER REQUIRED — (1) 440 volts, three-wire, three-phase, 60 Hz, 1.75 kva, operating current 3 amps per phase, starting current 3 times operating current. (2) 120 volts, one-phase, 60 Hz, 0.07 kva. (3) Convenience outlets; 120 v, one-phase, 60 Hz, 8 amps.

j. OPERATING TEMPERATURE — Minus 28° to +95°C.

k. INPUT CONNECTOR — Type C.

l. INSTALLATION REQUIREMENTS — The antenna assembly should be placed as high as possible on the ship, normally at the very top of the mainmast. It must be above all objects in its immediate vicinity. Objects within 15 feet or less should not be higher than the lower edge of the pedestal for optimum TACAN performance. The installing activity must supply a base plate on which to mount the pedestal. The "N" and "S" marks on the pedestal must be lined up parallel to the ship's centerline to an accuracy of 0.1 degree, with the "N" mark facing forward. The system requires an input from the ship's stable element to provide roll information for antenna stabilization.

m. SPECIAL CONSIDERATIONS — The cable lengths between the antenna and amplifier group must not exceed 200 feet.

n. REFERENCES — (1) Technical Manual, Antenna Group OA-7203E/URN-20(V), NAVSEA 0976-041-0130. (2) NAVSEA Drawing RE-D2695710.

ANTENNA AS-3240/URN

a. TYPE/USE — All the TACAN antennas discussed to this point share a common singular feature in that they rely upon an electro-mechanical device (a stationary central array encircled by one or two rotating concentric cylinders) to modulate and form the radiation pattern. Though satisfactory in function, these antennas are always extremely large and heavy, due mainly to the motorized assemblies required for spinning the cylinders and roll-stabilizing the antenna. Over a twenty-five year period, from the AS-677/URN-3 to the AS-1747/URN-20, the antenna weight in fact has been halved, from approximately 1000 pounds to near 500 pounds — a significant reduction, but still leaving a massive UHF antenna seeking placement at, or near, the top of the masts. It is, of course, highly desirable to install the TACAN antenna at the ship's highest point, normally atop the mainmast, free of all obstructions, for optimum performance in navigating pilots safely home.

Moreover, because the size and weight of these antennas would be absolutely intolerable if they were full range, they are necessarily limited to portions of the TACAN spectrum such as low-band or high-band. Thus it was a major breakthrough (of most welcome import to the Navy) when the new lightweight all-band AS-3240 TACAN antenna was developed and approved for fleetwide service.

What makes the AS-3240 lighter in weight than any shipboard TACAN antenna heretofore, is the use of electronic scanning rather than mechanical rotation, and the elimination of the need for roll stabilization.[13]

As part of the OE-273(V)/URN antenna group available for use with the Navy's new all-band (252-channel) AN/URN-25 TACAN, but readily adaptable to any of the older systems such as the AN/URN-20, the AS-3240 is ideally suited for mast-top installation.

The antenna assembly has three major components: the RF sub-assembly, pedestal, and pedestal electronics. The sub-assembly consists of replaceable parasitic modules arranged in a circular pattern on an aluminum honeycomb sandwich counterpoise. An inner ring contains 15 Hz modules encircling the central monopole radiator. An outer ring comprises the 135 Hz parasitic modules. Each module is a printed circuit board with divider, resistors, and a means for electrical connections. The modules are each supported by high temperature urethane foam bonded to the printed circuit board.

Two RF monitor probes, located on the counterpoise, sample the radiated output of the antenna for continuous analysis by the control monitor circuits. The electronic assembly mounted in the antenna pedestal contains printed circuit boards for digital antenna pattern generation, diode modulator drive, and monitoring of signals and drive. The entire active portion of the antenna assembly is enclosed by a fiberglass honeycomb radome to protect it from the sea environment. A detachable, RF-transparent, lightning arrestor is supplied also.

In essence, rather than forming the so-called "cogwheel" composite TACAN radiation pattern by mechanical rotation (modulation), the AS-3240 achieves the same far-field effect by digital switching of parasitic elements arranged in concentric arrays about the central radiator. Twelve inner parasitic elements provide the 15 Hz modulation, replacing the single phase rotating parasite in the mechanically rotating antenna. Thirty-two outer parasitic elements provide 135 Hz modulation, serving the same purpose as the nine outer elements of the rotating antenna. Beacon RF energy of the AS-3240 is radiated from a monopole at the counterpoise center. The 15 Hz and 135 Hz pattern modulation is generated by binary switching of the diodes in each of the parasitic units, in prescribed time sequence which is repeated once in each 15 Hz interval. In effect, the active parasites are rotated electrically rather than mechanically. Furthermore the basic bandwidth limitation of mechanical rotating antennas is overcome in the electronically scanned AS-3240 by employing three individual concentric rings of elements in the 135 Hz parasitic array. Each ring is positioned and adjusted to meet the necessary constraints for correct modulation over the entire TACAN range. The appropriate ring for a given frequency segment is activated by a fast electronic switch based on information from the beacon frequency synthesizer. The technique allows instantaneous band switching, resulting in an all-band operation.

b. PHYSICAL CONFIGURATION — [See Figure 3-23].

c. DIMENSIONS — (1) Height: 22 inches (antenna); 75 inches (lightning rod); (2) Diameter: 53 inches, (3) Weight: 107 pounds (including lightning rod).

d. FREQUENCY RANGE: All-band (252 channels); Transmit: 962-1024 and 1151-1213 MHz; Receive: 1025-1150 MHz.

e. INPUT IMPEDANCE/VSWR — 50 ohms, VSWR less than 2:1.

f. RF POWER RATING — 10 kW peak.

g. POLARIZATION — Vertical.

h. TYPE OF FEED — Unbalanced coaxial cable.

i. PRIMARY POWER REQUIRED — Operates directly from ac line power sources 60 to 400 Hz.

j. OPERATING TEMPERATURE — Minus 28°C to +95°C.

k. INPUT CONNECTOR — UG-30E/U.

l. INSTALLATION REQUIREMENTS — The antenna is designed and intended for top of the mainmast installation. When unable to place atop the mast, but as high as possible on the ship, other objects within 15 feet should not be higher than 10 feet below the pedestal mounting flange. It is preferable that objects beyond 15 feet be also below this line. If a thin object, such as a whip or steel mast, must project above this line, it should be a minimum of 50 feet away. Mounting of the AS-3240 is accomplished by eight mounting holes $^{11}/_{16}$ inches in diameter spaced 45 degrees apart on a 14-inch diameter aluminum pedestal flange.

m. SPECIAL CONSIDERATION — Ship's dry air at 1/10 psi must be supplied or optional desiccant agent provided. Cable lengths must not exceed 250 feet.

n. REFERENCES — (1) NAVELEX Installation Control Drawing, Antenna AS-3240/URN. (2) ITT Avionics Division Pamphlet, OE-273(V)URN, Solid State Tacan Antenna.

REFERENCES

1. L.S. Howeth, *History of Communications-Electronics in the United States Navy* (Washington, D.C.: U.S. Government Printing Office, 1963), p. 261.
2. *Ibid.*, p. 263.
3. *Ibid.*, p. 264.
4. *Ibid.*
5. Louis A. Gebhard, *Evolution of Naval Radio-Electronics and Contributions of the Naval Research Laboratory* (Washington, D.C.: Naval Research Laboratory, Report 8300, 1979), p. 264.
6. *Ibid.*, p. 267.
7. James J. Heigl, Jr., "The Quiet Revolution in Radio Navigation," *Signal*, March 1972, pp. 22-24.
8. Henry Jasik, *Antenna Engineering Handbook* (New York: McGraw-Hill Book Company, 1961), pp. 28-21 through 28-24.
9. Leonard J. Jacobson, "GPS Is On The Ground," *Signal*, March 1979, p. 47.
10. Gebhard, *op. cit.*, p. 271.
11. *Ibid.*, p. 274.
12. Robert I. Colin and Sven H. Dodington, "Principles of Tacan," *Electrical Communication*, March 1956, p. 16.
13. G.P. Scherer, "Electronically Scanned Shipboard Tacan Antenna," *Electrical Communication*, Vol. 50, No. 4, 1975, pp. 253-257.

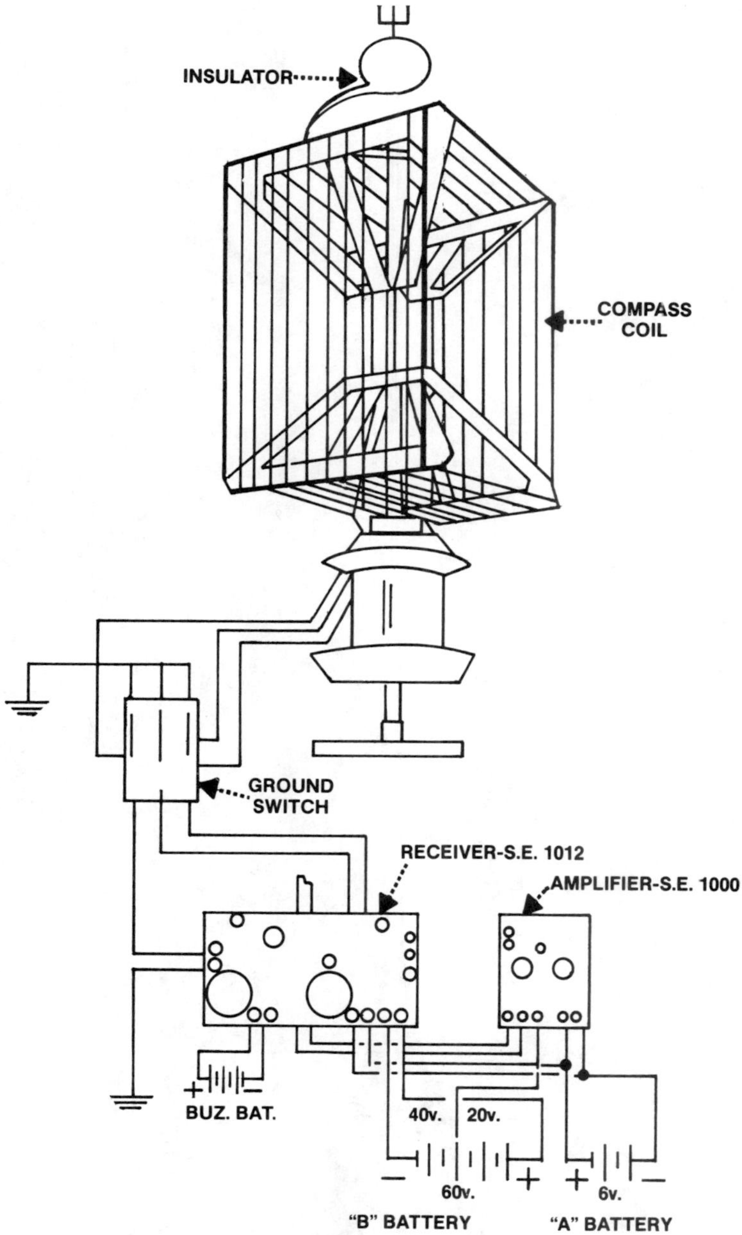

FIG. 3-1 Early U.S. Navy Radio Direction Finder Components and Connections

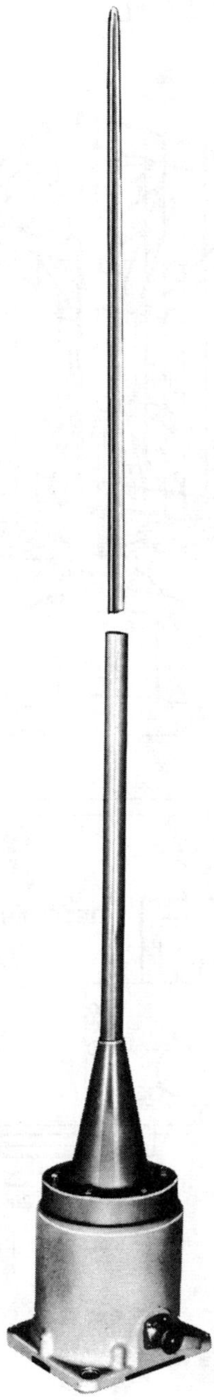

FIG. 3-2 AS-2283 VLF Whip OMEGA Navigation Antenna

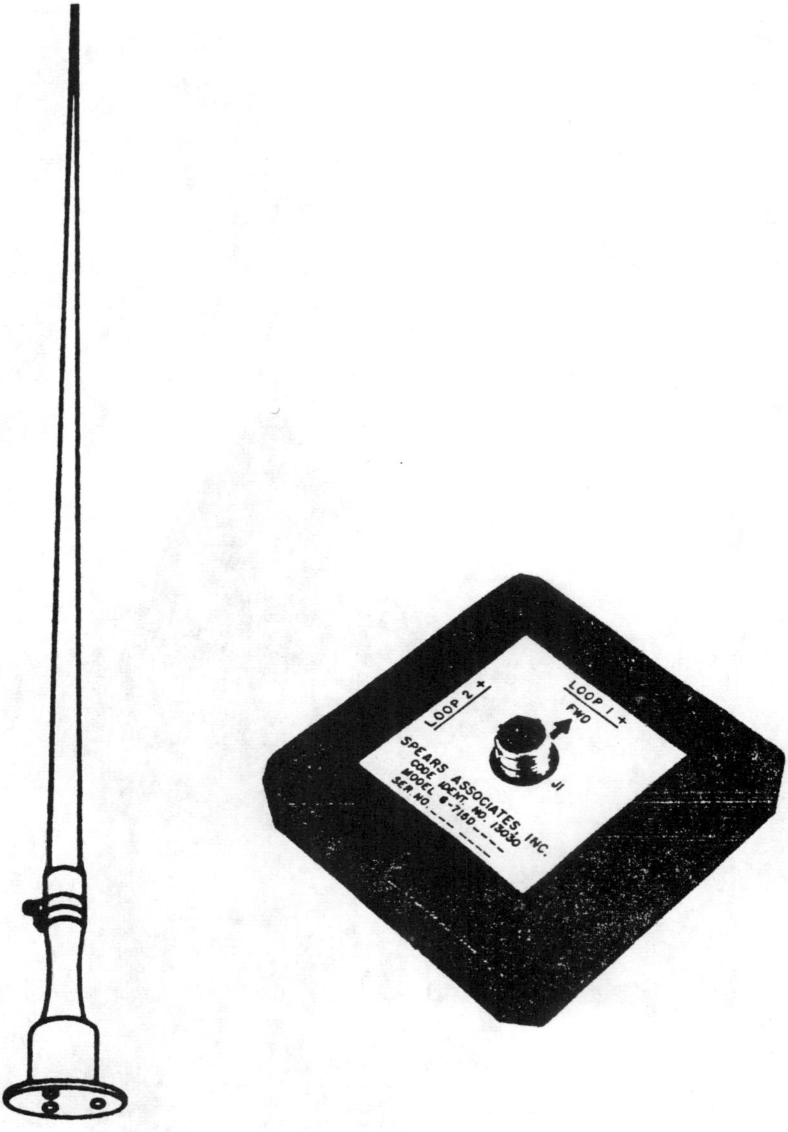

(a) (b)

Fig. 3-3 (a) AS-2271 VLF Whip OMEGA Antenna; (b) LTN-211 Loop Antenna

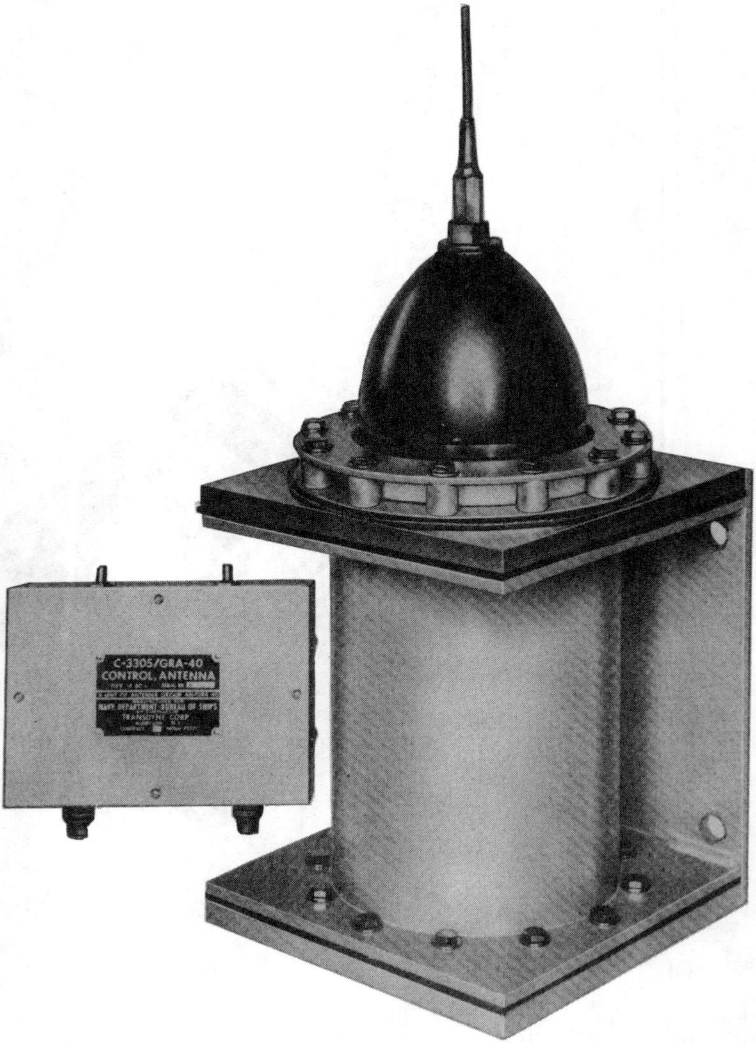

FIG. 3-4 AT-979/U LORAN Whip Antenna

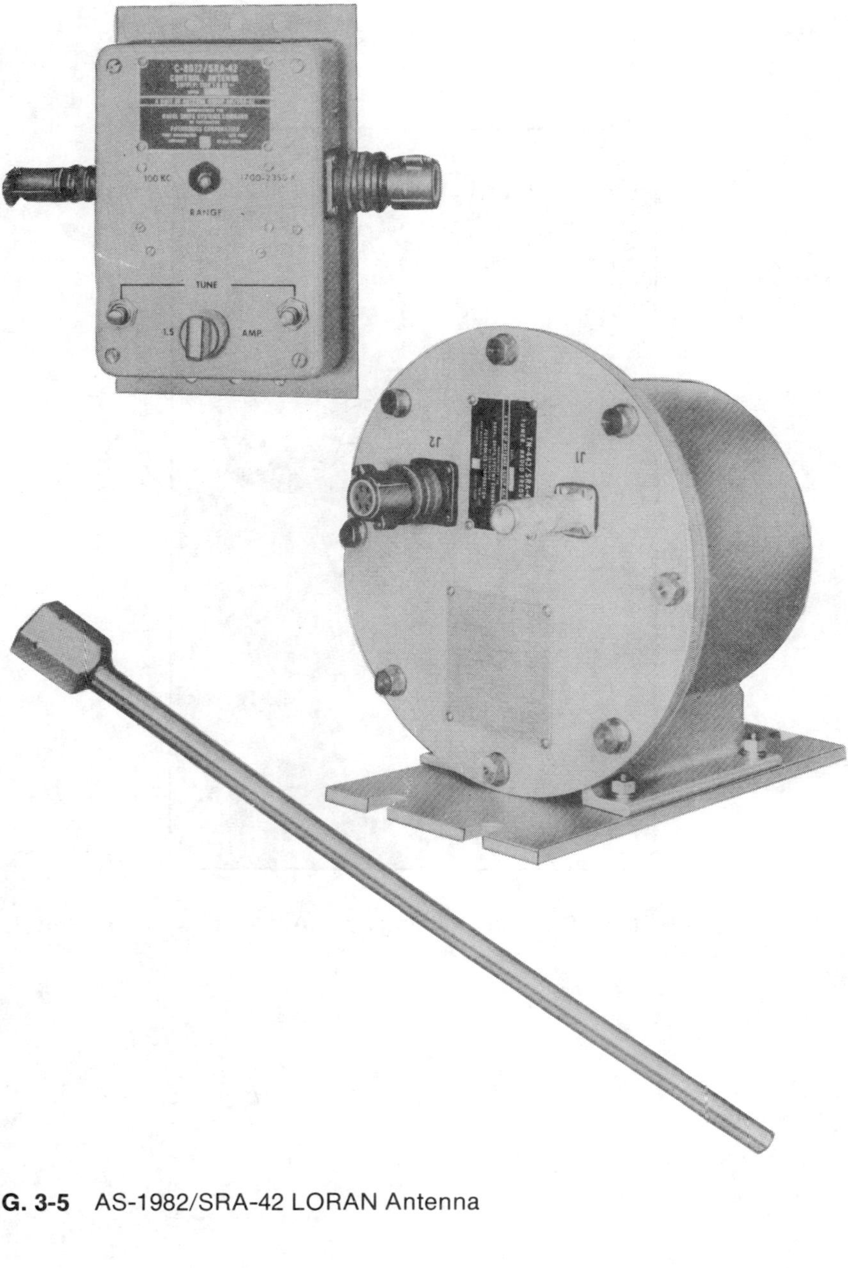

FIG. 3-5 AS-1982/SRA-42 LORAN Antenna

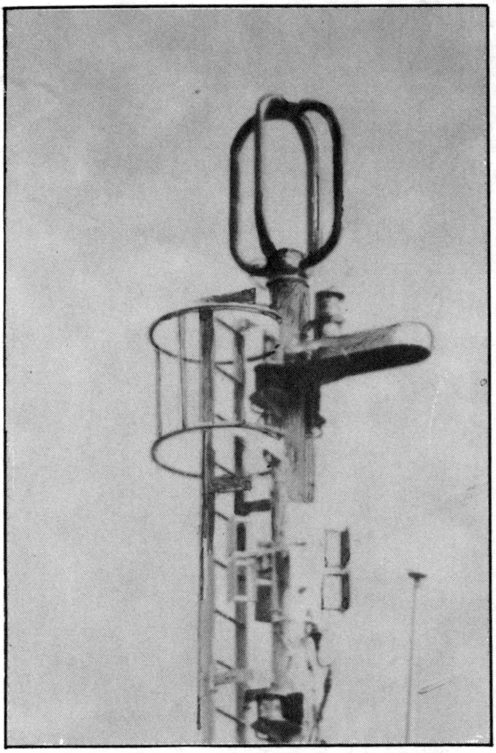

FIG. 3-6 LF/MF Radio Direction Finding Loop Antenna

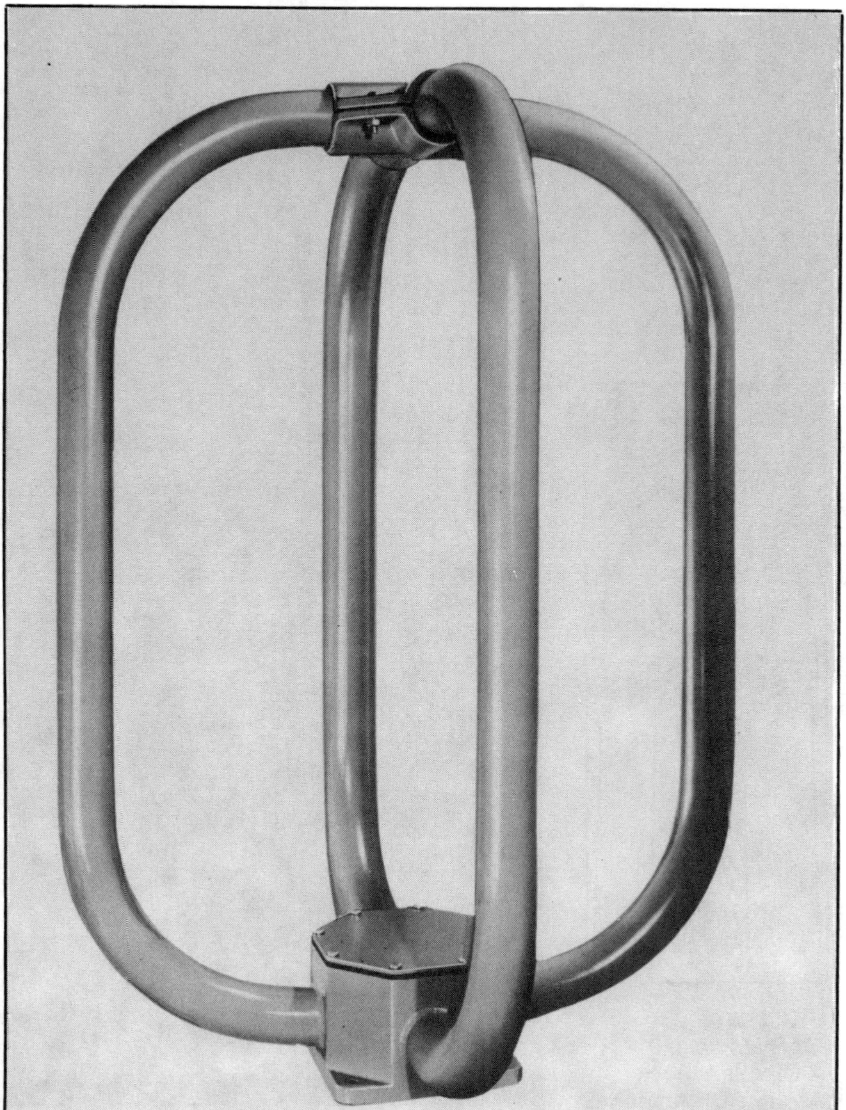

FIG. 3-7 4005A Crossed Loop DF Antenna
(Photo courtesy of ITT Mackay Marine)

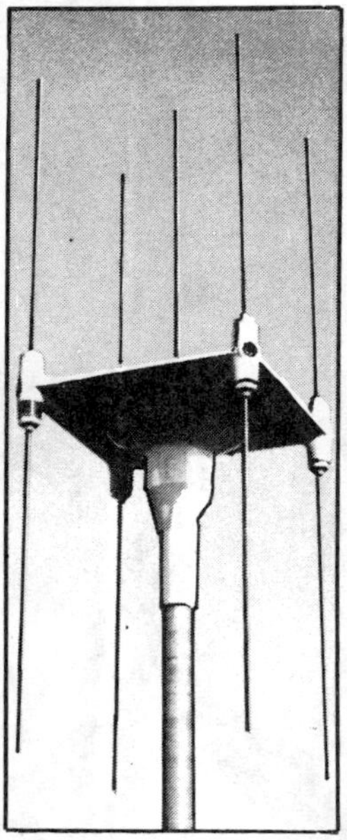

FIG. 3-8(a)
OAR Model 320
4-Dipole Adcock
Arrary RDF Antenna

FIG. 3-8(b) OAR Model 320 Adcock Array Antenna Installed (Photo courtesy of OAR)

FIG. 3-9 AS-514A UHF Direction Finder Antenna Installed

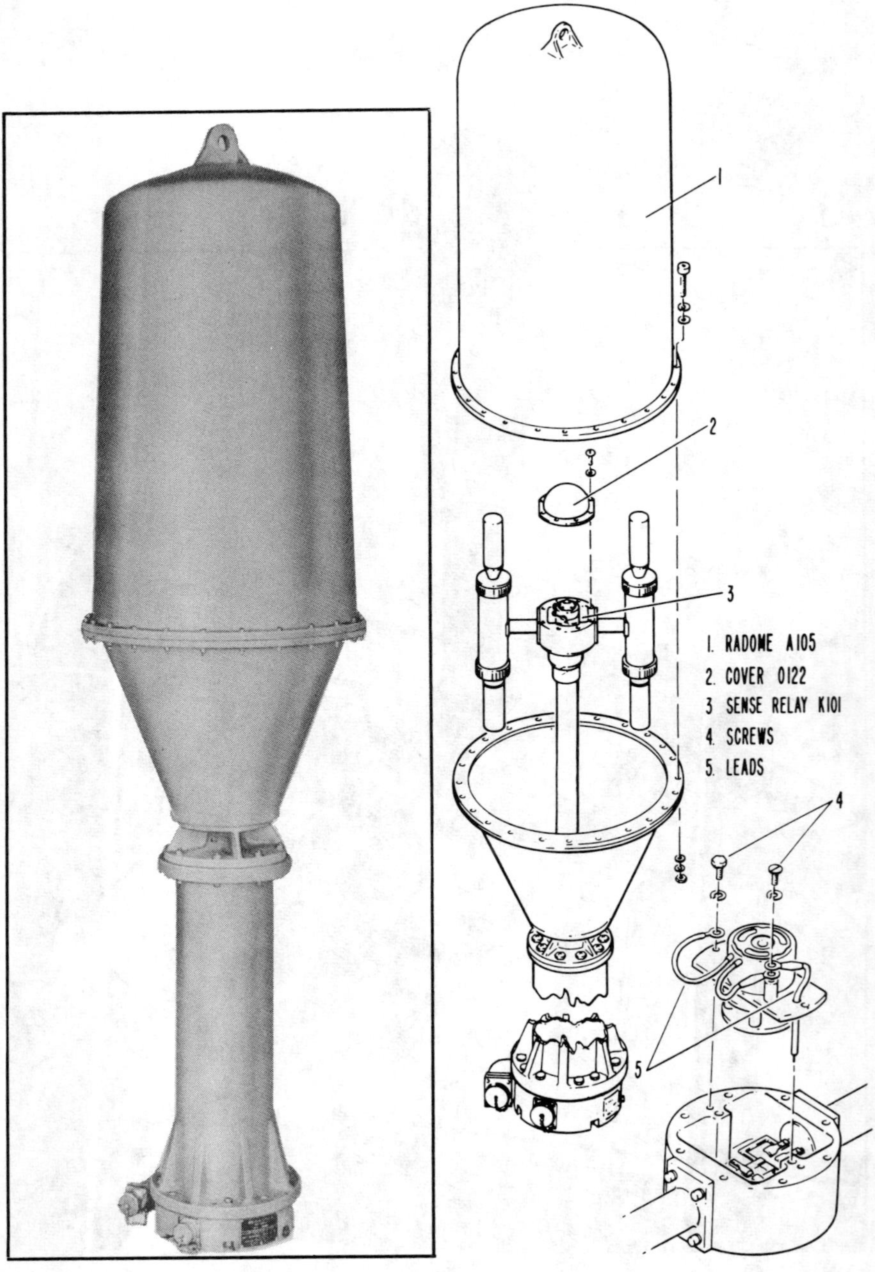

FIG. 3-10 AS-514A UHF Direction Finder Antenna

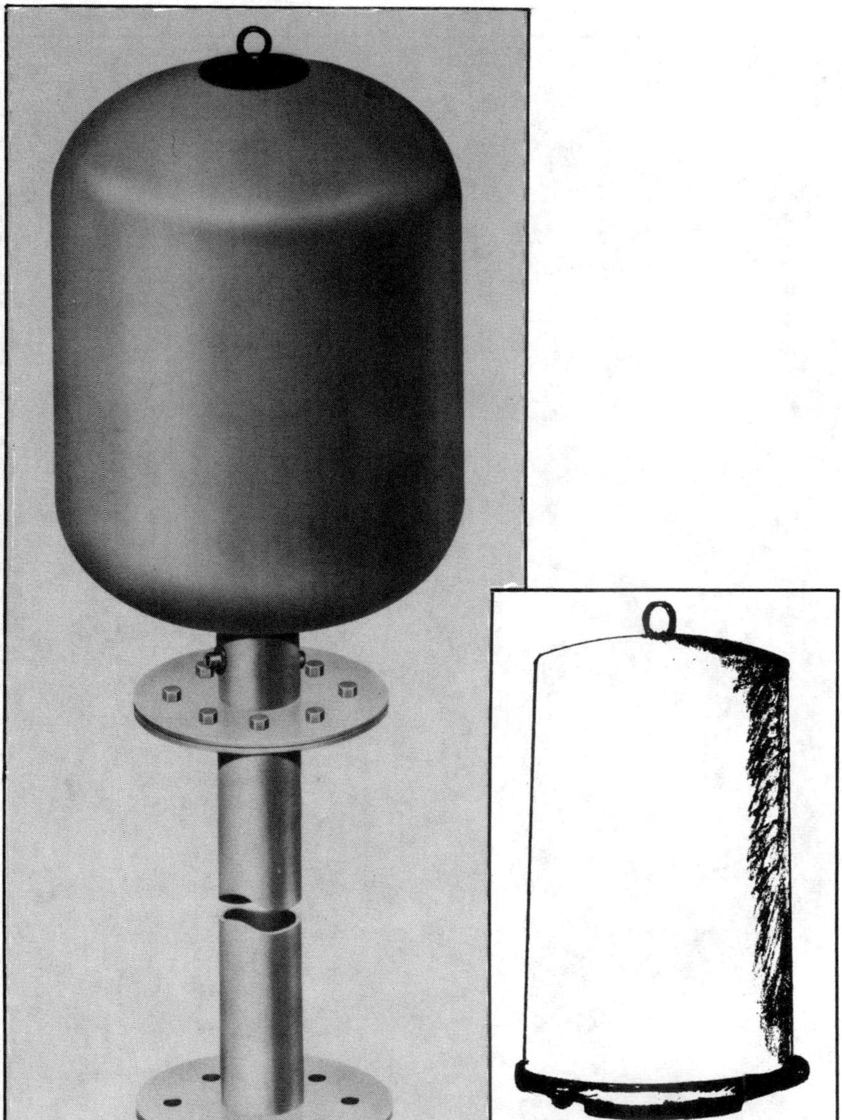

FIG. 3-11 AS-3386 UHF Direction Finder Antenna

FIG. 3-12 AS-2227 NAVSAT Antenna

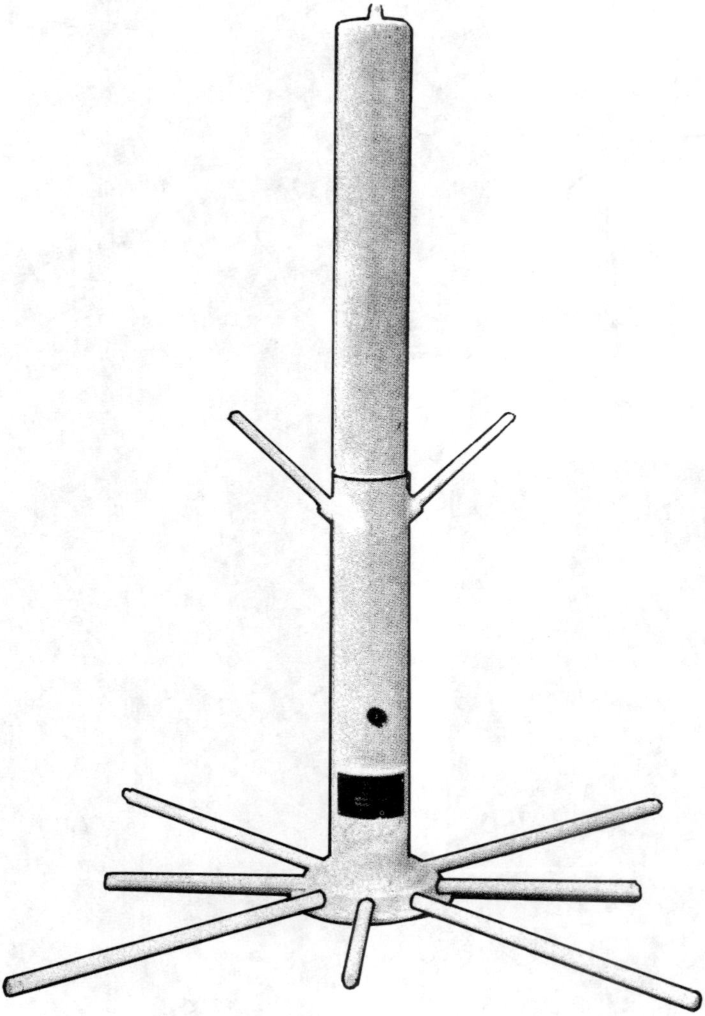

FIG. 3-13 CA-3086 NAVSAT Antenna
(Photo courtesy of Chu Associates)

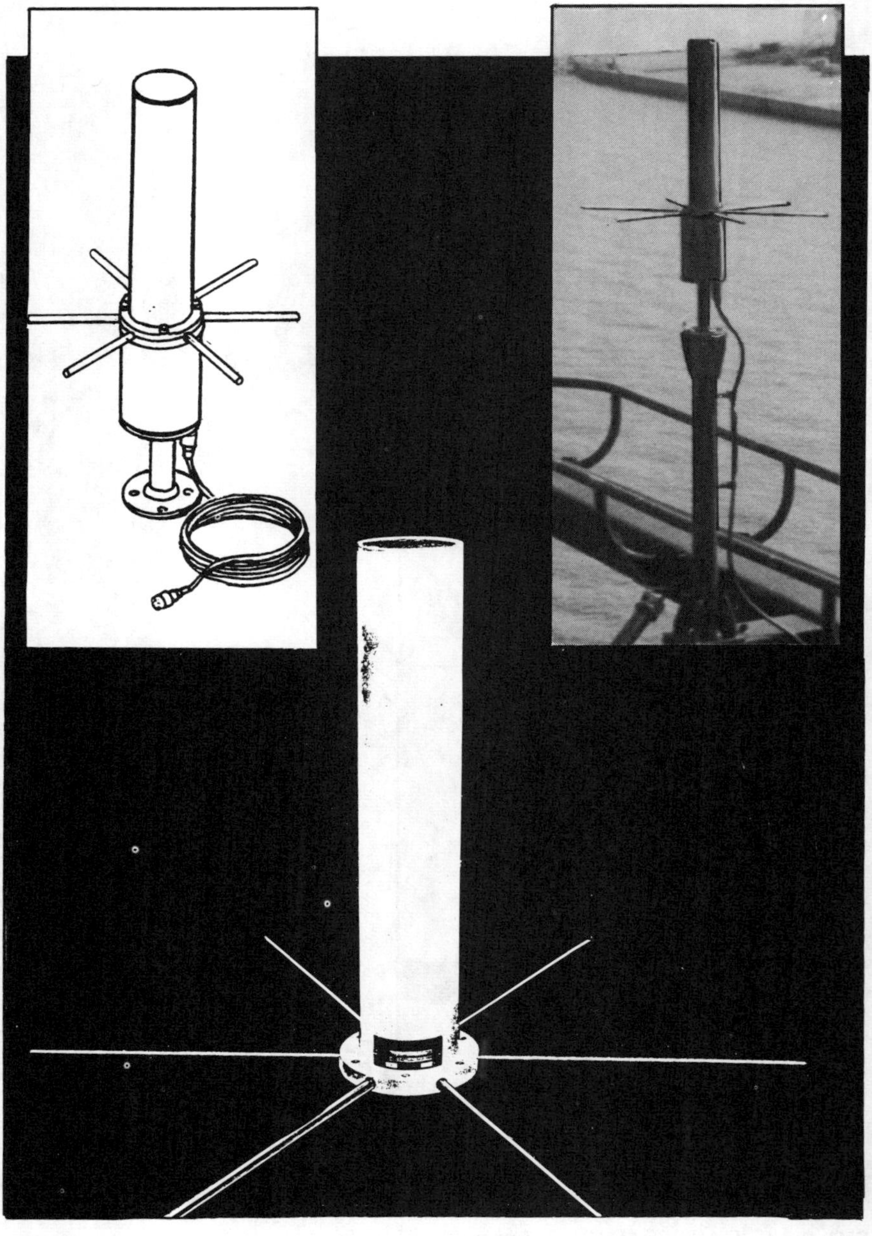

FIG. 3-14 AS-3330 UHF NAVSAT Antenna

FIG. 3-15 AS-677, AS-777, AS-678 TACAN Antenna

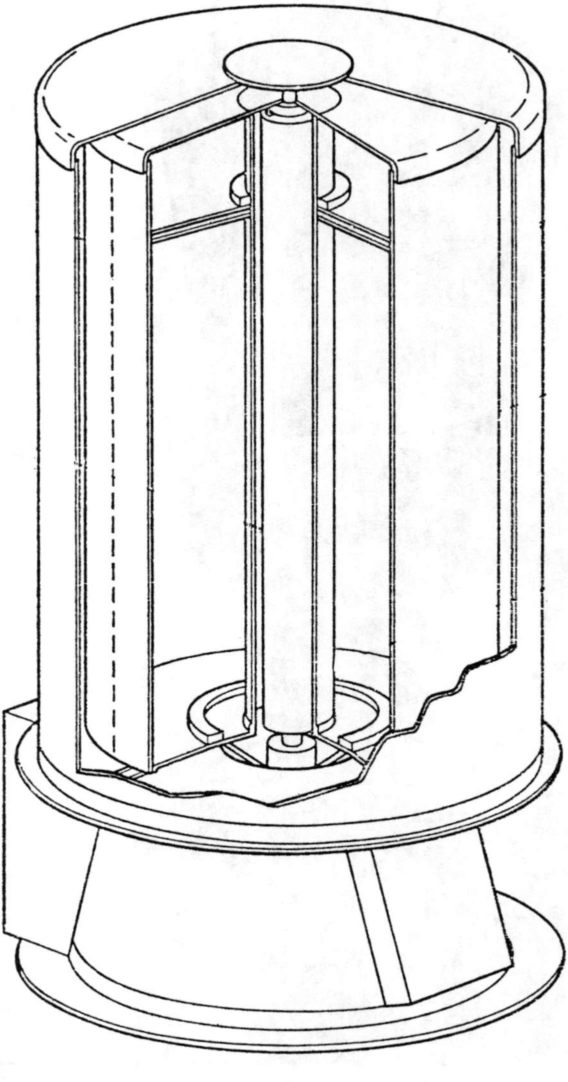

FIG. 3-16 AS-677, AS-777, AS-678 TACAN Antenna Physical Configuration

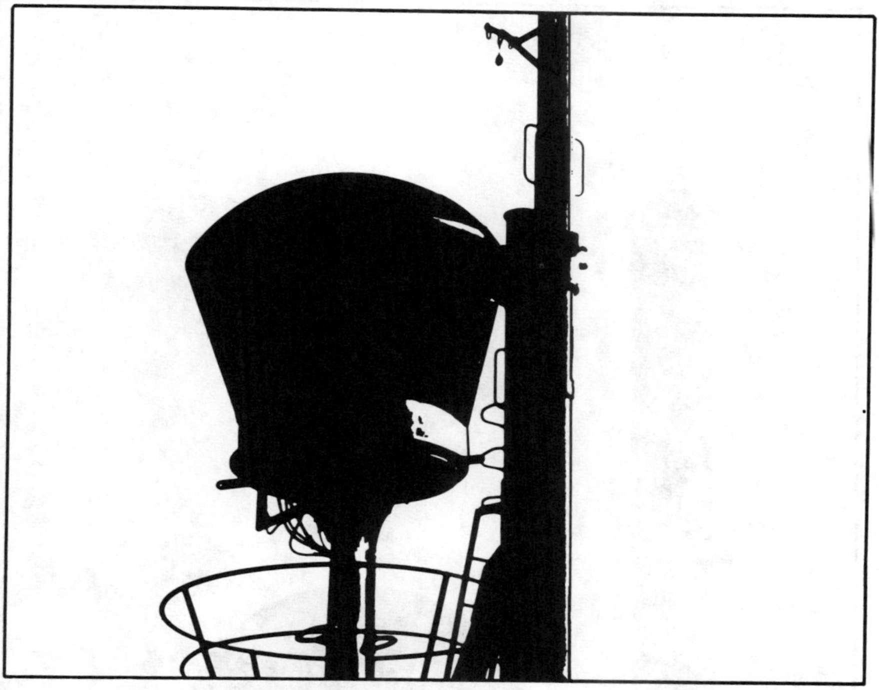

Fig. 3-17 AS-889, AS-890 TACAN Antenna

Fig. 3-18 AS-1075 TACAN Antenna

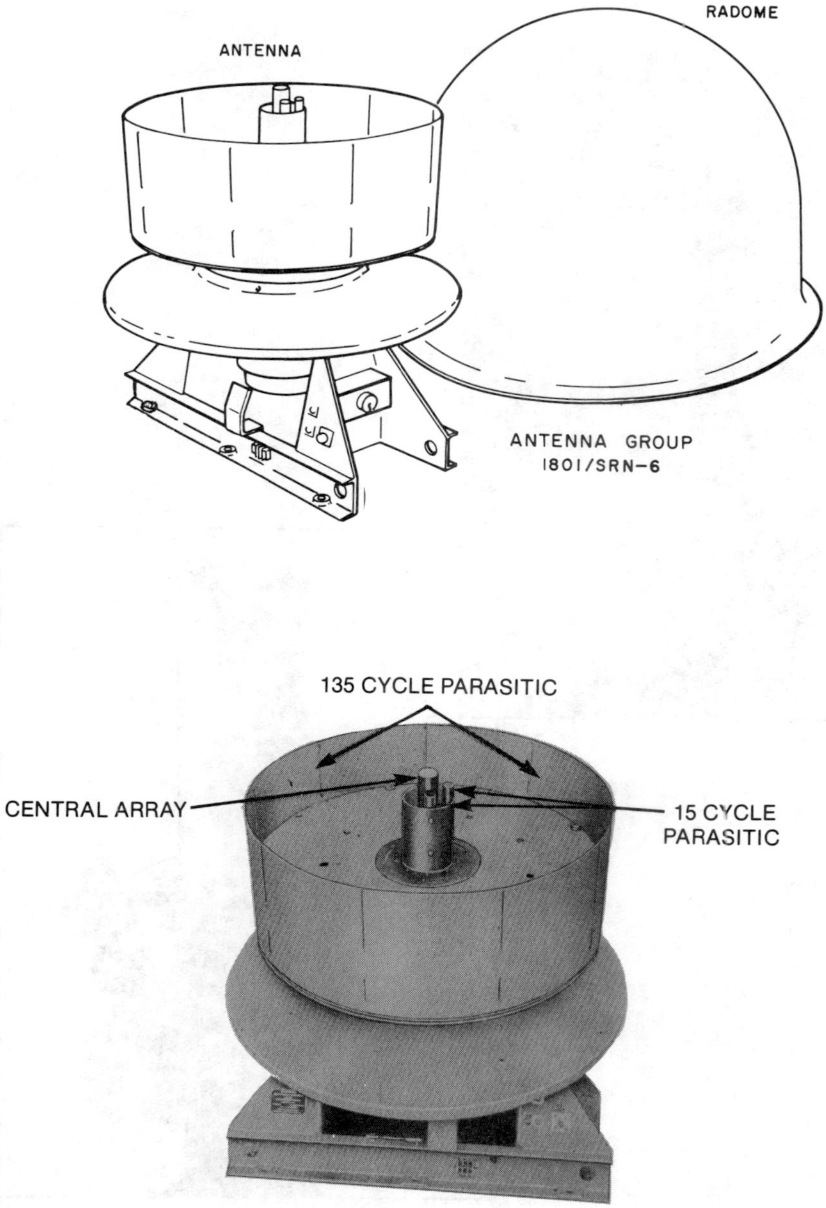

FIG. 3-19 AS-1075 TACAN Antenna Components

FIG. 3-20 AS-2822 Beacon-Transponder Antenna

FIG. 3-21 AS-4036 TACAN Antenna

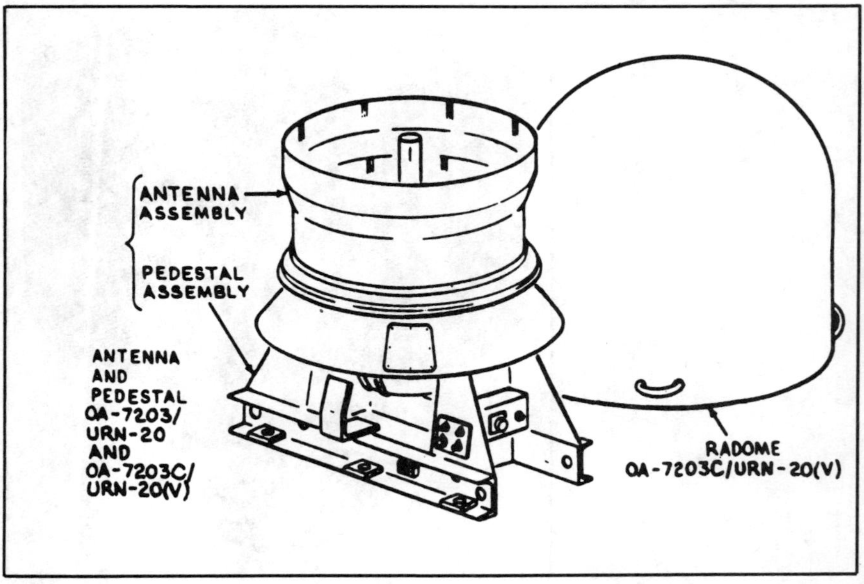

FIG. 3-22 AS-1747 TACAN Antenna

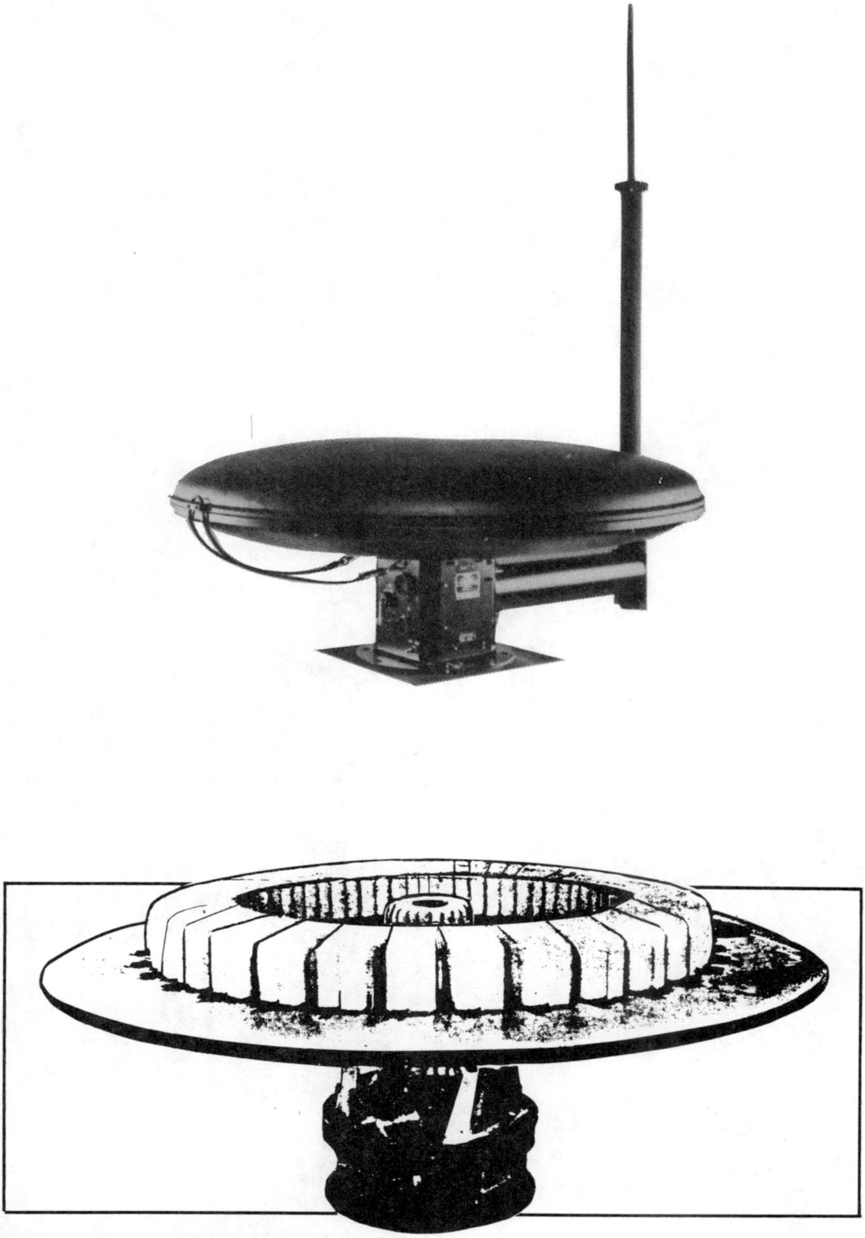

FIG. 3-23 AS-3240 TACAN Antenna

Chapter 4

Shipboard Radar Antennas

4-0 INTRODUCTION

> Yet, if the atomic bomb was not the decisive weapon of World War II in the Pacific, was there one at all, and what was it? What was it in the Atlantic? In Europe? . . . I am still of the opinion that the answer in all three theaters was one and the same: radar. It was radar that warned of the approach of the enemy, evolved and guided our tactics in both pursuit and evasion — in the air, on the sea, on the land, and under the sea — and aimed our weapons. Intelligence made a terrific difference too, but the big thing was radar.[1]

The above quotation by Edward L. Beach, Captain USN (Retired), a highly decorated World War II submariner and author of the book *Run Silent, Run Deep,* expresses unequivocally one veteran's opinion of the crucial role of radar in the 1940s. Now, forty years later, could anyone observing its progress underestimate the importance of radar today? Whether to control flight patterns of commercial airlines, prevent collisions at sea in bad weather, alert of approaching storms, or warn us to obey highway speed laws, radar has become essential to our daily lives. We have come to depend upon it, and take it for granted, as another of the silent, yet critical, elements of modern technology.

Like so many other leaps in technology, it was war, sadly enough, that spurred on the rapid development and immediate application of radar. Thus, unlike many other new technical devices of science and engineering, radar was originally developed specifically for military use. As the derivation of its name denotes, radar was thought of as a way to use radio for detecting and determining the location of enemy ships and planes. It is to the credit of man that its uses have been so widely and wisely adapted for peaceful purposes since World War II.

The U.S. Navy, as reviewed in Chapter Two, was one of the earliest proponents and pioneer users of radio communications. The U.S. Navy considers itself one of the very first developers, if not the original inventor, of radar. Without question it conducted some of the earliest experiments for practical application of radar. And two wartime U.S. Naval officers, Lieutenant Commanders F.R. Furth and S.M. Tucker, in an effort to keep secret its purpose, coined the very word "RADAR" as an acronym for radio detection and ranging equipment.[2] Both Furth and Tucker, incidentally, later served as directors of the Naval Research Laboratory (NRL) and rose to the rank of Rear Admiral.

Yet the principles of radio detection are about as old as the study of electromagnetism itself.[3] Heinrich Hertz, in the last part of the Nineteenth century, demonstrated that both metallic and dielectric bodies would reflect radio waves. Then, in 1903, Christian Hulsmeyer of Germany experimentally observed the detection of radio waves reflected from ships. A year later he patented his method as an aid for ship navigation and collision avoidance. A demonstration to the German Navy resulted in no great enthusiasm, however, due mainly to Hulsmeyer's radio results being limited to ranges of only a mile or so, and therefore being accepted as little better than visual lookouts.

Nearly twenty years elapsed before the subject was publicly proposed again, when, in 1922, Marconi suggested in a technical speech that a radio beam radiating upon an object such as a ship should reflect back to the sender information revealing the presence and bearing of the ship in fog or bad weather.[4] That very year, two Naval Research Laboratory scientists, A. Hoyt Taylor and L.C. Young, substantiated Marconi's speculations. Conducting experiments with high-frequency, continuous-wave radio transmissions across the Potomac River they noticed that a ship passing by interfered with their operations. More than likely perturbed by such interruptions at first, Taylor and Young, noting that the ship could be detected at ranges up to three miles, soon realized that here was a method permitting the detection of enemy ships at sea, irrespective of fog, darkness, or such man-made deceptions as smoke screens. Nevertheless, the Navy at this point in time was too busy satisfying pressing commitments for radio communication installations to sponsor further work in radio detection.[5]

Despite the lack of appreciation and large-scale funding for the development of radar *per se*, work continued in many related fields. One of the most important events was the application of pulsed transmission to radio techniques. In a cooperative effort with NRL, Gregory Breit and Merle Tuve of the Carnegie Insitute used radio pulses to investigate properties of the ionosphere, deriving height characteristics by calculating the time between initial transmission and return of reflected energy. The quiet intervals between pulses allowed clear reception of the echoes, a process that would be of fundamental importance to radar for determining range to target. Concurrently, there were other timely advances in radio engineering being made during this period: the cathode ray tube as an information display device, higher power transmitting tubes operating at higher frequencies, and much improved receiving tubes.[6] With the availability of these components and approaching threats of war, the development of radar began to accelerate to its fever pitch of the 1930s and early '40s.

In June 1935, Sir Robert Watson-Watt of England succeeded in receiving pulsed echoes from a flying boat at a range of 15 miles. By September, using a frequency of 12 MHz, ranges over 40 miles were attained on bomber aircraft. Then, on 28 April 1936, NRL's R.M. Page and L.C. Young demonstrated America's first truly effective pulse radar, obtaining high resolution detection of aircraft out to ten miles. This equipment, using a 28.6 MHz frequency with a 5 microsecond pulse width, was detecting aircraft at twenty-five miles by early June 1936. Meantime, the U.S. Army Signal Corps, after a visit to NRL in January, began developing its own pulsed radar equipment, resulting in tests by December 1936 of what would become the SCR-268, the first operational anti-aircraft gun control radar.

One of the biggest problems perplexing the Navy at the time was that radar antennas were too large for ordinary shipboard installation. NRL had overcome one major obstacle by inventing the duplexer, a switching device that allowed both transmitting and receiving from a single antenna. Still, because of the low frequencies being used, even one antenna was too large for the limited topside space aboard ship. It was realized, of course, that the solution to smaller antennas was to use higher frequency equipment. However, high power components simply did not yet exist for the very high (microwave) frequencies. The extensive Chain Home land-based radar system, for example, operated by the British twenty-four hours a day from September 1938 throughout the Battle of Britain and to the very end of the war, was designed for the HF band between 22 and 28 MHz. It was not that this was an optimum radar band but simply that it was the highest frequency at which reliable high power components were available.[7]

Nevertheless, in an effort to reduce the antenna size to something at least reasonable for shipboard use, the Navy decided in May 1936 to develop a 200 MHz VHF radar. This experimental equipment was the first radar to be installed on a U.S. Navy ship, the destroyer *USS Leary*, and, using a Yagi-type antenna mounted on a five-inch gun barrel for training in azimuth (see Figure 4-1), became the first radar to provide bearing as well as range of the target.[8] From this experimental 200 MHz model came the Navy's XAF, its first radar to be employed for evaluation during full-fledged fleet operations. Installed on the *USS New York* in December 1938, the 15 kW Model XAF proved conclusively, during exercises in the Caribbean, the importance of radar to the Navy. Aircraft were detected out to 100 nautical miles, ships out to 15 nautical miles, 14-inch shells in flight and fall-of-shot splashes at 7 nautical miles, and even birds at 5 miles away.[9] With perhaps a touch of humor, Louis Gebhard records, "The installation of its rather large planar antenna (20.5 by 23.5 feet overall; see Figure 4-2) caused considerable consternation, which disappeared when the military value of the equipment was proven."[10]

So enthused by the excellent performance of the XAF was the *New York's* commanding officer, Captain R.M. Griffin, that he wrote a letter of endorsement recommending:

a. The XAF be installed at once on all aircraft carriers and, as soon as possible, on other ships.
b. Development be continued, but that installation not be delayed while awaiting perfection.
c. Plans be included for installation on all new construction ships larger than destroyers, and space reservations be allowed for smaller versions of the XAF on destroyers.[11]

Thus, the success of the XAF resulted in Navy-wide interest in shipboard radar. Using the XAF as a prototype, RCA produced twenty CXAM models which were installed on large warships immediately prior to World War II, all giving long and reliable service. The 200 MHz frequency remained in use for many years thereafter for long-range air search.

In mid-1940, with war now raging in Europe, there occurred one of the most important events in the development of radar, particularly in the quest for higher frequency operation. In all the years before, the United States and Britain had worked independently in radar research. Then, in a joint mission to exchange technology, the design data received on England's cavity magnetron, capable of producing 1 kW at 3 GHz, provided a great boost to radar Research and Development in the United States.[12] From that point on it became the responsibility of the newly organized M.I.T. Radiation Laboratory in Cambridge, Massachusetts to explore microwave radar application, while the Armed Forces laboratories continued to concentrate on lower frequency development, principally at VHF.[13]

By the time America entered World War II, radar systems were being rushed into service. It is a matter of record that radar was there and operating at the very outset of hostilities; an Army SCR-270 working so well on 7 December 1941, that its display scope, indicating an unbelievable number of incoming aircraft, was mistrusted and shrugged off — with the tragic results of Pearl Harbor. Yet radar went on to play the mighty role which would win the singular praise of such an appreciative user as Captain Beach who was quoted at the beginning of this chapter. So significant were its achievements that the Congressional Record of July 2, 1943 called radar, "the super weapon — the most revolutionary device of this war . . . It represents the greatest technical advance in warfare since the original evolution of the military uses of aircraft."[14] As the war progressed, with most of the major technological obstacles now surmounted, the combined British and American work resulted in tremendous quantities of radar equipment being produced for the Allies. Thousands of models from VHF through SHF came out during the 1940s bearing the nomenclature SA, SC-1, SD, SE, SF, SG, SH, SJ, SK, SL, SM, SN, SO, SP, SQ, SR, SU, and SV.

From these beginnings has evolved, in a myriad of specialized forms, what many believe to be the primary sensors aboard ships and planes today, the heart of many navigation, surveillance, and weapons control systems.[15] Modern man has come to rely upon radar as an indispensable supplemental aid in extending his senses beyond human limits — seeing through darkness, fog, rain, and snow, *and* measuring the distance out to an object — doing far more than the human eye can do alone.[16] Moreover, during the forty years since becoming fully operational, radar development has never decreased in intensity, its design and application ever more sophisticated and complex. Higher transmitter power along with greatly improved antenna characteristics (higher gain, narrower beamwidths, lower sidelobes) allow vastly better resolution and accuracy than the now primitive equipment of World War II. But the tasks are tougher as the requirements levied upon radar in today's scenario have increased enormously. Whether it is civilian air traffic to be controlled or potential incoming anti-ship missiles to fend off, the demands for simultaneous detecting and tracking of closely-spaced targets in all environments, from very great distances away to close in, strains the limits of imagination and technology. So fast must present-day reaction times be that in some applications radar sensors automatically activate other control systems, defensive weapons for example, completely by-passing human operators.

The U.S. Navy has over 1500 shipboard radars of nearly two dozen types in service, including long range 2-D and 3-D air search, surface search, aircraft control, harbor navigation, and weapons firing. By the mid-1980s the total number will increase to perhaps 2000, but with a refreshing decrease in the types, as projected by Figure 4-3.[17] As might be expected, there is a wide variety of radar antennas installed aboard Navy ships. Some are outmoded but still in use; many are of current inventory and some, such as the new AEGIS fixed, planar, phased-arrays, are representative of the latest antenna technology. Within the limits of non-classified information, examples of those types currently in service will be described.

4-1 SURVEILLANCE RADAR ANTENNAS

During the war years, as radar technology rapidly progressed, the varied equipment types resulted in categorization by primary function. There were radars used for long range air search, others for very close-in targets, some for covering high elevation angles, some for surface search, and still others needed principally for marine navigation. Gradually over the years, naval surveillance radars became segregated by purpose: two-dimensional air search, three-dimensional air search, surface search, and navigation.

4-1.1 2-D Air Search Radar Antennas

AN/SPS-6

The first of the improved, post-war long-range air search radars was the AN/SPS-6, a direct descendant of the models SR-3 and SR-6 which had been designed especially for high-angle detection of Kamikaze attacks in the Pacific theater.[18] Operating in the mid-UHF region (L-band), and originally procured in 1950, many of the later versions of the SPS-6 are still in service on older ships. A series of antennas has been used with equipment models of the SPS-6, beginning with the AS-402 (Figure 4-4), and followed by the AS-429 (Figure 4-5) and the AS-430A (Figure 4-6). All of these have since been replaced by the AS-430B shown in Figure 4-7 and described in detail below:

ANTENNA AS-430B/SPS-6C

a. TYPE/USE — The AS-430B is a directional transmit and receive antenna used with long-range air and surface search radar set AN/SPS-6C designed for shipboard operation to obtain target bearing and range information. Consisting of a dual-frequency feedhorn radiator, a wind balancing vane, and parabolic section reflector, the antenna has a 30° vertical beamwidth and a horizontal beamwidth of three and one-half degrees. The reflector assembly is a stainless steel frame which is ribbed and braced with fabricated stainless steel channel members and has a reflecting surface made of wire mesh. The dual feedhorn (radar and IFF) is silver plated stainless steel and has a fiberglass cover protecting the orifice. The entire assembly is supported by a pedestal which rotates the antenna at varying rpm's to supply target bearing data.
b. PHYSICAL CONFIGURATION — [See Figure 4-7].
c. DIMENSIONS — (1) Height: 95-⅞ inches (113-⅞ with pedestal); (2) Width: 204 inches; (3) Depth: 153-⅝ inches; (4) Weight: 924 pounds (with pedestal).
d. FREQUENCY RANGE — 1250 to 1350 MHz (L-Band).
e. VSWR — Not to exceed 1.5:1 at the RF input.
f. POWER RATING — 750 kW peak.
g. POLARIZATION — Horizontal.
h. GAIN — 27 dB.
i. TYPE OF FEED — RG-132/W waveguide.
j. PRIMARY POWER REQUIRED — 115/230 vac, 60Hz.
k. REFERENCE — Technical Manual, Radar Set AN/SPS-6, NAVSEA 0967-114-8010.

AN/SPS-12

Shortly after the AN/SPS-6 equipment was purchased in the early 1950s, a second type of 2-D air search radar, the AN/SPS-12, was procured for fleet use. Operating at the same L-Band frequencies for long range target surveillance (out to 200 miles), the SPS-12, too, is now considered obsolete and being superseded by newer systems (refer to Fig. 4-3). Nevertheless, it is still found on a small number of naval ships. Its antenna, the AS-603 shown in shipboard configuration in Figure 4-8, has the following characteristics:

ANTENNA AS-603/SPS-12

a. TYPE/USE — The AS-603 is a directional, rotating, high gain antenna used with long range air and surface search radar set AN/SPS-12; designed for shipboard operation to obtain target bearing and range information. Consisting of a feedhorn exciter and parabolic section reflector, the antenna has a 30° cosecant-squared beam in the vertical plane and a three-degree beamwidth in the horizontal plane. The reflector assembly is an all-welded stainless steel frame of tubular construction with a reflector surface of welded wire mesh screening. The assembly rests on a tubular truss and drum support which bolts to the top of the rotary antenna pedestal. A vertically polarized antenna, the AT-388/SPS-12, is mounted in the mouth of the feedhorn for IFF use. To protect the dipole and horn interior from salt air and exhaust gas contaminates, a fiberglass cover is fitted over the front of the horn.
b. PHYSICAL CONFIGURATION — [See Figure 4-8].
c. DIMENSIONS — (1) Height: 92 inches (116 with pedestal); (2) Width: 205 inches; (3) Depth: 121 inches; (4) Weight: 990 pounds (with pedestal).
d. FREQUENCY RANGE — 1250 to 1350 MHz.
e. VSWR — Not to exceed 1.5:1 at RF input.
f. POWER RATING — 500 kW peak.
g. POLARIZATION — Horizontal.
h. GAIN — 25 dB.
i. TYPE OF FEED — CG-1027/U waveguide.
j. PRIMARY POWER REQUIRED — 440vac, 60Hz, three-phase.
k. REFERENCE — Technical Manual, Radar Set AN/SPS-12, NAVSEA-0967-069-2010.

AN/SPS-29

As reveiwed at the beginning of this chapter, the Navy, in its need for shipboard radar sets of manageable size (antennas in particular), had early centered its major research and development efforts around 200 MHz. Its first shipborne model, the XAF (Figure 4-2), operated at this frequency and from this experimental equipment came a long series of wartime successors including the CXAM, XAR, SA, SC-1, SK, and SR. Immediately following the war, the Navy was, for a time, required to give up the 200 MHz segment of the VHF band to the newly developing commercial television interests. By the mid-1950s, however, upon recognition of the merits of VHF radar for long range, high altitude detection, and the need for diversity of operation, the Navy won reassignment of the 200 MHz band for its uses.[19] Thereafter followed the development of the AN/SPS-17, which was superseded in turn by the AN/SPS-29 described below. The AN/SPS-29 is itself fast becoming obsolete (refer to Fig. 4-3).

ANTENNA AS-943/SPS-29

a. TYPE/USE — The AS-943 is a rotating planar array used with the shipborne AN/SPS-29 very long range air-search radar designed to detect the distance and bearing of remote targets. Arranged in four rows of seven horizontal radiators, there are 28 folded dipoles spaced one-half wavelength apart vertically and horizontally, and approximately 1/4 wavelength in front of a flat "bedspring" reflector, fabricated on an aluminum frame. The feed array leads from the rotating joint of the pedestal up to a tee assembly, where the power is divided equally to feed the top and bottom halves of the antenna. The power for each half of the antenna is further divided at the remainder of the dipoles on that branch of the feed array. The antenna input is tuned by a stub assembly between the input feed line and tee section. Internal construction of the dipoles provides for balun action and the impedance transformation necessary to obtain the proper current distribution for a highly directional beam. The antenna assembly is mounted on a rotatable pedestal and forms a radiation pattern having a 25.5° vertical beamwidth and 19° horizontal beamwidth. Provisions are made for installing a type AT-352/UPA IFF antenna above the radar reflector.
b. PHYSICAL CONFIGURATION — [See Figure 4-9].
c. DIMENSIONS — (1) Height: 102 inches (140 with pedestal); (2) Width: 210 inches; (3) Depth: 55 inches; (4) Weight: 1275 pounds (with pedestal).
d. FREQUENCY RANGE — 200 MHz VHF band.
e. INPUT IMPEDANCE — 50 ohms.
f. POWER RATING — 750 kW peak.
g. POLARIZATION — Horizontal.
h. GAIN — 18 dB.

i. TRANSMISSION LINE TYPE — (1) Radar: RG-154/U; (2) IFF: RG-18/U.
j. PRIMARY POWER REQUIRED — 440 volt, three-phase, 60 Hz power for antenna drive motor. 115 volt, one-phase, 60 Hz for synchro reference and monitoring voltages.
k. INSTALLATION REQUIREMENTS — The antenna is secured to the turntable on the antenna pedestal, which is mounted to the ship's structure by four mounting pads each drilled for three ¾-inch bolts. The radar transmission line is pressurized with dry air from the system dehydrator unit.
l. SPECIAL CONSIDERATIONS — The area selected for the antenna installation must have 360° unobstructed clearance from ship's structures and rigging to accommodate the antenna swing circle. The location for the antenna should have maximum unobstructed azimuth coverage to limit shading or blind spots. Since the synchro system and ship's heading marker are adjustable through 360° the orientation of the antenna pedestal is not critical.
m. REFERENCES — (1) Technical Manual, AN/SPS-29 Radar Set, NAVSEA 0967-LP-057-0010. (2) NAVSEA Drawings RE-H2690237 and RE-H2690714. (3) AS-943/SPS-29, NAVSEA 0967-052-8280.

AN/SPS-37 and AN/SPS-43

Improvements in radar technology gave rise to development of the AN/SPS-37 and AN/SPS-43 as derivatives of the AN/SPS-29. Using a much wider pulse width, the SPS-37 and SPS-43 achieve a greater maximum detection range with one-third the peak RF power of the SPS-29. The antenna used with the SPS-37 and SPS-43 on most ships is the same as that used for the SPS-29, i.e., the AS-943 planar dipole array (broadbanded and redesignated the AS-1091 for the SPS-43, but otherwise identical to the AS-943). For very large ships, such as heavy cruisers and aircraft carriers, the SPS-37A and SPS-43A versions are used, both of which employ the larger (40-ft. long) AS-1092/SP shown in Figure 4-10 and described below:

ANTENNA AS-1092/SP

a. TYPE/USE — The AS-1092 is a rotating dipole array used with either the AN/SPS-27A or the AN/SPS-43A very long range air-search radars designed to detect the distance and bearing of remote targets. The antenna has a double corner reflector, one stacked upon the other, to form a W shaped cross-section. In front of the aluminum reflector, arranged in an array of two rows of ten evenly spaced pairs, the 20 dipoles, made of brass, are fed from a common coaxial transmission line branching out through various power dividers to feed each pair. Producing a radiaton pattern with a 20° vertical beamwidth and 7°

horizontal beamwidth, the antenna assembly bolts to the turntable portion of the antenna pedestal. Provisions are made for mounting IFF antenna AT-352/UPA-22 atop the reflector.

b. PHYSICAL CONFIGURATION — [See Figures 4-10 (a) and (b)].
c. DIMENSIONS — (1) Height: 121 inches (161 with pedestal); (2) Width: 495 inches; (3) Depth: 63 inches; (4) Weight: 5720 pounds (with pedestal).
d. FREQUENCY RANGE — 200 MHz VHF band.
e. INPUT IMPEDANCE — 50 ohms.
f. POWER RATING — 180 kW peak.
g. POLARIZATION — Horizontal.
h. GAIN — 23 dB.
i. TRANSMISSION LINE TYPE — RG-154/U.
j. PRIMARY POWER REQUIRED — 440 volts, three-phase, 60 Hz.
k. REFERENCES — (1) Technical Manual, AN/SPS-37, NAVSEA 0967-052-7010. (2) AN/SPS-43, NAVSEA 0967-052-8010. (3) AS-1092/SPS-43A, NAVSEA 0967-LP-221-1010.

AN/SPS-40

While the 200 MHz AN/SPS-17 spawned the several VHF air search radars (AN/SPS-29, 37, 43) whose antennas are described above, a 400 MHz UHF version of the AN/SPS-17 was also produced. This resulted in an experimental AN/SPS-31, which was then followed by the procurement of a large number of AN/SPS-40 UHF air search radars. The earlier AN/SPS-40 equipment used an AS/1138/SP parabolic reflector antenna. Later models (i.e., the SPS-40B, C and D) use an AS-2782 model which is essentially identical to the AS-1138. Figure 4-11(a) is a shipboard view of the AS-2782/SPS-40B. As seen in Figure 4-3 and by Reference [20], the AN/SPS-40 is the current planned standard for naval long-range 2-D air search.

ANTENNA AS-2782/SPS-40B

a. TYPE/USE — The AS-2782 is a rotating, directional, high gain antenna used with the AN/SPS-40B long-range air-search radar designed for ships to detect distance and bearing of remote targets. A dual feedhorn and truncated paraboloid reflector, covered with a wire screen to reduce weight and wind resistance, comprises the antenna. The dual feed includes the radar section and an integral IFF. A slot type, the radar feed has a tuned cavity and flared shape to ensure proper illumination of the reflector. The reflector then forms the RF energy into a fan shaped beam with a 19° vertical beamwidth and 11° horizontal beamwidth.

The integral IFF feed is composed of two full-wave dipoles inserted in the feedhorn to simulate a corner reflector. This arrangement radiates vertically polarized IFF signals to effectively eliminate any interference with the horizontally polarized radar energy. A fiberglass cover is placed over the feed assembly to protect the opening from weather and corrosive contaminates.

b. PHYSICAL CONFIGURATION — [See Figure 4-11(b)].
c. DIMENSIONS — (1) Height: 140 inches (191 with pedestal); (2) Width: 213½ inches; (3) Depth: 108½ inches; (4) Weight: 1725 pounds (with pedestal).
d. FREQUENCY RANGE — 400 MHz UHF band.
e. VSWR — 1.5:1 or less into 50 ohms.
f. RF POWER RATING — 200 kW peak.
g. POLARIZATION — Horizontal.
h. GAIN - 21 dB.
i. TRANSMISSION LINE TYPE — (1) RG-153/U (radar); (2) RG-219/U (IFF).
j. PRIMARY POWER REQUIRED — 400 volts, three-phase, 60 Hz power for antenna drive motor. 115 volts, one-phase, 60Hz for synchro reference, and monitoring voltages.
k. INSTALLATION REQUIREMENTS — The base of the antenna pedestal has 16 bolt holes drilled for clearance for ¾-inch bolts spaced 20° apart on a 33-inch diameter bolt circle. The mounting base must provide a cutout access to permit removal of the rotary joint. The radar transmission line is pressurized with dry air by the radar dehydrator unit. The pedestal must be oriented fore and aft along the ship's centerline in accordance with the marking on the pedestal base.
l. SPECIAL CONSIDERATIONS — The area selected for the antenna installation must have a 360° unobstructed clearance to accommodate the antenna swing circle. The location of the antenna should have maximum unobstructed azimuth coverage to limit shading or blind spots.
m. REFERENCES — (1) Technical Manual, AN/SPS-40 Radar Set, NAVSEA 0967-LP-441-9010. (2) NAVSEA Drawing RE-D2699234.

AN/SPS-49

As a means of diversifying its radar operation frequencies, the Navy began production of an 800 MHz system, its modern-day, high performance, very long range, 2-D air-search radar equipment: the AN/SPS-49. A medium power system, the SPS-49 provides high resolution surveillance in varying environments, including severe clutter and jamming. Its antenna, the AS-3263/SPS-49, is pictured installed in Figure 4-12, and is described in detail below:

ANTENNA AS-3263/SPS-49

a. TYPE/USE — The AS-3263 is a high gain, rotating antenna used with the AN/SPS-49 radar system to provide target range and bearing information. Three major sections comprise the antenna: the feedhorn and its support boom; the reflector; and the antenna pedestal assembly. The reflector is 24 feet wide and has a double curved parabolic section composed of horizontal members to form the horizontally polarized RF energy into a 9° vertical and 3.3° horizontal beamwidth. RF energy is routed to the feedhorn via waveguide through elevation and azimuth rotary joints located within the pedestal.

The antenna is provided with two azimuth drive speeds. Optimum long range is obtained at the lower speed, 6 rpm. A higher speed, 12 rpm, is used if the operator desires to increase the data rate (at the expense of long range performance). To hold the radar beam near the horizon for stabilized long range detection as the ship rolls and pitches, the antenna is mounted on a trunion so that the elevation angle can be controlled by means of a jackscrew located behind the reflector. The jackscrew is turned by a motor-driven gearbox assembly whereby rotating threads cause the antenna to tilt up or down with varying ship attitude.

Four omnidirectional auxiliary antennas are used with special versions of the SPS-49 to provide coherent sidelobe cancelling (CSLC) in the presence of jamming. The antennas, types AS-3077/SP, AS-3078/SP, and AS-3079/SP shown in Figure 4-13, are mounted in an approximate 20-foot diameter circle. Two of the antennas are placed above the radar antenna on platforms attached to booms extending from the adjacent mast, and the other two are mounted below the radar antennas. The sidelobe canceller circuitry generates a signal equal in amplitude but opposite in phase to the jamming energy in order to effectively cancel it.

Provisions are made for mounting an IFF antenna at the extreme end of the feedhorn support. However, the IFF antenna is not part of the SPS-49 radar set.

b. PHYSICAL CONFIGURATION — [See Figure 4-12(b)].

c. DIMENSIONS — (1) Height: 171 inches (with pedestal); (2) Width: 288 inches; (3) Depth: 189 inches; (4) Weight: 3040 pounds (with pedestal). NOTE — The CSLC antennas are 17 inches high, 8 inches in diameter, and weigh 10 pounds each.

d. FREQUENCY RANGE — 851 to 942 MHz.

e. VSWR — Not greater than 2.0:1.

f. RF POWER RATING — 280 kW peak.

g. POLARIZATION — Horizontal.

h. GAIN — 29 dB.

i. TRANSMISSION LINE TYPE — (1) 9-¾ x 2.4-inch waveguide P/N WR975 (for radar antenna); (2) MS22931 Coax (for CSLC antennas).
j. PRIMARY POWER REQUIRED — 440 volts, three-phase, 60 Hz; 115 volts, one-phase, 60 Hz.
k. REFERENCE — Technical Manual, AN/SPS-49(V), NAVSEA 0967-LP-584-8010.

AN/SPS-58, AN/SPS-62, AN/SPS-65 Series

The AN/SPS-58 is the air surveillance portion of the shipboard acquisition and designation system known as the Basic Point Defense Surface Missile System (BPDSMS). This equipment is intended for rapid designation of high speed, low altitude radar-detected targets, at very short range, to the ship weapons control systems even in a severe clutter-environment. Early models of the SPS-58 used the AS-2607 antenna depicted in Figure 4-14. However, subsequent versions (i.e., the AN/SPS-58A and then the AN/SPS-62), incorporated a modification to the ship's SPS-10 antenna, the AS-936, by replacing the single feedhorn of the SPS-10 with the dual-feed AS-4022/SPS-58A waveguide horn. Since the SPS-10 is basically a C-band surface search radar, while the SPS-58 is an L-band air search acquisition system with target designation capability, this unified dual antenna affords an overall ship sensor system featuring independent, but simultaneous, operation of both air and surface surveillance within a 25-mile range.

A further improved variant of the SPS-58 is the AN/SPS-65, which offers enhanced performance in an ECM environment, an automatic target detection interface, and higher reliability.[21] The SPS-65 also uses the modified SPS-10 antenna with the dual-feed waveguide horn, redesignated the AS-936C as shown in Figure 4-15 and described below. (Only the L-band air search characteristics will be detailed here. The C-band portion is given later under the Surface Search radar section of this chapter.)

ANTENNA AS-936C/SPS-10B

a. TYPE/USE — The AS-936C is an antenna assembly consisting of an SPS-10 antenna altered to operate with the very short range SPS-65 air surveillance radar for BPDSMS and NATO Sea Sparrow systems. The SPS-10 antenna modification is accomplished by replacing the existing C-band horn of an AS-936B antenna with the dual-feed AS-4022/SPS-58A waveguide horn. The resultant integrated L and C band antenna is fed from a rotary joint incorporated within the existing SPS-10 pedestal, and, for the air search portion, has a 16° vertical beamwidth and a 6° horizontal beamwidth. The 16° elevation beam ensures horizon detection capability for all but extreme ship roll conditions with no antenna stabilization feature.

b. PHYSICAL CONFIGURATION — [See Figure 4-15].
c. DIMENSIONS — (1) Height: 76 inches (with pedestal); (2) Width: 126 inches (swing circle); (3) Depth: 126 inches (swing circle); (4) Weight: 440 pounds (with pedestal).
d. FREQUENCY RANGE — L-band.
e. POWER RATING — 20 kW peak.
f. POLARIZATION — Vertical.
g. GAIN — 23 dB minimum.
h. TRANSMISSION LINE — Waveguide.
i. REFERENCE — Technical Manual, Radar Set AN/SPS-65, NAVSEA 0967-609-5010.

4-1.2 3-D AIR SEARCH RADAR ANTENNAS

Toward the end of World War II, as radar devices capable of handling much greater power at increasingly higher operating frequencies were developed, several S-band (2 to 4 GHz) radar systems were manufactured and rushed into service. Among these were the Navy's 700 kW models SM and SP, the first radars designed specifically for three-dimension detection and interception of figher planes. Using parabolic dish antennas having an extremely narrow conical beam to provide simultaneous azimuth and elevation information, these radars were, therefore, restricted to covering only one airplane at a time. This necessitated continual switching back and forth between an enemy aircraft and the Allied plane being guided to intercept.

To overcome this limitation, the model SX multiple-intercept 3-D radar was introduced into the fleet at the end of the war and used extensively thereafter. As is evident in Figure 4-16, the SX antenna was actually composed of two separate antennas, one for routine air search and the other working independently as a height-finder. That is, while the total assembly rotated 360° in azimuth, the height-finder scanned angular sectors in elevation to achieve three-dimensional coverage.

By the early 1950s, however, the 3-D fighter director function was being accomplished by use of a single integrated antenna. Initially, this had been done at L-band, on a limited basis aboard ship, with the AN/SPS-2. Having an output of 7 megawatts to provide very long range detection (out to 300 miles) at altitudes of 100,000 feet, this was the world's most powerful radar. The AS-627 antenna unit of the SPS-2 is shown in Figure 4-17. It had a diamond shaped parabolic reflector and a vertical stack of seven feedhorns to produce a single composite cosecant squared radiation pattern. Weighing five tons, with dimensions of 40 feet wide by 30 feet deep and 27 feet long, the AS-627 obviously was meant for use only on the largest of warships.

Operation at S-band frequencies allowed the employment of comparatively smaller, lighter, 3-D radar antennas, beginning with the AN/SPS-8. Its antenna was the AS-434 which employed a very narrow vertical beam of 1.1 degrees to scan 11° angular sectors. By changing the elevation of the reflector, the sector itself could be varied between 0° and 36° vertically. Moreover, the antenna was stabilized for pitch and roll of the ship. Figure 4-18 shows two slightly different variations of the AS-434/SPS-8 antenna, displaying its distinctive features: the Robinson feedhorn (i.e., a fixed horn aperture having an internally rotating input feeder to produce vertical scanning.), a bifocal parabolic reflector, and a large counter balance wind sail. The later edition SPS-8B used the AS-828 seen in Figure 4-19 with its large elliptical parabolic reflector, its "organ pipe" scanner (consisting of 20 fixed, vertically stacked waveguide feddhorns to provide elevation scanning), and its small balance sail. These antennas, both types of which are very large and heavy (weighing more than two tons) are classified as electro-mechanical in nature (in terms of affording 3-D detection) and are now obsolete. Along with the SPS-8 they have been superseded by the AN/SPS-30, a 3-D radar system which itself is quickly becoming a relic of by-gone days.

AN/SPS-30

The SPS-30 is a very high power (2 megawatt) long range S-band shipboard radar system designed for integrated 3-D air search and height-finding. Where particularly close surveillance of a special sector in space is desired, the search pattern can be varied to provide segment coverage. This is accomplished through a major component of the SPS-30, its very large, high gain microwave antenna, the AS-1167 described below:

ANTENNA AS-1167/SPS-30

a. TYPE/USE - The AS-1167 is an S-band radar antenna used for 3-D air search to provide range, bearing, and height information for separate, multiple targets. Its principal elements are an elliptical parabolic reflector, an unusually long feed boom, and an "organ pipe" scanner feed assembly. The antenna is stabilized for ship pitch and roll, and is, for the most part, fabricated of aluminum. RF energy is routed from the radar transmitter (or conversely, to the radar receiver) via waveguide to a rotating spinner-feed, and then into twenty vertically stacked organ pipe feedhorns. The reflector is thereby excited at varying angles to elevation scanning. The entire antenna assembly rotates continously through 360° at five speeds ranging from 1 to 10 rpm. To scan a selected sector the reflector may be tilted at any angle between 6 and 30 degrees above horizontal. For high resolution sweeping of the vertical sectors the radiated beam is made as narrow as possible, approximately 1.2° in elevation, and 1.5° in the azimuth plane, and scanned at rates variable from 240 to 2400 per minute.

b. PHYSICAL CONFIGURATION - **[See Figure 4-20]**.
c. DIMENSIONS - (1) Height: 320 inches (with pedestal); (2) Width: 203 inches (swing circle); (3) Depth: 203 inches (swing circle); (4) Weight: 6000 pounds (with pedestal).
d. FREQUENCY RANGE - 3400 to 3600 MHz.
e. RF POWER RATING - 2500 kW peak.
f. POLARIZATION - Vertical.
g. GAIN - 40 dB minimum.
h. TRANSMISSION LINE TYPE - Waveguide.
i. REFERENCE - (1) Technical Manual, AN/SPS-30, NAVSEA 0967-445-2010,

AN/SPS-39 and AN/SPS-42

With the advent of the AN/SPS-39 and AN/SPS-42 air search radar equipment, shipboard 3-D radar antennas evolved to a dramatic new look and mode of operation. The height-finding function changed from electromechanical to electronic in nature, the forerunner of phased arrays, as the geometric shape eventually changed from parabolic to a planar surface. Whereas azimuth searching would continue to be performed by continuous mechanical rotation of the antenna, elevation scanning would now be accomplished electronically, by varying the frequency of the transmitted energy in discrete programmed steps. The first of these antennas, the AS-999A (later the AS-1109) is shown in Figure 4-21 and described below.

ANTENNA AS-999A/SPS-39 (later AS-1109/SP)

a. TYPE/USE - The AS-999A antenna is part of antenna group AN/SPA-45 (later AN/SPA-54) used with radar sets AN/SPS-39 and AN/SPS-42 designed for shipboard use to provide 3-D surface and airborne target position data, under all weather conditions, at ranges up to 160 miles in azimuth and 75,000 feet in altitude. Consisting of a vertical linear array of slot radiators and a parabolic cylinder reflector, the antenna forms a pencil beam having a 3° vertical beamwidth and 2.4° horizontal beamwidth. As a result of the physical and electrical spacing of the array slots and feed, the beam of energy radiated by the linear array changes direction electronically, scanning through 48° in elevation in 26 discrete steps. The antenna reflector is constructed of metal-faced honeycomb panels bolted together to form a smooth, parabolic surface, and is fixed at an incline of 25° with respect fo the ship's deck. Mounting on a rotating base pedestal permits continuous scanning of the beam through 360° in azimuth.

The vertical array of radiators is shaped by folding the waveguide back and forth upon itself to form a sinuous feed of 44 elements. At 3000 MHz all slots are fed in phase and radiate simultaneously to produce a broadside beam. By changing the applied frequency the phase of the energy fed to the slots is changed, thereby causing them to radiate at different moments in time. A decrease in frequency causes the beam to move downward. Thus, by altering the frequency in steps, the vertical position of the beam can be made to step in angular increments relative to the broadside position. Moreover, mechanical stabilization of the antenna is not required; additional frequencies are available for electronically positioning the beam above 48° and below horizon during ship pitch and roll conditions.

For suppression of sidelobe energy a dual horn SLS antenna is mounted atop the AS-999A. Provision is also made for installing an AS-1026 or AT-352 IFF antenna at the reflector base.

b. PHYSICAL CONFIGURATION - [See Figure 4-21].
c. DIMENSIONS - (1) Height: 184 inches (with pedestal); (2) Width: 110 inches; (3) Depth: 94 inches; (4) Weight: 2475 pounds (with pedestal).
d. FREQUENCY RANGE - 2910 to 3100.5 MHz (S-band).
e. VSWR - less than 1.7:1 (except 4.0:1 at 3000 ±6 Mhz).
f. RF POWER RATING - 1000 kW peak.
g. POLARIZATION - Horizontal.
h. GAIN - 33.5 dB.
i. TRANSMISSION LINE TYPE - Waveguide.
j. PRIMARY POWER REQUIRED - 115 vac; 60Hz; one-phase.
k. REFERENCES - (1) Technical Manual, AN/SPS-39, NAVSEA 0967-074-4010. (2) Technical Manual, AN/SPS-42, NAVSEA 0967-074-6010. (3) Directory of Radar Equipment, AN/SPA-45 and AN/SPA-54, NAVSEA 0967-420-0020, Section 2.

ANTENNA AS-1838/SPA-72

a. TYPE/USE - The AS-1838 is part of antenna group AN/SPA-72 used with the upgraded AN/SPS-39A for the same purposes as described above. In contrast to the AS-999 and AS-1109 parabolic cylinders earlier employed with the SPS-39, the AS-1838 is a planar array. About 6dB higher in gain, this antenna uses a matrix of waveguide slot antennas to produce a very narrow pencil beam having beamwidths of 2.25° in elevation and 1.1° in azimuth. Rapid elevation scanning is achieved electronically by changing the transmitted frequency as the antenna rotates in azimuth at speeds of 7.5, 10, or 15 rpm.

The antenna consists of a sinuous delay line assembly stacked vertically to feed a matrix of 60 horizontal linear arrays. Each of the linear arrays is itself an identical waveguide subsection containing 98 inclined radiating slots cut into one narrow wall. It is sheathed in radome material to protect the total array from the weather and other environmental contaminates.

The planar antenna is inclined from the vertical such that a line through the broadside axis makes an angle of 24.65 degrees with respect to the ship's deck. To permit variable positioning of the radiated beam in elevation, a stepped-frequency scanning technique is used. That is, the vertical angle of the beam is determined by the frequency applied, and a series of differing frequencies results in elevation scanning so that the antenna does not mechanically move in the vertical plane. The overall effect is that an increase in frequency causes the beam angle to decrease with the respect to the horizon, and a decrease in frequency will raise the beam upwards.

Antenna AS-1838 is normally configured with an AS-1666/SPA-64 sidelobe suppression (SLS) horn antenna attached integrally to the underside. Provision is also made for mounting an IFF antenna atop the AS-1838 when required.

b. PHYSICAL CONFIGURATION - [See Figure 4-22].

c. DIMENSIONS - (1) Height: 167 inches with pedestal); (2) Width: 165 inches; (3) Depth: 55 inches; (4) Weight: 3200 pounds (with pedestal.

d. FREQUENCY RANGE - 2910 to 3100.5 MHz.

e. VSWR - less than 1.5:1 (except 3.5:1 at 3000 ±6 MHz).

f. RF POWER RATING - 1000 kW peak.

g. POLARIZATION - Horizontal.

h. GAIN - 39.5 dB.

i. TRANSMISSION LINE TYPE - Waveguide.

j. PRIMARY POWER REQUIRED - 115vac, 60Hz, one-phase.

k. REFERENCES - (1) Technical Manual, AN/SPS-39A, NAVSEA 0967-311-2010. (2) Technical Manual, AN/SPS-39A, NAVSEA 0967-

074-5310. (3) Technical Manual, AN/SPA-72, NAVSEA 0967-004-1020 and 0967-004-1030.

AN/SPS-48

By the early 1970s, the Navy was procuring and installing its modern era 3-D radar systems, the AN/SPS-48 and AN/SPS-52. These two systems are the planned standard for the fleet. The AN/SPS-48 is a long range, very high power air search and height-finding radar operating at S-band to provide airborne target range, azimuth, and elevation data to shipboard weapons systems and displays. The radar set is computer programmed for virtually simultaneous, multiple-beam elevation scanning, using the AS-1686 high-gain planar array described below:

ANTENNA AS-1686/SPS-48

a. TYPE/USE — **The AS-1686 is a major part of antenna group** OA-7028/SPS-48 used with the AN/SPS-48 for 3-D air surveillance. A planar array, it is composed of 76 horizontally positioned linear slotted array sections stacked one atop the other and fed from a vertical serpentine assembly. The antenna is tilted back 15° and is attached to a two-section support of horizontal and vertical beams mounted on a vertical tubular structure. Ladders, one inside and one outside the support structure, allow maintenance access. Because the radar is used for precision height-finding, the scan coverage is stabilized ±20° for pitch and roll movements of the ship and for varying environmental conditions. The entire array is covered with radome material and dry-air pressurized.

   Elevation scanning from horizon up to 45° is accomplished by changing the frequency of RF energy applied to the antenna's horizontal arrays. Simultaneous with the elevation scanning, the antenna is rotated continuously at either 7 1/2 or 15 rpm to allow searching in azimuth. A pencil beam 1.6 by 1.5 degrees in width is radiated, with the elevation angle being inversely related to frequency, i.e., as the frequency is increased the elevation angle decreases.

   Integral to the AS-1686 main antenna are two small ancillary antennas, the AS-2263 reference antenna, and, either the AS-1688 or the AS-1065 IFF antenna. The AS-2263 reference antenna is an omnidirectional receiving antenna, a small 8 1/2 inch square by 35 inches high section of S-band waveguide shaped like an inverted "L", used to provide sidelobe blanking to the radar set.

b. **PHYSICAL CONFIGURATION — [See Figure 4-23].**

c. DIMENSIONS — (1) Height: 228 inches (with pedestal); (2) Width: 194

inches (swing circle); (3) Depth: 108 inches; (4) Weight: 4500 pounds (with pedestal).

d. FREQUENCY RANGE — 2900 to 3100 MHz.
e. VSWR — 3.5:1 maximum at beam normal.
f. RF POWER RATING — 2200 kW peak.
g. POLARIZATION — Horizontal.
h. GAIN — 38.5 dB.
i. TRANSMISSION LINE TYPE — Waveguide.
j. PRIMARY POWER REQUIRED — (1) 440 volt, three-phase, 60Hz. (2) 115 volt, three-phase, 60Hz. (3)440 volt, three-phase, 400Hz.
k. REFERENCES — (1) Technical Manual, AN/SPS-48(V), NAVSEA 0967-186-1010. (2) Technical Manual, Antenna Group OA-7028/SPS-48(V), NAVSEA 0967-443-0010. (3) NAVSEA Drawing RE-E2698987.

AN/SPS-52

Radar Set AN/SPS-52 is an S-band frequency-scanning surveillance system which provides 3-D target position data for naval ships under all weather conditions. It is in reality an improved version of the earlier SPS-39 and 42 equipment and the latest model, i.e., the SPS-52C, is to be one of the two planned standard 3-D air search systems for the U.S. Navy (the alternative system being the SPS-48 discussed above). This long range radar has a maximum range of 240 nautical miles, elevation coverage of 42° with respect to the horizon, a height capability of 100,000 feet, and continuous 360° azimuthal scanning at variable rpm's. It is capable of operating in any one of ten different modes.

Current installations of the SPS-52 employ the AS-1838/ SPA-72 antenna described above. However, some of the earlier AS-1667/ SPA-64 antennas are still in use. Identical in outward appearance to the SPA-72, the SPA-64 differs only slightly internally.

ANTENNA AS-1667/SPA-64

a. TYPE/USE — The AS-1667 is part of antenna group AN/SPA-64 initially used with the AN/SPS-52 radar system. (This antenna is now being replaced by the AS-1838/SPA-72.) In essence the antenna is a matrix of waveguide slot antennas which forms a very slender beam of RF energy radiating at a discrete angle with respect to the plane containing the slots. The 60 identical, horizontal, linear waveguide sections stacked one above the other, each have 63 radiating slots cut into one narrow wall. (Compare this to the 98 slots of each section of the SPA-72 array.) The RF energy from the transmitter is fed to the 60 linear subarrays from a sinuous vertical feedline running up the side of the stack. A progressive change in the inclination of the radiating slots results in the formation of a pencil beam having a 2.25° vertical and 1.9° horizontal beamwidth.

The AS-1667 antenna is installed at a fixed tilt such that the broadside frequency causes the main beam to form at 24.65 degrees with respect to the ship's deck. Stepped increases in the frequency of the applied energy lower the beam elevation; conversely, a decrease in frequency causes the beam to rise. Vertical scanning, therefore, is achieved at a very rapid rate while the antenna rotates continuously in azimuth.

Mounted on the underside of the AS-1667 array, and functioning integrally with it, is a small horn-type S-band sidelobe suppressor (SLS) antenna, the AS-1666. Provisions are also made for accomodating the AT-352/UPA-22 type IFF antenna when required.

b. PHYSICAL CONFIGURATION — [See Figure 4-24].
c. DIMENSIONS — (1) Height: 160 inches (with pedestal); (2) Width: 165 inches; (3) Depth: 55 inches; (4) Weight: 2950 pounds (with pedestal).
d. FREQUENCY RANGE — 1920 to 3100.5 MHz.
e. VSWR — less that 1.5:1 (except 4.0:1 at 3000 ±6 MHz).
f. RF POWER RATING — 1000 kW.
g. POLARIZATION — Horizontal.
h. GAIN — 37 dB.
i. TRANSMISSION LINE TYPE — Waveguide.
j. PRIMARY POWER REQUIRED — 115vac, 60Hz, one-phase
k. REFERENCES — (1) Technical Manual, AN/SPS-52, NAVSEA 0967-003-8010. (2) Technical Manual, AN/SPA-64, NAVSEA 0967-004-0000.

4-1.3 Surface Search Radar Antennas

Almost from the very beginning, in fact just as soon as advanced technology allowed it, the surface search radar function aboard naval vessels was moved up to microwave frequencies. This move was brought about largely in an effort to make radar equipment as small as possible and, therefore, feasible for installation on all types of craft. As early as 1942, the Model SG (see Figure 4-25) was introduced into the fleet as the Navy's first successful S-band shipboard radar. Nearly one thousand were produced, many seeing extensive wartime use and remaining in service for twenty years.[22] Intended for combatant ships, the SG provided a better surface search performance than lower frequency radars, for navigation as well as detection of surface targets and low-flying aircraft. Evidently well pleased with the results, the Navy decided to jump up another notch, doubling the operating frequency again to reach C-band. Accordingly, it was in this band that two of its most extensively employed surface search radars were developed, the AN/SPS-5 and the AN/SPS-10. These two workhorses, modified many times over, would serve faithfully for more than 30 years.

AN/SPS-5

The first postwar C-band radar to be widely installed in the fleet was the AN/SPS-5, a medium power, lightweight surface search system designed especially for use on light and intermediate size craft. Undergoing a series of revisions, from SPS-5A to SPS-5D, it initially operated between 6275 and 6576 MHz, but with the SPS-5C version began using the 5450 to 5825 MHz range. Moreover, three antenna variations have been used. The early SPS-5 set had the AS-511 shown in Figure 4-26. Then the SPS-5A and 5B used the AS-651 in Figure 4-27. Finally the later models, that is the SPS-5C and 5D, use the AS-1002 described below:

ANTENNA AS-1002/SPS-5C

a. TYPE/USE — The AS-1002 is a rotating, directional, high gain radar antenna designed for use with the AN/SPS-5C and 5D to provide surface target range and bearing to ship displays. Consisting of a sectoral feedhorn and slatted parabolic section reflector, the AS-1002 forms a concentrated beam having a 15° vertical and 1.7° horizontal beamwidth. The antenna rotates at a nominal speed of 17 rpm.
b. PHYSICAL CONFIGURATION — [See Figure 4-28].
c. DIMENSIONS — (1) Height: 52½ inches (with pedestal); (2) Width: 101 inches (swing circle); (3) Depth: 101 inches (swing circle); (4) Weight: 205 pounds (with pedestal).
d. FREQUENCY RANGE — 5450 to 5825 MHz.
e. RF POWER RATING — 350 kW peak.
f. POLARIZATION — Horizontal.
g. GAIN — 29 dB.
h. TRANSMISSION LINE TYPE — RG-95/U waveguide.
i. PRIMARY POWER REQUIRED — 115 vac, 60 Hz, one-phase.
j. REFERENCE — Technical Manual, AN/SPS-5D, NAVSEA 0967-032-7000

AN/SPS-10

The AN/SPS-10 is a high power radar set designed primarily for the detection, ranging, and tracking of surface targets at sea. It has been widely used on large naval vessels for more than thirty years, being several times revised and improved from the original SPS-10 through the SPS-10G model. Early versions of the system used the AS-615 antenna. Later sets employed the AS-936A and then the

AS-936B. There is also a small, lightweight AS-1161 employed occasionally. However, because the AS-1161 is identical to the AS-1002 shown in Figure 4-28 and described under the preceding section on the AN/SPS-5C, its characteristics will not be repeated below.

ANTENNA AS-615/SPS-10

a. TYPE/USE — The AS-615 antenna is a unit of the AN/SPS-10 shipboard radar system. It consists of a feedhorn, a truncated slatted reflector, and an integral IFF function comprised of dipole and corner reflectors mounted on the horn aperture. The antenna is designed to produce a shaped beam having a cosecant-squared vertical pattern of 16° beamwidth and a sharp 1.5° horizontal directivity. Additionally, the elevation pattern provides limited air search capability in the range of 7 to 22 degrees above the horizon.
b. PHYSICAL CONFIGURATION — [See Figure 4-29].
c. DIMENSIONS — (1) Height: 76 inches (with pedestal); (2) Width: 126 inches (swing circle); (3) Depth: 126 inches (swing circle); (4) Weight: 324 pounds (with pedestal).
d. FREQUENCY RANGE — 5450 to 5825 MHz.
e. RF POWER RATING — 285 kW peak.
f. POLARIZATION — Horizontal.
g. GAIN — 30 dB.
h. TRANSMISSION LINE TYPE — (1) RG-49/U waveguide (radar); (2) RG-18/U (IFF).
i. PRIMARY POWER REQUIRED — 115 vac, 60 Hz, one-phase.
j. REFERENCE — Technical Manual, AN/SPS-10, NAVSEA SE211-FO-MMA-010/SPS-10.

ANTENNA AS-936A/SPS-10B and AS-936B/SPS-10B

a. TYPE/USE — The AS-936A is essentially the same antenna as the AS-615 described above, except that the drive assembly is 440 V ac, three-phase, rather than 110 V single phase. The AS-936B, however, is a different antenna. This is readily apparent by comparing Figure 4-30 (a) and (b), where it can be seen that the AS-936B has an extension added to the bottom of the reflector and "blinders" on the side of the feedhorn. This modification improves the antenna characteristics by lowering the sidelobe levels and reducing minor lobe interference with other ship radars.
b. PHYSICAL CONFIGURATION — [See Figure 4-30].
c. DIMENSIONS — (1) Height: 74 inches (with pedestal); (2) Width: 126 inches (swing circle); (3) Depth: 126 inches (swing circle); (4) Weight: 442 pounds (with pedestal).
d. FREQUENCY RANGE — 5450 to 5825 MHz.
e. RF POWER RATING — 285 kW peak.
f. POLARIZATION — Horizontal.

g. GAIN — 30 dB.
h. TRANSMISSION LINE TYPE — Waveguide.
i. PRIMARY POWER REQUIRED — 440 vac, 60 Hz, three-phase.
j. REFERENCE — Technical Manual, AN/SPS-10, NAVSEA SE211-FO-MMA-010/SPS-10.

NOTE: A third version of the AS-936, i.e., the AS-936C, exists with modified feedhorn assembly to be compatible with the 2-D air search radar series AN/SPS-58A/62/65, as described earlier in Section 4-1.1.

AN/SPS-53

The principal purpose of a shipboard surface search radar is to detect and present accurate range and bearing of surface targets while maintaining 360° surveillance for all such targets within line of sight of the radar antenna. Since the maximum range requirement of a surface search radar is limited primarily by the radar horizon, super high frequencies (X-band) are now most often used to allow maximum reflection from small targets such as ship masts and submarine periscopes. Narrow pulse widths and high pulse repetition rates afford high range resolution, range accuracy, and maximum illumination of targets. Compensation for ship pitch and roll and detection of low-flying aircraft may be realized by designing for wide vertical beamwidths, while employing very narrow horizontal beamwidths greatly improves bearing accuracy.

Recognizing these factors the Navy selected the AN/SPS-53 radar as a compact, lightweight, medium power, surface search system for its smaller vessels. Operating at X-band, the SPS-53 is capable of detecting targets between a 25-yard minimum and 32-nautical mile maximum range. Many models of the SPS-53 have been produced over the years, from the original in the mid-1960s to the present-day 53M version. Depending on the model employed, four different antennas are available. Two of the types are five feet in length and the other two types are eight feet long. The five-foot versions have a 1.6° horizontal beamwidth, while the eight-foot models have a 1° horizontal beamwidth. Both types have a 20° vertical beamwidth.

ANTENNAS AS-1652/SPS-53, AS-2321/SPS-53, AS-2322/SPS-53E, and AS-2645/SPS-53J

a. TYPE/USE — These antennas are virtually identical in function and description, varying only in size and weight and, accordingly, horizontal beamwidth. The AS-1652 is used with the original model SPS-53 and the A, B, and H versions; the AS-2321 with the D, F, G, K, and M models; the AS-2322 with the E and L versions; and the AS-2645 is employed with the SPS-53J. Variations exist as well with the particular pedestal type used in each case. All versions rotate at 15 rmp.

Each of the antenna types consists of a feedhorn and slotted waveguide radiating element. Being small and light, they are ideally suited for mounting in high, clear locations aboard ship, preferably at or near the top of a mast. However, the waveguide run should not exceed 100 feet in length.

b. PHYSICAL CONFIGURATION — [See Figure 4-31].
c. DIMENSIONS — (1) Height: 22⅜ in. (AS-1652 and AS-2645, with pedestal), 21¼ in. (AS-2321 and AS-2322, with pedestal); (2) Width: 59⅞ in. (AS-1652 and AS-2321), 96¼ in (AS-2322), 96 in. (AS-2645); (3) Depth: [swing circles are same as width dimensions given above], 5⅜ in. (AS-1652 and AS-2645), 8¼ in. (AS-2321 and AS-2322); (4) Weight: 100 pounds (AS-1652 and AS-2645 with pedestal), 87 pounds (AS-2321 with pedestal), 97 pounds (AS-2322 with pedestal).
d. FREQUENCY RANGE — 9345 to 9405 MHz (X-band).
e. RF POWER RATING — 35 kW peak.
f. POLARIZATION — Horizontal.
g. GAIN — 30 dB average.
h. TRANSMISSION LINE TYPE — RG-51/U waveguide.
i. PRIMARY POWER REQUIRED — 115 vac, 60 Hz, one-phase.
j. REFERENCES — (1) Technical Manual, AN/SPS-53A, NAVSEA 0967-212-6010. (2) Technical Manual, AN/SPS-53A(MOD), NAVSEA 0967-312-4010. (3) Technical Manual, AN/SPS-53 Models B through M, NAVSEA 0967-LP-603-8010.

AN/SPS-55

The AN/SPS-55 is a solid-state shipboard surface search and navigation radar capable of detecting targets as close in as 50 yards and out to 50 miles distant. A high power X-band system, it is designed for installation on naval combatants and large military vessels and it is the current standard equipment of its type. It offers selection of a very short, 0.12 microsecond pulse for high resolution at close range detection, or a wider, 1.0 microsecond pulse for longe range surveillance. Antenna AS-2953 is used with the system and is described below:

ANTENNA AS-2953/SPS-55

a. TYPE/USE — The AS-2953 is part of antenna group OE-172 composed principally of two end-fed slotted waveguide arrays mounted in a back-to-back configuration. One array provides linear horizontal polarization to be used under normal weather conditions, while the other array may be selected by the operator to provide circular polarization when experiencing high precipitation clutter. The advantage of circular polarization is that a circularly polarized wave incident on an approximately spherical object (raindrops, mist and fog particles) is reflected as a circularly polarized wave but with opposite sense rotation. Since the same antenna is used for both transmitting and receiving, the antenna is not responsive to the opposite sense of rotation and the echo energy from precipitation will not be accepted by the receiver. In addition, normal radar targets are rather complex, unsymmetrical objects so they will return some energy with the correct polarization as well as energy with the incorrect polarization. In this manner, rain clutter is considerably reduced while sacrificing only a few dB of real target echoes. Imperfections in the circular polarizer and slight ellipticity of falling raindrops prevent complete cancellation of rain echo.

The antenna functions in a manner to form the RF energy into a narrow fan beam having a 1.5° horizontal beamwidth and a vertical beamwidth of 20°, i.e., -10° to +10° centered on the horizon. During transmission the RF signal enters the antenna through a feed waveguide and is distributed to the 80 periodic, narrow-wall slots. Skewed in angle and alternated in direction of skew, the slots in this feed manifold are separated by approximately one-half wavelength, resulting in broadside radiation at the vertex of the elevation-flared sectoral horn. The beam is then rotated at a continuous rate of 16 rpm.

Due to the slotted array construction, target bearing data of the antenna is automatically frequency compensated for squint (that is to correct the angular offset in azimuth plane between the center of the radiated beam and a perpendicular to the antenna face) as the transmitter is tuned.

b. PHYSICAL CONFIGURATION — [See Figure 4-32].
c. DIMENSIONS — (1) Height: 28 inches (with pedestal); (2) Width: 80 inches (swing circle); (3) Depth: 21 inches; (4) Weight: 195 pounds (with pedestal).
d. FREQUENCY RANGE — 9.05 to 10.0 GHz (X-band).
e. RF POWER RATING — 130 kW peak.
f. POLARIZATION — Selectable: horizontal or circular.
g. GAIN — 31 dB maximum.
h. TRANSMISSION LINE TYPE — RG-68/U waveguide.
i. PRIMARY POWER REQUIRED — 115 vac, 60 Hz, one-phase.
j. REFERENCE — Technical Manual, Radar Set AN/SPS-55, NAVSEA 0967-LP-531-5010.

AN/SPS-67

As the 1980s begin, the Navy is making plans to procure the new AN/SPS-67 equipment as its primary surface search and navigational radar.[23,24] Designed as a replacement for the 30-year old SPS-10 series, the SPS-67 is a medium range, C-band, high definition system employing solid-state standard electronic modules (SEMs) for greater reliability. SEMs are solid-state plug-in modules developed to replace the old vacuum tube circuits of the SPS-10. As a consequence the SPS-67 requires less installation space. Moreover, it has a microprocessor built-in test equipment (BITE) system to signal the operator immediately of any troubles, thus greatly facilitating maintenance.

Performance of the SPS-67 is further improved over the SPS-10 by the incorporation of a very narrow, 0.1 microsecond pulse mode which affords better resolution and detection of buoys and small craft, at short range, for harbor navigation. In open seas, for longer range targets, wider pulse modes (0.25 and 1.0 microseconds) are used. A further advancement in performance over the SPS-10 is realized by improved noise suppression from such unwanted sources of interference as nearby radars and wave action clutter.

Replacement of the SPS-10s with the SPS-67, however, involves below-decks equipment only. The SPS-10 antennas, i.e., the AS-936 series described above in Section 4-1.3, will continue to be used with the new SPS-67 system.

4-1.4 Surface Navigation Radar Antennas

It was noted in the introduction to this chapter that radar was developed chiefly as a military device. Yet it must have occurred quite early to the pioneers in radar science that although it was being produced for detection, early warning, and tracking of enemy ships and planes, and guiding of friendly interceptors — radar must surely have peacetime applications as well. Certainly one of the most obvious uses would be for safe navigation at sea during periods of poor visibility; that is, for collision avoidance, coastal steaming, and harbor piloting under all weather conditions including fog, rain, snow, and night.

The utility of radar for such purposes became especially evident by the mid-1940s when the move up to microwave frequencies became feasible, making possible the detection of close and medium range objects on the surface as small as a submarine periscope.[25] No doubt naval ships then began using their surface surveillance radars to perform secondary roles in navigation and by the war's end specialized ship radar systems such as the AN/SPN-4, AN/SPN-5, and AN/SPN-11 were being developed and procured to be used primarily as an aid to close-in navigation. In the years since, an impressive array of surface navigation radars has been manufactured and installed on virtually every sea-going vessel of any size, commercial and military.

For many years the traditional parabolic section reflector antenna was used almost exclusively with U.S. Navy surface navigation radar sets. However, these antennas have in turn given way to the slotted waveguide type. The slotted array eliminates the need of a separate feed horn and reflector, being in essence the last few feet of waveguide mounted horizontally, with one of the narrow edges facing forward and having thin slots from which the energy is radiated. Moreover, an additional desirable feature of the slotted array is its excellent sidelobe suppression.[26]

AN/SPS-21

The AN/SPS-21 is a short range, compact, lightweight C-band surface scanning radar designed especially for small ships where space is at a premium, or for use as a second radar on larger vessels requiring an auxiliary navigation system. In recent years it has been employed principally as a surface navigation radar and is now considered obsolete, to be replaced by more modern equipment. The SPS-21 has been produced in several models, up to and including the SPS-21D. All models use the AS-710 antenna or its derivatives, the AS-710B or AS-710C. Each of these versions is identical electrically and mechanically except that the AS-710C is eight inches higher overall than its predecessors.

ANTENNA AS-710B/SPS-21

a. TYPE/USE — The AS-710 is a rotatable, directional radar antenna comprised of a parabolic reflector, a pyramidal feedhorn, and drive assembly. As part of the AN/SPS-21 surface navigation radar system, the antenna concentrates the RF energy into a sharp, fan beam having a 15° vertical beamwidth and a 2.0° horizontal beamwidth. This wide vertical beamwidth compensates for ship pitch and roll, and, rotating at 20 rpm, provides surveillance of surface and low-flying airborne targets.
b. PHYSICAL CONFIGURATION — [See Figure 4-33].
c. DIMENSIONS — (1) Height: 53 inches (AS-710, 710A, 710B with pedestal), 61 inches (AS-710C, with pedestal); (2) Width: 98 inches (swing circle); (3) Depth: 48 inches; (4) Weight: 169 lbs. (with pedestal).
d. FREQUENCY RANGE — 5500 to 5600 MHz.
e. RF POWER RATING — 10 kW peak.
f. POLARIZATION — Horizontal.

g. GAIN — 28 dB.
h. TRANSMISSION LINE TYPE — RG-49/U waveguide.
i. PRIMARY POWER REQUIRED — 115 vac, 60 Hz, one-phase.
j. REFERENCES — (1) Technical Manual, Radar Set AN/SPS-21, NAVSEA 0967-953-5010. (2) Instruction Book, Radar Sets AN/SPS-21C and AN/SPS-21D, NAVSEA 0967-317-6010. (3) Technical Manual, Radar Set AN/SPS-21C, NAVSEA 0967-952-9010.

AN/SPS-35

The AN/SPS-35 is a very compact, lightweight X-band surface navigation radar set designed especially for small craft where space is a a premium. It is the military equivalent of the Raytheon Model 1500 "Pathfinder" used extensively on commercial and recreational vessels. An unusual feature of the SPS-35 is that the radar transmitter-receiver is housed in the pedestal mount above deck, so that the only below-decks equipment is the display unit. The antenna used with the SPS-35 is the AS-923 described below.

ANTENNA AS-923/SPS-35

a. TYPE/USE — The AS-923 is a rotating, directional radar antenna comprised of a feedhorn, small parabolic section reflector, and transmitter-receiver integral with the pedestal mount. Designed for use with the AN/SPS-35 surface navigation radar system, it forms a fan-shaped beam having a 15° vertical beamwidth and a 2° horizontal beamwidth.
b. PHYSICAL CONFIGURATION — [See Figure 4-34].
c. DIMENSIONS — (1) **Height: 42$^{7}/_{16}$ inches (with pedestal)**; (2) Width: 49 inches (swing circle); (3) Depth: 24 inches; (4) Weight: 150 pounds (with pedestal).
d. FREQUENCY RANGE — 9335 to 9405 MHz.
e. RF POWER RATING — 7 kW peak.
f. POLARIZATION — Horizontal.
g. GAIN — 29 dB nominal.
h. TRANSMISSION LINE TYPE — [No external transmission line used except for very short piece of waveguide between feedhorn and pedestal housing, furnished as part of set].
i. PRIMARY POWER REQUIRED — 115 vac, 60 Hz, one-phase; (115 vdc and 24 vdc versions also available).
j. REFERENCES — (1) Technical Manual, AN/SPS-35, NAVSEA 0967-161-1010. (2) Field Service Manual, Pathfinder Radar Model 1500, NAVSEA 0967-185-9010.

AN/SPS-36

The AN/SPS-36 is a very compact, lightweight X-band surface navigation radar set designed especially for small craft, such as naval minesweepers, to detect surface and land targets from 25 yards minimum to a maximum of 16 miles. It has a very unusual feature in that the transmitter-receiver is mounted on the back side of the antenna and rotates along with it. Therefore, the only below-decks equipment is the display unit. The antenna used with the SPS-36 system is the AS-925 described below.

ANTENNA AS-925/SPS-36

a. TYPE/USE — The AS-925 is a rotating, directional, slotted waveguide antenna designed for use with the AN/SPS-36 surface navigation radar system. Rotating at 17 rpm, this small, lightweight antenna forms a fan-shaped beam having a 28° vertical and 2° horizontal beamwidth.
b. PHYSICAL CONFIGURATION — [See Figure 4-35].
c. DIMENSIONS — (1) Height: 19 inches (with pedestal); (2) Width: 53 inches (swing circle); (3) Depth: 20⁵⁄₁₆ inches; (4) Weight: 100 pounds.
d. FREQUENCY RANGE — 9350 to 9400 MHz.
e. RF POWER RATING — 10 kW peak.
f. POLARIZATION — Horizontal.
g. GAIN — 29 dB.
h. TRANSMISSION LINE TYPE — None. [Transmitter-receiver is integral part of antenna so no waveguide is used].
i. PRIMARY POWER REQUIRED — 24 vdc.
j. REFERENCE — Technical Manual, Radar Set AN/SPS-36, NAVSEA 0967-953-8010.

AN/SPS-41

The AN/SPS-41 is very small, lightweight X-band surface navigation radar set designed for small seagoing vessels to detect such objects as buoys, shorelines, and other ships within its operating range of 25 yards to 32 nautical miles. The transmitter-receiver is housed integrally with the base pedestal above deck, thus eliminating lengthy waveguide runs and resulting in the display unit and power supply being the only equipment requiring below deck installation space. The antenna used with the SPS-41 system is the AS-1004 described below.

ANTENNA AS-1004/SPS-41

a. TYPE/USE - The AS-1004 is a rotating, directional, radar antenna comprised of a feedhorn, small parabolic section reflector, and transmitter-receiver integral with the pedestal mount. Designed for use with the AN/SPS-41 surface navigation radar system, it forms a fan beam having a 20° vertical and 2° horizontal beamwidth and rotates at 21 rpm.
b. PHYSICAL CONFIGURATION — [See Figure 4-36].
c. DIMENSIONS — (1) Height: 33½ inches (with pedestal); (2) Width: 37½ inches; (3) Depth: 31¾ inches; (4) Weight: 119 pounds (with pedestal).
d. FREQUENCY RANGE — 9375 MHz.
e. RF POWER RATING — 10 kW peak.
f. POLARIZATION — Horizontal.
g. GAIN — 27 dB.
h. TRANSMISSION LINE TYPE — [No external transmission line used except for very short section of waveguide between feedhorn and pedestal housing, furnished as part of set].
i. REFERENCE — Technical Manual, AN/SPS-41, NAVSEA 0967-072-6010.

AN/SPS-46

The AN/SPS-46 is a small, X-band surface navigation radar set designed for seagoing vessels where space is at a premium. It has a range of 25 yards minimum out to 32 nautical miles and uses the AS-1066 antenna described below.

ANTENNA AS-1066/SPS-46

a. TYPE/USE — The AS-1066 is a rotating, directional radar antenna consisting of a feedhorn and small parabolic section reflector. Designed for use with the AN/SPS-46 surface navigation radar set, it forms a fan-shaped beam having a 15° vertical and 2.2° horizontal beamwidth, and rotates at 18 ± 3 rpm.
b. PHYSICAL CONFIGURATION — [See Figure 4-37].
c. DIMENSIONS — (1) Height: 27¾ inches (with pedestal); (2) Width: 48½ inches (swing circle); (3) Depth: 19 9/16 inches; (4) Weight: 85 pounds (with pedestal).
d. FREQUENCY RANGE — 9375 ± 30 MHz.
e. RF POWER RATING — 7 kW peak.
f. POLARIZATION — Horizontal.
g. GAIN — 20 dB.
h. TRANSMISSION LINE TYPE — Waveguide (not to exceed 25 feet).
i. PRIMARY POWER REQUIRED — 115 vac, 60 Hz, one-phase (24 vdc model also available).

j. REFERENCE — Technical Manual, Radar Set AN/SPS-46, NAVSEA 0967-085-2010.

AN/SPS-63

The AN/SPS-63 is a lightweight, solid-state, X-band shipboard radar set intended for installation on small craft for both narrow-water and open-sea navigation. It is representative of modern technology, using modular construction and plug-in printed cards for ease of maintenance and high reliability. Moreover the equipment is designed to be especially rugged at high speeds in the severe sea environment. Emphasis is placed on high resolution at short range, the minimum detectable range being as close as 10 to 20 yards, with maximum displayed range of 40 nautical miles. Transmission power is optional at either 7 kW or 20 kW. This system is installed on the Navy's hydrofoil patrol craft, the PHM.

ANTENNA AS-4052/SPS-63

a. TYPE/USE — The AS-4052 is a rotating directional antenna comprised of a slotted waveguide illuminator and parabolic cylinder reflector having a 20° vertical and 1.2° horizontal beamwidth rotating at 25 rpm.
b. PHYSICAL CONFIGURATION — [See Figure 4-38].
c. DIMENSIONS — (1) Height: 29½ inches (with pedestal); (2) Width: 74 inches; (3) Depth: 16½ inches; (4) Weight: 90.5 pounds (with pedestal).
d. FREQUENCY RANGE — 9375 ± 30 MHz.
e. RF POWER RATING — 17 kW nominal.
f. POLARIZATION — Horizontal.
g. GAIN — 28 dB.
h. VSWR — less than 1.3:1.
i. PRIMARY POWER REQUIRED — 28 vdc.
j. REFERENCE — Technical Manual, Navigation Radar AN/SPS-63, NAVSEA 0967-LP-583-7010.

AN/SPS-64

The AN/SPS-64 is a lightweight, solid-state, modular-constructed X-band shipboard radar set intended primarily for close-in navigation of harbor channels and other congested waterways. It features a bright daylight display and versatile, high performance, interference rejection under severely adverse weather conditions for collision avoidance and surface surveillance. The standard SPS-64 system comes equipped with a six-foot AS-3194 antenna, and 20 kW transceiver. However, a dual X-band/S-band system is often employed to meet longer range requirements, and the many varying conditions at sea. This requires a second, higher power transceiver and an additional twelve-foot antenna installed as illustrated in Figure 4-39(a).

ANTENNA AS-3194/SPS-64 and AS-3195/SPS-64

a. TYPE/USE — The AS-3194 (6-foot X-band) and AS-3195 (12-foot S-band) antennas are rotating, end-fed, slotted waveguide arrays with very low (30 dB sidelobe levels and sharp main beams having a 22° vertical beamwidth and 1.25° (for AS-3194) or 1.90° (for AS-3195) horizontal beamwidths. Rapid target updating, a desirable advantage for close-in navigation situations, is achieved by rotating the antennas at 33 rpm.
b. PHYSICAL CONFIGURATION — [See Figure 4-39].
c. DIMENSIONS — (1) Height: 24 inches (AS-3194); 49 inches (AS-3195); (2) Width: 80 inches (AS-3194 swing circle), 148 inches (AS-3195 swing circle); (3) Depth: swing circles as above; (4) Weight: 140 pounds (AS-3194), 332 pounds (AS-3195).
d. FREQUENCY RANGE — 9375 ± 30 MHz (AS-3194); 3050 ± 25 MHz (AS-3195).
e. RF POWER RATING — 20 kW (AS-3194), with 50 kW option; 60 kW (AS-3195).
f. POLARIZATION — Horizontal.
g. GAIN — 28 dB nominal.
h. TRANSMISSION LINE TYPE — Waveguide.
i. PRIMARY POWER REQUIRED — 115 vac, 60 Hz, one-phase.
j. REFERENCES — (1) Technical Brochure, AN/SPS-64(V), Raytheon Marine Company. (2) Technical Manual, Radar Set AN/SPS-64(V), SE211-R9-MMC-010 and SE211-R9-MMC-020.

AN/SPS-66A

The AN/SPS-66A is a very compact, lightweight, low power, solid-state X-band navigational radar system designed for marine use on small vessels. Meant especially to detect small, close-in targets such as boats, buoys, and shorelines, the SPS-66A has a minimum range of 25 yards, but is capable of surface surveillance out to 32 nautical miles in six selectable range scales. The solid-state circuitry, plug-in printed circuit boards, and modular construction ensure quick, easy maintenance when required. The four-foot AS-4072 antenna used with the SPS-66A is described below:

ANTENNA AS-4072/SPS-66A

a. TYPE/USE — The AS-4072 is a rotating, directional, end-fed, slotted array having very low sidelobes of -27dB, and a radiation pattern having a 23° vertical and 2° horizontal beamwidth rotating at 30 rpm. The four-foot antenna described here is shown in Figure 4-40(a). However, a smaller (33½″) antenna is also available housed with transceiver under a dome as seen in Figure 4-40(b). The dome covered version is used extensively by the Coast Guard on its patrol craft.

b. PHYSICAL CONFIGURATION — [See Figure 4-40].
c. DIMENSIONS — (1) Height: 22½ inches; (2) Swing Circle: 56 inches; (3) Weight: 100 pounds.
d. FREQUENCY RANGE — 9375 ± 30 MHz.
e. RF POWER RATING — 7 kW.
f. POLARIZATION — Horizontal.
g. GAIN — 28 dB.
h. TRANSMISSION LINE TYPE — Virtually no waveguide required as transceiver is coupled to antenna directly by waveguide flange plate at antenna base.
i. PRIMARY POWER REQUIRED — 115 vac, 60 Hz, one-phase.
j. REFERENCES — (1) Technical Manual, Radar Sets AN/SPS-66 and AN/SPS-66A. (2) Technical Brochure, AN/SPS-66, Raytheon Marine Company.

Commercial Navigation Marine Radars

A large variety of modern, compact, lightweight marine navigation radars are currently available commercially from a small number of U.S. and overseas manufacturers. Among these is equipment marketed by such well-known firms as Bendix, Raytheon, Decca, Marconi, Selenia, and SMA. Commercial sets are purchased from time to time by the Navy to be installed on fleet vessels for back-up service or test and evaluation purposes. Typical of such equipment is one that has come into fairly widespread use aboard naval ships, the LN-66 Seafarer produced by Kaar Electronics of Canadian Marconi. Because of its popularity as an auxiliary navigation system, the two antenna versions (three-foot or five-foot options) used with the LN-66 will be described below:

ANTENNA LN-66

a. TYPE/USE — The LN-66 antenna is a rotating, directonal, slotted waveguide array having a 22° vertical and 2.5° horizontal beamwidth for the three-foot antenna (rotating at 22 rpm), and a 11° vertical by 1.5° horizontal beamwidth for the five-foot version (rotating at 24 rpm).
b. PHYSICAL CONFIGURATION — [See Figure 4-41].
c. DIMENSIONS — (1) Height: 24 inches; (2) Width: 36 in. or 60 in., optional; (3) Weight: 21½ pounds (3-ft. version), 41½ pounds (5-ft. version).

d. FREQUENCY RANGE — 9345 to 9405 MHz.
e. RF POWER RATING — 10 kW (higher power versions available as option).
f. POLARIZATION — Horizontal.
g. GAIN — 30 dB nominal.
h. TRANSMISSION LINE TYPE — RG-52 waveguide.
i. PRIMARY POWER REQUIRED — 115 vac, 60 Hz, one-phase.
j. REFERENCE — Technical Manual, Navigation Radar Set LN-66, NAVSEA 0967-LP-522-5010, NAVSEA 0967-LP-583-5010.

4-2 AIR CONTROL RADAR ANTENNAS

The aircraft carrier has about it an aura of power and suspense, and the excitement of carrier air operations at sea never fails to capture the observer's fascination. Highly dramatized news scenes and Hollywood movies have, over the years, left vivid impressions of carriers in action. Fighter planes hurriedly thrust one after the other into the air on attack missions return later out of dark clouds, triumphantly limping and sputtering, fuel tanks nearly empty, to be flagged and waved to a jolting touchdown upon a tilting deck strewn with smoldering debris. In admiration we wonder time and again, how do they do it? Through the years carrier-borne aircraft have become increasingly sophisticated and so complex that even in the calmest waters, under clear blue skies, air operations remain the most critical and delicate of naval activities at sea. Were it not for the marvel of radar electronics appearing on the historical scene contemporaneously with the aircraft carrier — were it not for radar navigational and control approach systems in particular — the landing of modern, high velocity, heavy jet aircraft onto rolling, pitching flight decks would be extremely hazardous at best and, at worst, impossible.

The problem of carrier air operations is of course aggravated and magnified tenfold by real life conditions in an ocean environment, where, most often, seas are not calm nor skies blue. The requirement in real time is to ensure an immediate readiness for controlling and landing aircraft under the most adverse weather conditions, i.e., during poor visibility caused by fog, rain, or darkness, and in rough seas. The need was recognized early, and it was brought urgently to the fore in the numerous Pacific Theater battles of World War II. As a result a concentrated effort went into developing carrier controlled approach (CCA) radar equipment. By the end of the war CCA systems were being tested on the *USS Solomons* and operationally installed, in the late 1940s, on the *Valley Forge* and *Philippine Sea*.

A report of early test results is worth repeating here:

> In the demonstrations on the *USS Solomons*, the system was used to control aircraft on 141 approaches. Fifty-seven of these were made with the pilot in a hooded cockpit, 19 were made at night, and 15 were made with the carrier and escorts totally darkened until the aircraft was less than 200 yards astern of the carrier, in landing position. In a final demonstration, under conditions previously considered impossible, on a night when no horizon was visible, in rain, with seas roughened by 43 knots of wind and all ships completely blacked out, an F6F plane was landed on the carrier, repeatedly and successfully. The unique features of the radar system (X-band) included an antenna mounted on the after edge and slightly below the top of the flight deck. This antenna provided a beam, broad in elevation but sharp in azimuth, which horizontally scanned the area astern the carrier. Another feature of the system was an indicator which clearly displayed echoes from the approaching plane as it entered the glide path. This indicator gave the Landing Signal Officer (LSO) the position data required to keep the aircraft accurately approaching the carrier, and the timing data needed to determine when the aircraft would pass over the end of the flight deck; at that moment, the LSO's "cut" signal would drop the aircraft precisely onto the carrier's arresting gear. The plane was first picked up by search radar and identified by CIC, and the pilot guided by voice communication into the sector scanned by the new radar system.[27]

These promising results, and a succession of continued improvements, led to the eventual production by the mid-1950s of the AN/SPN-6, 8, 10, and 12 equipment that in turn gave way, a decade later, to the AN/SPN-35, 41, 42, 43 and 44 air control radar systems in use today. It should be noted, however, that not all CCA radars are installed aboard aircraft carriers. Many are also on special helicopter-bearing ships such as the LPH and LHA.

4-2.1 AN/SPN-6

One of the older postwar carrier control approach systems still in active service is the AN/SPN-6, a high-power, S-band equipment originally designed for installation on medium and large aircraft carriers. Now considered obsolete (having been replaced by the AN/SPN-43 on all carriers), it is nevertheless included here because it is still to be found on numerous LPH ships in the fleet.

ANTENNA AS-538/SPN-6

a. TYPE/USE — The AS-538 is a rotating, directional, high gain antenna designed for use with the AN/SPN-6 CCA radar system. Comprised of a pyramidal feedhorn, a parabolic system reflector, a ground plane, and a tilting reflector mechanism, the AS-538 assembly is mounted atop a four-axis stabilized base pedestal. The antenna forms a sharp pencil beam having a 2½° vertical and 2° horizontal beamwidth which rotates in azimuth at either 15 or 30 rpm.
b. PHYSICAL CONFIGURATION — [See Figure 4-42].
c. DIMENSIONS — (1) Height: 180 inches (with base); (2) Width: 145 inches (219 inches swing circle); (3) Depth: 82 inches; (4) Weight: 2585 pounds.
d. FREQUENCY RANGE — 3550 to 3700 MHz.
e. RF POWER RATING - 500 kW.
f. POLARIZATION — Horizontal.
g. GAIN — 37 dB.
h. TRANSMISSION LINE TYPE — RG-48/U waveguide.
i. PRIMARY POWER REQUIRED — All power and control voltages are supplied from C-851/SPN-6 antenna control.
j. INSTALLATION REQUIREMENTS — The antenna base has eight equally spaced 13/16-inch bolt holes on a 16½-inch bolt circle for mounting the antenna. The "Right-Left" line on the base should be located perpendicular to the centerline of the ship. This is a stabilized antenna and four points of maximum displacement symmetrically located, one in each quadrant, indicate that a swing contour on a diameter of 224 inches is required when operating in stabilized mode.
k. SPECIAL CONSIDERATIONS — The antenna must be mounted so that its view of the flight deck toward the stern of the ship is completely unobstructed.
l. REFERENCES — (1) Instruction Book, Radar Set AN/SPN-6, NAVSEA 0967-193-0010. (2) NAVSEA Drawing RE-F2689809.

4-2.2 AN/SPN-35

The AN/SPS-35 radar set is a carrier-controlled approach system used in conjunction with optical landing for precision approaches to an aircraft carrier during adverse weather conditions. Both azimuth and elevation data are displayed on a control-indicator unit enabling the radar operator to direct the pilot along a predetermined glidepath and azimuth courseline to a point one mile from the ship.

The SPN-35 employs a very large, distinctive "igloo" radome (shown in Figure 4-43) which houses a rather elaborate antenna system consisting of two separate radiators (the AS-1292 azimuth antenna and the AS-1669 elevation antenna) with individual drive units and a stabilized base pedestal. The radome itself is so unusual as to warrant further description:

> Radome CW-910/SPN-35 is a domed circular structure of 220¼ inches outside diameter having cylindrical walls seven feet high merging into a hemispherical dome. The total height is 16 feet 2 inches. The radome is secured to the ship by attachment to an angle iron ring which is welded or otherwise attached to the deck. The complete unit consists of 12 panels, each covering 30° of circumference, 4 curved galvanized iron angles which form a circular base ring, and a circular dome cap six feet in diameter which forms the apex of the structure. The wall panels consist of a fiberglass woven structure impregnated with polyester resin with fluted ducts running vertically between the outer and center skin. The fiberglass fluted material, being woven, will not delaminate. The lower 9 inches of each panel is reinforced with dense foam to provide a solid fastening area. The space between the center and inner skins is filled with polyurethane foam. Warm air may be circulated through the flutes next to the outer skin to prevent icing, and a 1½-inch thick layer of polyurethane foam is bonded to the inner surface of the panels for additional thermal installation.
>
> The base ring is 2 x 3 x ⅜-inch galvanized angle iron rolled to the radius of the dome such that the two-inch leg lies on the deck directed toward the center. The upright leg (the three-inch leg) has holes for ½-inch bolts spaced to provide three bolts per panel. The base ring is supplied in four identical sections. Manifolds are installed to permit the use of heaters and blowers to force warm air up through the flutes of the panels and out through small apertures in the top, exhausting into the interior of the radome chamber. This arrangement serves the dual purpose of de-icing the exterior of the radome and maintaining the inside temperature at the designed value. The blowers may operate continuously but the heaters should be controlled thermostatically. The radome may be air conditioned by cutting the appropriate aperture in a panel in an area not considered "window area." Otherwise, ship's air conditioning may be used by piping through the deck. Window area is that area through which radar energy is transmitted during operation. It extends from 36 to 144 inches above the deck.

ANTENNA AS-1292/TPN-8

a. TYPE/USE — The AS-1292 is a rotatable, directional X-band radar antenna used to provide azimuthal information to the SPN-35. It consists of a truncated paraboloid reflector with an offset feedhorn, and polarized assembly which allows remote controlled selection of either horizontal or circular polarization. The narrow beam formed by the antenna has a 3.5° vertical and 1.1° horizontal beamwidth, and is capable of being tilted from -1° to +25°. An azimuth drive system functions to rotate the antenna in either 360° or limited sector modes in the horizontal plane.
b. PHYSICAL CONFIGURATION — [See Figure 4-44].
c. DIMENSIONS — (1) Height: 159 inches (note radome height is 194 inches); (2) Width: 78 inches (note radome diameter is 220¼ inches); (3) Weight: 3500 pounds (total with radome).
d. FREQUENCY RANGE — 9000 to 9160 MHz.
e. RF POWER RATING — 200 kW peak.
f. POLARIZATION — Horizontal or circular (selectable).
g. GAIN — 36.5 dB.
h. VSWR — 1.5:1 maximum.
i. TRANSMISSION LINE TYPE — Waveguide.
j. PRIMARY POWER REQUIRED — 120/208 vac, 400 Hz, three-phase.
k. REFERENCES — (1) Technical Manual, Aircraft Control Approach Central, AN/SPN-35A, NAVSEA 0967-254-6010. (2) Technical Manual, Radome CW-910/SPN-35, NAVSEA 0967-204-2010.

ANTENNA AS-1669/SPN-35

a. TYPE/USE — The AS-1669 is a rotatable, directional X-band radar antenna used to provide monopulse-type radiation and reception for elevation information to the SPN-35. It consists of a truncated paraboloid reflector with a dual-channel feedhorn and polarizer assembly fixed in circular polarization by an external grid device. The narrow cosecant-squared beam formed by the antenna has a 1.15° vertical and 3° horizontal beamwidth which is scanned in elevation in either 10° or 35° sectors. The antenna is also capable of being rotated ± 15° horizontally.
b. PHYSICAL CONFIGURATION — [See Figure 4-44].
c. DIMENSIONS — (1) Height: 152 inches (note radome height is 194 inches); (2) Width: 30 inches (note radome diameter is 220¼ inches); (3) Weight: 3500 pounds (total with radome).
d. FREQUENCY RANGE — 9000 to 9160 MHz.
e. RF POWER RATING — 200 kW peak.
f. POLARIZATION — Circular.
g. GAIN — 34 dB.
h. VSWR — 2.0:1 maximum.
i. TRANSMISSION LINE TYPE — Waveguide.

j. PRIMARY POWER REQUIRED — 120/208 vac, 400 Hz, three-phase.
k. REFERENCES — (1) Technical Manual, Air Control Approach Central, AN/SPN-35A, NAVSEA 0967-254-6010. (2) Technical Manual, Radome CW-910/SPN-35, NAVSEA 0967-204-2010.

4-2.3 AN/SPN-41

The AN/SPN-41 is the shipboard microwave transmission portion of the Aircraft Approach Control System (ACCS) designed to provide proper flightpath data to an approaching aircraft as it flies into range of the AN/SPN-42 automatic radar landing system, or visual contact with the Fresnel Lens Optical Landing System. Though not technically a radar system, the SPN-41 does operate in a pulse mode at Ku-band and is usually included with the carrier complement of air traffic control radar equipment.

Two separate transmitters with individual antennas are used for sector scanning. One, the azimuth set, is installed on the ship's stern at the centerline of the landing deck runway. The other, the elevation transmitter, is located on the flight deck in the vicinity of the arresting cables and provides the glide slope signal. The SPN-41 uses one-way transmission, ship to air, to a receiver in the aircraft where the angular information is decoded and displayed on a cross-pointer indicator. With a vertical needle for azimuth and horizontal needle for vertical guidance, the aircraft becomes perfectly aligned with the runway centerline, on a standard glide path to the carrier deck, when both needles are precisely centered. The two SPN-41 antennas used to accomplish the required scanning are described below:

ANTENNA AS-2579/UPN

a. TYPE/USE — The AS-2579 is the elevation antenna portion of transmitter group OT-33/SPN-41 used to transmit vertical angle data to approaching aircraft. A scannable, directional microwave radiator, it is comprised of an end-fed slotted waveguide and drive assembly to scan the antenna from 0° to 10° above horizon at a mechanical rate of 2.5 Hz. The radiation pattern has a 1.3° vertical and 20° horizontal beamwidth. The antenna (and entire OT-33 transmitter group) is protected by a large radome box and is installed on the flight deck near the arresting cable.
b. PHYSICAL CONFIGURATION — [See Figure 4-45].
c. DIMENSIONS — (1) Height: 60 inches including base support assembly (note radome box is 66½ inches high); (2) Width: 49 inches including base assembly (note radome box is 90 inches wide); (3) Depth: 36 inches including base assembly (note radome box is 81 inches deep); (4) Weight: 1275 pounds (total with radome).

d. FREQUENCY RANGE — 15.4 to 15.7 GHz (Ku-band).
e. RF POWER RATING — 2.2 kW peak.
f. POLARIZATION — Horizontal.
g. GAIN — 28 dB.
h. TRANSMISSION LINE TYPE — Waveguide.
i. PRIMARY POWER REQUIRED — 115 vac, 400 Hz, three-phase.
j. REFERENCES — (1) Technical Manual, Transmitting Set AN/SPN-41, NAVELEX 0967-384-6010. (2) Handbook for Operating and Maintenance, Transmitting Set AN/SPN-41 (XN-1), NAVSEA 0967-171-6010.

ANTENNA AS-2580/UPN

a. TYPE/USE — The AS-2580 is the azimuth antenna portion of transmitter group OT-32/SPN-41 used to transmit azimuth information to approaching aircraft. A scannable, directional microwave radiator, it is comprised of a parabolic section reflector, feedhorn, and drive assembly to scan the 2° pencil beam at 10° on either side of the centerline at a mechanical rate of 2.5 Hz. The antenna (and entire OT-32 transmitter group) is protected by a large box-type radome and is installed on the ship stern at the runway centerline.
b. PHYSICAL CONFIGURATION — [See Figure 4-46].
c. DIMENSIONS — (1) Height: 56¾ inches including base support assembly (note radome box is 68½ inches high); (2) Width: 48¼ inches including base assembly (note radome box is 102 inches wide); (3) Depth: 30 inches including base assembly (note radome box is 82¼ inches deep); (4) Weight: 1350 pounds (total with radome).
d. FREQUENCY RANGE — 15.4 to 15.7 GHz (Ku-band).
e. RF POWER RATING — 2.2. kW peak.
f. POLARIZATION — Horizontal.
g. GAIN — 28 dB.
h. TRANSMISSION LINE TYPE — Waveguide.
i. PRIMARY POWER REQUIRED — 115 vac, 400 Hz, three-phase.
j. REFERENCES — (1) Technical Manual, Transmitting Set AN/SPN-41, NAVELEX 0967-384-6010. (2) Handbook for Operating and Maintenance, Transmitting Set AN/SPN-41(XN-1), NAVSEA 0967-171-6010. NOTE — Both the AS-2579 elevation antenna and the AS-2580 azimuth antenna use a small waveguide horn, type AS-2581, for signal monitoring purposes. Each of the two horns are mounted in a special manner prescribed by Reference (1).

4-2.4 AN/SPN-42

The AN/SPN-42 is a dual-channel aircraft carrier radar system integrated with the Navy Tactical Data System (NTDS) to provide precise control of landing aircraft during the final approach and landing sequence. It is an updated replacement of the old AN/SPN-

10 equipment. Each of the two identical and independent channels is capable of controlling one suitably equipped plane per minute to actual touchdown on a moving carrier deck, thus allowing a landing rate of two aircraft per minute. Functionally, the SPN-42 system operates as follows:

Returning aircraft are assigned to Landing Control Central AN/SPN-42A (the radar system) by the NTDS computer. When an aircraft assignment is received the radar system computer programs the most desirable flight path and establishes a radar acquisition window. Each of the two channels of the radar system (channel A and channel B) can handle one aircraft assignment at a time. As the aircraft enters the acquisition window it is detected by the radar system and tracked to the exclusion of all other aircraft. The aircraft radar-derived position vectors are then stabilized to compensate for variations due to ship's motion and compared with the parameters of the programmed flight path. Aircraft flight correction signals are then generated to control the aircraft along the flight path to touchdown. If an unsafe flight or landing condition is probable, the radar system initiates a waveoff signal and control of the aircraft is returned to the NTDS computer. In addition to the automatic waveoff, the LSO or radar operator may initiate a waveoff command when, in their judgment, a safe landing cannot be accomplished. The pilot may also discontinue the landing sequence at any time.

Modes of Operation

Each of the two radar system channels may be operated in any one of three modes. The mode selection for channel A is independent of that for channel B. The three modes are as follows:

Mode I - Automatic
Mode II - Semi-automatic
Mode III - Manual (voice talkdown)

Mode I — During the automatic mode of operation, the aircraft is controlled by the radar system from the time it enters the acquisition window and the pilot signals he is ready for control commands until touchdown on the flight deck is accomplished. The pilot, the controlling channel radar operator, and the LSO all monitor the landing operation at their respective indicators. Any one of the three may terminate the landing sequence if he thinks an unsafe condition occurs.

Mode II — The semi-automatic mode of operation is the same as Mode I, except that the role of the pilot is more extensive. Observing a cross-pointer indicator in the cockpit, the pilot makes the required flight correction maneuvers to guide his aircraft along the desired flight path. The pilot's actions are based on flight correction signals received from the radar system.

Mode III — The manual mode is a talkdown mode of operation. The aircraft is acquired and tracked in the same manner as for Modes I and II. As in the other two modes of operation, precise aircraft position and flight path information is displayed on the radar operator's position indicator. The operator interprets the display and associated meter indications. He then verbally transmits precise flight correction instructions to the pilot.

To accomplish this function the SPN-42 employs two identical antenna systems, one for channel A, one for channel B. Each of the two antenna systems consists of two individual parabolic dish antennas, the larger one operating at Ka-band for radar transmission, the smaller one, riding "piggyback," used for beacon reception at X-band. These antennas are described below.

ANTENNAS AS-1347/SPN-42 and AS-2407/SPN-42

a. TYPE/USE — (1) The AS-1347 is a very high gain, rotatable, highly directive, automatic tracking radar antenna designed to locate and continuously track an assigned aircraft. It is part of antenna-receiver group OE-29A/SPN-42 and consists of an offset, rapid-scan, nutating feed and four-foot parabolic dish reflector mounted on a stabilized base. Employing an extremely thin pencil beam of only 0.57° (in both horizontal and vertical planes) which scans at a rate of 6000 rpm, the antenna is capable of being trained -30 to +75° in azimuth and ±32° from horizontal in the elevation axis.

During search or acquisition operations the AS-1347 scans sectors, or "windows," until the assigned aircraft is detected and locked onto. After lock-on the antenna continues to track the plane, at the same time transmiting beacon interrogations at Ka-band. When the aircraft transponder responds (at X-band) to the AS-2407 beacon antenna satisfactorily, the radar mode is automatically switched over to the beacon video mode unless manually overridden to remain on radar.

(2) The AS-2407/SPN-42 is a small X-band parabolic dish antenna mounted rigidly "piggy back" to the AS-1347 reflector and thus is trained in tandem with it. Beacon transmissions from interrogated aircraft are received by the AS-2407 at 9.3 MHz and, when of satisfactory level, cause the SPN-42 to automatically switch from the radar to

the beacon mode. If the beacon signal becomes unsatisfactory, the system automatically switches back to the radar mode. A manual override is provided for selecting either the radar or beacon mode when desired.

b. PHYSICAL CONFIGURATION — [See Figure 4-47].
c. DIMENSIONS — (1) Height: 77 inches (overall maximum); (2) Width: 100 inches (swing circle); (3) Depth: 75 inches (overall); (4) Weight: 906 pounds (total with pedestal).
d. FREQUENCY RANGE — 33.0 to 33.4 GHz (AS-1347); 8800 to 9500 MHz (AS-2407).
e. RF POWER RATING — 50 kW peak.
f. POLARIZATION — Circular or linear, selectable (AS-1347); linear (AS-2407).
g. GAIN — 48.7 dB (AS-1347); 25 dB (AS-2407).
h. VSWR — 1.3:1 maximum.
i. TRANSMISSION LINE TYPE — Waveguide.
j. PRIMARY POWER REQUIRED — 440 vac, 60 Hz, three-phase.
k. INSTALLATION REQUIREMENTS — The base of the pedestal has five bolt holes $^{15}/_{16}$ inches in diameter for mounting the unit to the platform. One bolt hole is on the centerline. The other bolt holes are on 16 x 20 centers. Each pedestal must be mechanically aligned parallel with the keel (roll axis) within 5 degrees.
l. SPECIAL CONSIDERATIONS — The waveguide sections in the antenna assembly require pressurization of 18 to 23 psig with leakage rates up to 20 cubic inches per minute. Peak surge pressures must not exceed 40 psig. Two antenna units are used in each installation to satisfy the operational requirements. The locations for the antennas must be selected in the island superstructure, facing aft, with an unobstructed view of the approach path and the touchdown point of the landing area.
m. REFERENCES — (1) Technical Manual, Landing Control Central AN/SPN-42, NAVSEA 0967-304-4010/NAVELEX 0967-304-4015. (2) NAVSEA Drawing RE-B2697186.

NOTE — The AN/SPN-42 system also uses three very small corner reflector antennas as point targets for calibrated reference tracking. These antennas are installed in a precise manner per NAVSEA drawing RE-F2696925. An example of one of the corner reflectors installed in place is shown in Figure 4-48.

4-2.5 AN/SPN-43A

The AN/SPN-43A is an S-band surveillance and air traffic control radar used on carriers to provide air navigational data for simultaneous control and identification of aircraft in the terminal area. It replaces the AN/SPN-6 and earlier version AN/SPN-43 equipment shown in Figure 4-49. Used in conjunction with the precision carrier controlled approach (CCA) system for landing operations, the SPN-43A has a range of 50 nautical miles, 360° coverage at altitudes from radar horizon to 30,000 feet, and can track low flying planes to a minimum of 250 yards. Azimuth and range data is displayed to an operator who interprets it and then directs control of the aircraft to the precision CCA transfer point. Positive identification of aircraft is achieved by means of an IFF system synchronized with the radar. The combined antenna assembly used to accomplish the SPN-43 function is described below:

ANTENNAS AS-2785/SPN-43 and AS-2188/U

a. TYPE/USE — (1) The AS-2785 is a part of antenna group OE-138/SPN-43A. It consists of a flared feedhorn and parabolic section reflector mounted on a gyro-stabilized pedestal to compensate for ship pitch and roll, and rotates at 15 rpm. The antenna forms a sharp beam having a 4.2° cosecant-squared elevation beamwidth and a 1.5° azimuth beamwidth. The radiating surface of the S-band reflector is composed of laminated sheets of fiberglass and foil over a resin impregnated core of honeycomb fiberglass. The top of the reflector structure is attached to the elevation tilt actuator which allows tilting the beam from 2.5 to 6.0 degrees.

(2) Antenna AS-2188/U (the IFF antenna) consists of a hybrid-ring duplexer and directional coupler, a ten-foot dual-aperture array with associated RF cables, and a supplementary backfill radiator. The antenna is physically attached to the antenna support structure above the S-band reflector, and thus rotates with it. A dual-aperture, longitudinal shunt-slot array designed for IFF signals (1030 MHz transmit and 1090 MHz receive), the antenna consists of two independent radiating elements divided by a septum. When in-phase signals are coupled from the sum port of the hybrid coupler to the antenna feed cables, the radiating element functions as an integral element and the antenna propagates a conventional pattern with the power concentrated at beam center (sum pattern). When the two elements are excited 180 degrees out of phase with respect to each other, a pattern is produced which is characterized by a deep null along the centerline of propagation (difference pattern).

The sum pattern is transmitted at 1030 MHz and is vertically polarized with a 45° vertical beamwidth and 9° horizontal beamwidth measured at the 3 dB (half-power) points. The antenna is tilted +10° in the elevation plane with respect to ship's reference ordnance level when the co-mounted S-band radar antenna level bracket is at zero degrees.

The difference pattern is also transmitted at 1030 MHz with the same polarization and elevation beamwidth as the sum pattern. The null depth of the difference pattern, when measured at boresight with respect to the sum pattern main lobe, is a minimum of 25 dB.

The hybrid ring duplexer and directional coupler is a stripline device which performs the power dividing and phase shifting functions necessary to obtain the sum and difference pattern outputs from the IFF antenna.

Radiation of the backfill IFF antenna provides sufficient difference pattern power in the back direction to prevent nearby reflections of the sum pattern from causing spurious responses from aircraft transponders in the rear hemishpere area. The amount of power propagated by the backfill IFF antenna is regulated by the value of the ship's directional coupler. Variation of radiation in the horizontal plane is approximately 0.5 dB and the transmitted frequency is the same as for the IFF antenna. Polarization is vertical.

b. PHYSICAL CONFIGURATION — [See Figure 4-50].
c. DIMENSIONS — (1) Height: 150 inches, overall max (tilted); (2) Width: 120 inches; (3) Depth: 182 inches, overall max; (4) Weight: 3400 pounds (total, with pedestal).
d. FREQUENCY RANGE — (1) 3590 to 3700 MHz (S-band antenna); (2) 1000 to 1100 MHz (IFF antenna).
e. RF POWER RATING — 850 kW peak.
f. POLARIZATION — (1) Horizontal (S-band antenna); (2) Vertical (IFF antenna).
g. GAIN — (1) 32 dB (S-band); (2) 18 dB (IFF antenna).
h. VSWR — (1) 1.5:1 (S-band antenna); (2) 1.8:1 to 4.0:1 (IFF antenna).
i. TRANSMISSION LINE TYPE — Waveguide.
j. REFERENCE — Technical Manual, Radar Set AN/SPN-43A, NAVELEX 0967-436-3010.

4-2.6 AN/SPN-44

Radar Set AN/SPN-44 is a navigational, speed-detecting radar set that provides an aircraft carrier's Landing Signal Officer (LSO) with an accurate indication of the true and relative air speed of aircraft making a landing approach on the carrier. With this information, the LSO can accurately control the aircraft's approach, thus reducing the possibilities of damage to the aircraft and to arresting gear because of improper landing speed. Aircraft approach speeds from 40 to 200 knots can be measured to an accuracy of ± 2.5 knots within a 100 to 4000 yard range. The system computes, records, and indicates the true and relative speeds of the approaching aircraft, thereby allowing close control of the approach pattern.

To provide speed information, the radar set utilizes the doppler effect. When a fixed frequency CW carrier strikes a moving target, the reflected wave is shifted in frequency because of relative motion between the wave and target. A fixed frequency carrier is generated by the transmitter and conducted through an RF waveguide system to the antenna. The antenna radiates vertically polarized fixed frequency energy to the moving target. The energy received by the antenna is horizontally polarized and at a slightly altered frequency. The reflected frequency and a sample of the transmitted frequency are mixed in the receiver to produce the difference (doppler) frequency which is proportional to relative speed between the aircraft and the radar set. The difference frequency is amplified, converted to a dc analog, and applied to a servo system which positions a synchro transmitter, precision potentiometer, and relative speed dial. True air speed is obtained by adding wind speed from the ship's anemometer and the resultant synchro information is forwarded to the speed indicators.

The AN/SPN-44 is a replacement for the older AN/SPN-12 system. It is very similar in function and appearance, employing the same type of parabolic dish described below:

ANTENNA AN/SPN-44

a. The AN/SPN-44 antenna is part of the topside-mounted reciever-transmitter unit RT-989/SPN-44. A highly directive X-band microwave radiator, the antenna consists of a parabolic dish reflector with a feedhorn attached at the center extending along the axis. The feedhorn itself is a length of dielectric-filled waveguide terminated in a radiating element which directs its energy back toward the parabolic reflector. The sharp pencil beam thus formed has a beamwidth of only 3.3 degrees. The entire unit is supported on a two-axis pivot and is manually trained by the operator.
b. PHYSICAL CONFIGURATION — [See Figure 4-51]. For example of installed antenna see Figure 4-47(a) (upper left).
c. DIMENSIONS — (1) Height: 35 inches (overall with base); (2) Width: 44 inches (max swing circle); (3) Depth: 35¼ inches; (4) Weight: 233 pounds (total with base).
d. FREQUENCY RANGE — 10,000 to 10,250 MHz (X-band).
e. RF POWER RATING — 10 watts CW.
f. POLARIZATION — Horizontal.
g. GAIN — 31 dB.
h. VSWR — 1.5:1.
i. TRANSMISSION LINE TYPE — None. [Antenna is connected directly to transmitter-receiver unit].

j. PRIMARY POWER REQUIRED — 115 vac, 60 Hz, one-phase.
k. SPECIAL CONSIDERATIONS — The RT-989/SPN-44 must be located aft of the carrier's island, about 20 to 30 feet above the flight deck, with a clear view over the stern of the carrier. It is designed for platform mounting with shock mounts attached to the base. The length of the cabling between the receiver-transmitter and computer is limited to 100 feet. The mounting platform must be of sufficient size to permit the operator to stand behind and on either side while using the gunsight to track the approaching aircraft.
l. REFERENCE — Technical Manual, Radar Set AN/SPN-44, NAVSEA 0967-318-1010.

4-3 WEAPONS CONTROL RADAR ANTENNAS

Naval demonstrations of radar performance at sea in the late 1930s, particularly in the ability to track projectiles and spot the point of shot fall, aroused enthusiasm about the possibilities of using radar for gunfire control.[28] Heretofore, by use of the traditional optical range finder, the procedure was to fire once over the target, once short of the target, observe the relative points of splash down, and then attempt to calculate the range and bearing of the target for a hit on the third try. With its phenomenal accuracy, radar theoretically promised to put an end to the rather wasteful and limited trial and error method by providing a direct hit on the first shot. This conclusion, and the imminent threat of war, led to the immediate development of automatic-track gun support radar systems.

Just prior to America's entry into World War II the first gunfire control (range-only) equipment, the Model FA (or Mark 1) was installed on the *USS Wichita*. Shortly thereafter, the first system to provide bearing as well as range, the Model FC (or Mark 3), was put aboard the *USS Philadelphia*, and over a hundred more FC's procured. That was followed almost immediately by the Model FD (or Mark 4), and the FM (Mark 12) radars which, using a sequential lobing antenna for azimuth train, and a separate nodding antenna for elevation data, were used for anti-aircraft defense. Several hundred of these models were bought by the Navy. All of the early fire control radars operated in the UHF frequency range (700 MHz) with increasing pulse power as improvements allowed. The Model FA was a 25 kW system, the FC and FD were 40 kW, and the FM was 250 kW.

As technology advanced, the operating frequency was quickly moved upwards and scanning capability added, first with the Model FH (Mark 8) at S-band (3000 MHz), then, before the end of the war, with the Mark 13 at X-band (8800 MHz). (Interestingly, more than two decades later, when the battleship *New Jersey* was reactivated for its 1968-69 Vietnam service, it was still using the Mark 13 for shore bombardment.*)[29]

One of the more important initial improvements to fire control radars occurred early in the 1940s with the introduction of conical-scan antennas.[30] The radar beam in this type antenna rotates around the central axis of the parabolic reflector, describing a cone of radiation. The first conical-scan antennas used a rotating dipole element; however, the resulting distortion caused by a continuously changing polarization as the dipole was rotated proved a hindrance. This problem was alleviated by employing a nutating feed rather than physical rotation, so that the transmitted energy remained of constant polarization.

Several hundred of the early S-band (3000 MHz) conical-scan fire control radars, Model FJ (or Mark 9) and Model FL (or Mark 10), were purchased during the war years. However, to realize the advantage of much smaller antennas having very narrow beams with far greater angular accuracy, fire control radar equipment soon moved up to X-band. Since then several thousand anti-aircraft conical-scan weapons control radars have been placed in naval service; e.g., the Mark 28, 29, 34, 35, 37, 39, and the AN/SPG-48, 50, and 53. Moreover, with the advent of missile technology, conical-scan radars were developed for missile tracking purposes. The AN/SPG-51, for example, was by 1951 being introduced for Tartar missile guidance.[31]

Following the war, continued research in fire control radar systems resulted in a further improvement; i.e., the development of monopulse type equipment for yet another remarkable increase in angular accuracy. In the monopulse method, a system of four feedhorns receive the return radar energy and apply it to a hybrid network where the four individual inputs are compared. The difference in signal characteristics is used for indicating angular error in train and the sum for computing range. In the transmit case, either all four horns are used in concert, or a fifth center horn is separately employed. Adapted immediately for missile guidance, the first shipboard monopulse radar was the AN/SPG-49 installed aboard the *USS Galveston* in 1957.[32] This was followed by such systems as the

*As of this writing, early in the 1980s, the Navy is again bringing the *New Jersey* back into active service, and she will come out equipped yet with the Mark 13 — after nearly 40 years!

AN/SPQ-5 and the AN/SPG-55, 56, and 59, many of which are still in use today.

Now in the 1980s, naval weapons control and missile tracking radars have become evermore sophisticated and are indispensable elements of shipboard combat systems. As will be seen in the following sections, a variety of modern weapon systems come equipped with fire control and tracking radars as an integral subset.

4-3.1 MARK 13

Radar equipment MK 13 is an X-band system used for main battery fire control on battleships and heavy cruisers. It is capable of measuring bearing and range with great accuracy to about 50,000 yards, and out to 80,000 yards with somewhat less accuracy. Introduced in 1944, the MK 13 is of course now considered obsolete. Nevertheless, it is still carried aboard the battleships and heavy cruisers currently in moth balls, and is, therefore, included here in the likely event that these ships are reactivated from time to time for use of their heavy fire power support in shore bombardment.

ANTENNA MK 13

a. TYPE/USE — The MK 13 radar antenna is a highly directive, scannable radiator consisting of an elliptical-paraboliod section reflector made of perforated sheet aluminum and fed by a waveguide horn. The antenna assembly is mounted on a framework which permits it to be rocked from side to side by an antenna scanning motor through an eccentric mechanism. Thus, by mechanically moving the paraboloid reflector, the narrow beam (0.9° azimuth and 3.6° elevation beamwidth) is made to scan through a horizontal angle of 11 degrees. Also furnished as part of the antenna system is a dummy load; essentially a section of waveguide equipped with a ceramic resistance dissipating unit, to permit tuning of the transmitter without radiating from the antenna. To protect the antenna from harmful effects of the sea environment, the entire assembly is covered by a streamline-shaped radome.
b. PHYSICAL CONFIGURATION — [See Figure 4-52].
c. DIMENSIONS — (1) Height: 45³⁄₁₆ inches (with cover); (2) Width: 126 inches (with cover); (3) Depth: 77¼ inches (with cover); (4) Weight: 1000 pounds (total with cover).
d. FREQUENCY RANGE — 8800 MHz.
e. RF POWER RATING — 50 kW peak.
f. POLARIZATION — Horizontal.
g. GAIN — 39 dB nominal.
h. TRANSMISSION LINE TYPE — Waveguide.
i. PRIMARY POWER REQUIRED — 220/440 vac, 60 Hz, three-phase.

j. REFERENCE — Instruction Book, Radar Equipment Mark 13, NAVSEA OP-1773.

4-3.2 MARK 25

Radar equipment MK 25 is an automatic tracking X-band fire control radar associated with the MK 67 Gun Fire Control System (GFCS). It provides a selection of conical, circular, and spiral antenna scans for use in target acquisition, tracking, and shell spotting with a minimum range of 300 yards and a maximum range out to 100,000 yards. The system may be operated either manually or automatically, in range, azimuth, and elevation. Furthermore, each coordinate may be operated manually or automatically, independent of the others. An old mainstay of naval fire control radars, the MK 25, first introduced in the 1940s and modified several times since, is still used aboard a wide range of ships in support of 5″ guns.

ANTENNA MK 25 (MOD 5)

a. TYPE/USE — The MK 25 radar antenna is a five-foot parabolic dish reflector (type AT-276/SPG) fed by a nutating waveguide rod to provide conical, circular, or spiral scanning. The MOD 5 reflector consists of metallic material embedded in fiberglass. An earlier version MOD 2 antenna, used for many years and fabricated of wire mesh over a metal frame, is shown in Figure 4-53.

The antenna's feed-rod scanner is enclosed in four housings: the shield, support tube, motor, and control housings. These housings are fastened together and constitute a completely waterproof unit. The shield housing encloses a gimbal and the pivoted end of the support tube. The support tube housing encloses the waveguide, waveguide clamp, waveguide position adjusting screws, and the support tube assembly. The control housing encloses the control transformer, mechanical gearing, springs, and electrical controls. The motor housing encloses the drive motor, reference generator, bell crank, and counterweight assembly. The drive motor is energized by the motor controller located below decks. The entire antenna assembly is supported by a mount to provide the mechanical means for elevating and stabilizing the antenna.

To increase the tactical utility of the radar, three types of scans are available: conical (small circle), circular (large circle), and spiral. Selection of scans is under control of the antenna scan selector switch at the radar set control.

During conical scan, the average cross section of the beam is about

1.6° and the angle that the axis of the beam makes with the gun director line of sight is 0.5°. The rate of scan is 30 Hz. If the target is in the beam but off the gun director line of sight axis, different reflecting intensities result for different positions of the beam axis in its conical path. The position of the received echo on the indicators shows the position of the target with respect to the gun director line of sight. When the antenna is pointed directly at the target, equal signals result for all positions of the beam and the received echo is centered in bearing and elevation in the form of a straight line of uniform intensity. This type of scan provides accurate bearing and elevation information and is used for automatic target tracking.

Circular scanning is the same as conical scanning except that the axis of the radar beam with respect to the gun director line of sight is approximately 5 degrees and the radar beam is approximately 2 degrees wide. If a target were centered on the gun director line of sight, there would be no received echo, as there is a cone with a vertex angle of approximately 8 degrees about the line of sight containing very little radiation. This type of scan is used for spotting shell splashes. Its advantage over conical and spiral scanning for this purpose is that it gives a flickerless pattern on the radar indicator screens.

Spiral scan offsets the radar beam from the conical position to the circle position and back to the conical position in a spiral form. The scan goes from conical to circular in approximately 8½ spirals and returns in approximately 6½ spirals. The time for a complete spiral cycle is one-half second. This type of scan covers the complete area of the conical and circular scan plus the area between the two and is used for target acquisition.

b. PHYSICAL CONFIGURATION — [See Figure 4-54].
c. DIMENSIONS — (1) Height: 102 inches (overall, with mount), note parabolic dish alone is 60 inches; (2) Width: 111 inches (base of mount); (3) Depth: 67 inches (with mount); (4) Weight: 1143 pounds (total, with mount).
d. FREQUENCY RANGE — 8500 to 9600 MHz.
e. RF POWER RATING — 500 kW peak.
f. POLARIZATION — Vertical.
g. GAIN — 39 dB.
h. TRANSMISSION LINE TYPE — Waveguide.
i. PRIMARY POWER REQUIRED — 115 vac, 60 Hz, three-phase.
j. REFERENCES — (1) Technical Manual, Radar Equipment Mark 25 MOD 5, NAVSEA 0P-3977. (2) Technical Manual, Radar Equipment Mark 25 MOD 3, NAVSEA OD-43741 and OD-43742. (3) Technical Manual, Radar Equipment Mark 25 MOD 2, NAVSEA OP-1651 and OP-1788A.

4-3.3 MARK 35

Radar Equipment MK 35 is an automatic acquisition and tracking X-band fire control radar mounted integrally with the MK 56 Gun Fire Control System. It provides a selection of either conical or spiral beam scanning of aerial, surface, and shore targets at a minimum range of 250 yards and out to a maximum range of 30,000 yards. The MK 56 GFCS is employed on several ships of the fleet for control of three-inch and five-inch guns.

ANTENNA AS-515A/SPG MK 35

a. TYPE/USE — The MK 35 AS-515A radar antenna consists of a four-foot metal-impregnated, fiberglass, parabolic dish reflector fed by a broadband, pressurized waveguide section terminated in a dielectric radiator. Three curved fiberglass rods are placed over the front of the reflector to protect the feed radiator. Included in the antenna assembly is a motor, gear box, and mechanisms by which the antenna feed is nutated for both conical and spiral scan. The assembly is mounted on a yoke to allow tilting the radar beam in elevation. During the search mode of operation the antenna employs a spiral nutation at 29.1 Hz to develop a 12.5° cone four times each second. In the track mode, a much narrower 3° cone is formed, also at 29.1 Hz.
b. PHYSICAL CONFIGURATION - [See Figure 4-55].
c. DIMENSIONS — The MK 35 dish antenna is four feet in diameter. However, as an integral part of the MK 56 director, it must be considered in the overall director dimensions, which are 111 inches high by 97 inches wide, and weighing 7550 pounds.
d. FREQUENCY RANGE — 8500 to 9600 MHz.
e. RF POWER RATING — 40 kW peak.
f. POLARIZATION — Vertical.
g. GAIN — 35.5. dB.
h. TRANSMISSION LINE TYPE — Waveguide.
i. PRIMARY POWER REQUIRED — 115 vac, 60 Hz, one-phase.
j. REFERENCE — Gun Fire Control System, Mark 56, NAVSEA OP-1600.

4-3.4 AN/SPG-34 (MK 34)

Radar Set AN/SPG-34 is part of the MK 63 Gun Fire Control System used to control the firing of shipboard 3″/50 and 40 mm antiaircraft guns. While principally supplying target range for the GFCS, the SPG-34 also provides director pointing information necessary to acquire a designated target and for tracking a target after acquisition. It should be noted, however, that the SPG-34 is not an automatic tracking radar. It requires the director operator to continuously position the director so that either the optical target image or the radar indication is kept in the center of the gunsight. In this manner the gun director operator manually positions the director to acquire the target in elevation and bearing while the radar operator tracks the target in range.

The SPG-34 radar system may be operated in any of three modes; i.e., designation, acquisition, or track. Switching from one mode to another is nearly instantaneous and is controlled by the radar operator. The designation mode is used to acquire a designated target which has been detected by search radars or lookout personnel. In this mode the antenna operates in spiral scan to provide information indicating in which direction the gun director must be moved to bring it and the radar antenna to bear on the designated target. When the gun director is then positioned so that the radar beam intercepts the target, the operator gates the desired target in range and switches the set to the acquisition mode.

The acquisition mode of operation supplies pointing information derived from the previously gated target signal so that the gun director can be accurately pointed on target. The antenna remains in spiral scan to sweep a broad cone-shaped area centering the target on line with the antenna axis.

The track mode is used for precisely tracking a gated target in preparation for firing. In this mode the antenna radiation is switched to conical scan to keep the target tightly centered in the gun sights. The director, meanwhile, through continuous computer updating, automatically trains the guns on the future (calculated) position of the target while the gunsights remain independently centered on target. The X-band microwave antenna used to accomplish these AN/SPG-34 missions is described below:

ANTENNA AS-515A MK 34 (AN/SPG-34)

a. TYPE/USE — The SPG-34 fire control radar antenna consists of a 40-inch diameter parabolic dish reflector (type AT-194A) fed by an AS-515A nutating horn radiator. The parabolic reflector is fabricated of laminated fiberglass embedded with wire mesh. Three curved fiberglass rods are fastened over the reflector face to protect the radiator horn. While the reflector concentrates the radiated energy into a narrow conical beam, a motor-driven mechanism nutates the feedhorn around the reflector axis. Thus, depending on the mode of operation, either a spiral or a conical beam formation may be selected. Spiral scanning is used initially to facilitate location of targets. Here the radiated beam, 2.6° in beamwidth, scans an enlarged 12.4 degree cone at a slow 2 Hz rate. During conical scan operation, the radiated beam, still only 2.6° in beamwidth, scans a tighter area encompassing a 4 degree cone, at a faster rate of 29 Hz. To permit the antenna to move in train and elevation so as to remain on target while tracking (as the guns point to the future, anticipated, target position), the entire antenna assembly is mounted on the gun in a Y-shaped yoke mechanism.

b. PHYSICAL CONFIGURATION — [See Figure 4-56].
c. DIMENSIONS — (1) Diameter: 40 inches (reflector); (2) Depth: 35⅛ inches (total antenna); (3) Weight: 126 pounds (antenna only).
d. FREQUENCY RANGE — X-band.
e. RF POWER RATING — 30 kW peak.
f. POLARIZATION — Vertical.
g. GAIN — 36 dB.
h. TRANSMISSION LINE TYPE — Waveguide.
i. PRIMARY POWER REQUIREMENTS — (1) 440 vac, 60 Hz, three-phase; (2) 115 vac, 60 Hz, one-phase.
j. REFERENCE — Radar Set AN/SPG-34, NAVSEA OP-3682.

4-3.5 AN/SPG-49

Radar Set AN/SPG-49 provides target tracking and illumination for the MK 77 shipboard guided missile fire control system. The equipment is installed principally on large missile cruisers for surface-to-air, surface-to-surface, and surface-to-shore operations. A major subsystem of the SPG-49 is its very large, high-power microwave lens antenna used for acquisition, automatic tracking, and illumination of a designated target.

In a typical surface-to-air engagement the SPG-49 is assigned a target by a weapons director operator, whereupon it commences an acquisition program until defined to the point that automatic tracking is initiated. Separate action is then taken to guide a missile to target. Once the missile-to-target range reaches a predetermined value, the SPG-49 begins CW radiation to illuminate the target so that the missile, sensing the energy reflected by the target from the CW illumination beam, homes in.

The SPG-49 is obsolescent and is being replaced by more modern guided missile fire control systems. Nevertheless the supporting documentation has yet to be declassified and, therefore, further details of the system philosophy and capabilities cannot be disclosed here.

ANTENNA UD 20 AN/SPG-49

a. TYPE/USE — The UD 20 antenna used with the AN/SPG-49 radar set is a very large, complex assembly. In addition to the antenna itself, the assembly consists of four major structures: the pedestal, train, elevation, and traverse members. This configuration affords full hemispherical coverage of high velocity targets despite the very heavy antenna structure. The train member is a U-shaped yoke that supports the elevation member and permits the antenna to tilt in elevation. The elevation member is an octagonal gimbal which pivots in elevation between the yoke arms and supports the traverse member on swivels within the gimbal. The traverse member is a radome-covered box structure, octagonal at the front and dome-shaped in the rear, which pivots traversely and houses all units of the antenna subsystem.

The antenna subsystem is an intricate assembly of several devices including: (1) an AT-828 microwave lens; (2) an AT-636 acquisition reflector; (3) an AS-782 acquisition antenna; and (4) a UD18 dual tracking radar horn antenna. The AT-828 microwave lens is an octagonal-shaped polyester, resin-impregnated, fiberglass, honeycombed structure of over 4000 cells used to focus the radar beams sharply into space. Each individual cell consists of an aluminum loading element and plastic support internally covered with a smooth, conductive, resin-bonded, silver coating. Mounted on the inside front of the traverse member, the microwave lens is protected from the weather by a spherical sectioned fiberglass radome.

The AT-636 acquisition reflector, located in the center of the traverse member, is composed of vertically strung wires on an octagonal frame, invisible to the horizontally polarized track radar energy, but which acts as a reflector to the vertically polarized acquisition radar microwave energy.

Located on the floor of the traverse member, the AS-782 acquisition antenna consists of a high-power folded reflector, waveguide wedge, high-power scanner, divergent lens, and multi-element horn radiator. In the acquisition mode of operation, microwave energy from the radar transmitter is dispersed into the reflector which acts as a collimation device. The wedge section directs the energy emerging from the folded reflector into the scanner to produce a symmetrical scan.

The tracking radar sensor assembly is located in the rear of the traverse member and contains the dual tracking antenna horns with ancillary equipment.

b. PHYSICAL CONFIGURATION — [See Figure 4-57].
c. DIMENSIONS — (1) Height: 209 inches (total assembly); (2) Width: 200 inches (total); (3) Depth: 216 inches (total); (4) Weight: 44,407 pounds (total assembly).
d. FREQUENCY RANGE — C-band.
e. RF POWER RATING — 3000 kW peak.
f. POLARIZATION — (1) Horizontal (tracking); (2) Vertical (acquisition and illumination).
g. TRANSMISSION LINE TYPE — Waveguide.
h. REFERENCE — Radar Set AN/SPG-49, NAVSEA OP 2819.
NOTE — In addition to the main UD 20 SPG-49 antenna, two small horn antennas are mounted on the outside of the traverse member box structure. These are used for sidelobe suppression and beam spoilage during anti-jam conditions.

4-3.6 AN/SPG-50

The AN/SPG-50 is an X-band radar system used on surface ships as part of antiaircraft Gunfire Control System MK 63. As such it has the same purpose and function as the AN/SPG-34, also meant for use with the MK63 GFCS as described in Section 4-3.4. However, there is one major and significant difference: During the track mode of operation, the capability for tracking in range is performed automatically, rather than manually, by the SPG-50. In all other aspects the two radar systems are virtually identical. In fact, the antenna used with the SPG-50 is the same one used with the SPG-34, i.e., the AS-515A with AT-194A reflector.Thus the technical characteristics are the same as those given under Sec. 4-3.4, and the antenna is mounted on the guns as shown in Figure 4-56.*

4-3.7 AN/SPG-51

Radar Set AN/SPG-51 is part of the surface ship MK 74 Guided Missile Fire Control System, an integral component of the medium range Tartar missile weapon system. The SPG-51 consists of a C-band computer-controlled, pulsed doppler, automatic acquisition and tracking radar and a high-power X-band, automatic or manual, continuous wave illuminator (CWI) transmitter for missile guidance.

The GMFCS MK 74 receives target designations from the weapons direction equipment after detection by air or surface search radars as primary sources, or NTDS acting from other sources. Upon target

---

*REFERENCE — Radar Set AN/SPG-50, NAVSEA OP 2166.

designation the director trains, elevates, and initiates a local search pattern at the range and position indicated by the external sensor. The GMFCS radar acquires the target and commences automatic tracking. The target is illuminated during the tracking phase and acts as a beacon for the guided missile during the homing phase.

Once the radar acquires the target, data are also transmitted to the GMFCS computer for processing and resolving launcher pointing position and time-of-fire as a function of missile intercept range capabilities. The radar continues to track the target to provide illumination for the missile until the calculated intercept time plus a short interval. Depending upon the results of intercept, the GMFCS may be reassigned to the same target or assigned to a new target.

Both the tracking radar and the CWI transmitter share the same antenna, but employ separate feeds and reflector surfaces. Moreover, the CWI operation makes use of two individual feedhorns: a main horn is used to illuminate the target and an ancillary feedhorn provides a wide angle, rear reference lock-on signal to the missile's rear reference antenna. These signals guide the missile in flight while the target is being tracked.

ANTENNA AS-2649/SPG-51C

a. TYPE/USE — The AS-2649 radar antenna is part of antenna group OE-119/SPG-51C designed for use with the MK 74 GMFCS. An essentially identical antenna, the AS-2234, is used with the SPG-51D model and an older version, somewhat different in appearance, was the AS-1483 employed with the SPG-51B system shown in Figure 4-58. The AS-2649 antenna consists of a 7½-foot dual-purpose parabolic dish reflector, common to both the track and CWI transmitters, and an offset, stationary, multi-element feed subassembly. While the antenna appears to be a single parabolic dish, in reality there are two electrically separate reflectors. That is, a vertically-polarized X-band reflector comprised of a wire grid imbedded in a dielectric material is molded to the front of the conventinal, horizontally polarized, solid C-band reflector.

The X-band antanna subassembly is made up of the following major components: (1) a main X-band CWI feedhorn having an extremely narrow pencil beamwidth of 0.9°; (2) an ancillary rear reference guidance signal horn located at the reflector center and having a very wide 70° vertical by 80° horizontal beamwidth and (3) a power divider mounted on the main feedhorn support which divides the X-band power between the main and ancillary feedhorns.

The C-band antenna subsystem consists of the following units: (1) a four-horn cluster, producing a sharp 1.6° beam, and (2) a comparator unit which is mounted adjacent to the four-horn feed, functioning to compare the return energy received by the horns, and sense the direction and magnitude of the angular displacement between the target and the radar beam.

Finally, in addition to the above subassemblies, there are various other units associated with the AS-2649, including a conical-scan-on-receive-only (COSRO) scanner used to modulate the difference output of the comparator unit.

b. PHYSICAL CONFIGURATION — [See Figure 4-58(b)].
c. DIMENSIONS — (1) Diameter: 91 inches; (2) Weight: 550 pounds (total: OE-119 Antenna Group).
d. FREQUENCY RANGE — C-band (tracking radar); X-band (CWI).
e. RF POWER RATING — 30 kW peak.
f. POLARIZATION — (1) Horizontal: (tracking radar); (2) Vertical (CWI).
g. GAIN — (1) 39.5 dB (tracking radar); (2) 45.0 dB (main CWI); (3) 12.5 dB (ancillary CWI).
h. TRANSMISSION TYPE — Waveguide.
i. PRIMARY POWER REQUIREMENTS — (1) 115 vac, 60 Hz, one-phase; (2) 115 vac, 400 Hz, one-phase.
j. REFERENCES — (1) Radar Set AN/SPG-51B, NAVSEA OP-3010. (2) Radar Set AN/SPG-51C, NAVSEA OP-3433. (3) Radar Set AN/SPG-51D, NAVSEA OP-3541.

4-3.8 AN/SPG-52

The AN/SPG-52 is a small, lightweight, Ku-band radar set designed as part of the shipboard three-inch/50 and five-inch/38 MK 70 Gun Fire Control System used principally for antiaircraft fire against medium-range (450 to 8850 yards) targets. With proper modifications the system also may be used to direct gunfire against surface targets at ranges from 1000 to 20,000 yards.

The SPG-52 normally operates in a continuous range-search mode at a fixed repetitive rate from 500 to 10,000 yards until a target is acquired, at which time it switches automatically to the range-track mode. It then supplies range and target range rate data to the computer for calculating the target motion lead angle used in proper firing action.

ANTENNA RT-266A/SPG-52

a. TYPE/USE — The SPG-52 radar antenna is an integral unit of the

RT-266A receiver-transmitter unit. It consists of a small, 11-inch aluminum parabolic dish reflector and waveguide feedhorn which forms a 5° conical beam capable of being scanned in elevation from -25 to +90 degrees (independent of the gun director) and 375° in azimuth (trained in concert with the director). To protect against the environment, a fiberglass radome shell covers the complete RT-266A assembly.

b. PHYSICAL CONFIGURATION — [See Figure 4-59].
c. DIMENSIONS — (1) Height: 32 inches (total RT-266A); (2) Width: 23⅛ inches (total); (3) Depth: 21⅞ inches (total); (4) Weight: 85 pounds (total).
d. FREQUENCY RANGE — 16,400 to 16,600 MHz (Ku-band).
e. RF POWER RATING — 40 kW.
f. GAIN — 30 dB.
g. TRANSMISSION LINE TYPE — None external. [Antenna is mounted integrally with the receiver-transmitter unit].
h. PRIMARY POWER REQUIREMENTS — 115 vac, 400 Hz, one-phase.
i. REFERENCES — (1) Gun Fire Control System, Mark 70, NAVSEA OP-1940A. (2) Radar Sets AN/SPG-52 and AN/SPG-52A, NAVSEA OP-1940F.

4-3.9 AN/SPG-53

Radar Set AN/SPG-53 is part of the MK 68 shipboard Gun Fire Control System used for directing orders to five-inch guns in engagement of air and surface targets. The engagement sequence in general consists of search, acquisition, and track, and in so doing the MK68 GFCS performs the required computations to provide the correct gun orders for engagement, i.e., directing the line of fire to the predicted point of target interception.

The SPG-53 X-band radar system is integrally associated with the MK68 to supply accurate train and elevation tracking signals to the gun director, to determine precise range to the target under all conditions of visibility and to transmit target range data to the associated computer. Several models of the SPG-53 have been produced over the years, up to and including the SPG-53F. Two important variants are the SPG-53B and SPG-53C, versions modified to have the additional capability of CW illumination required for guidance and control of Tartar missiles (also known as MK 68 Mod 9 and 10). The SPG-53B/C antenna, therefore, differs considerably from the conventional SPG-53 antenna, and will be described separately.

The classified nature of the AN/SPG-53 radar equipment precludes disclosing any further details of the system here.

ANTENNA AS-515A/SPG

a. TYPE/USE — The conventional AN/SPG-53 radar antenna consists of a type AT-276 five-foot parabolic dish reflector, with type AS-515A nutating feed rod extending through the reflector center to the focal point. Note that this reflector is the same one used with the MK 25 fire control radar equipment, and the AS-515A feed system is the same unit employed with the MK-35 and the SPG-34 radars described previously. The antenna forms a narrow 1.6° beam which it radiates in either a spiral (11.6° cone at a 2 Hz rate) or conical (3.1° cone at 30 Hz rate) scanning mode, as selected, to obtain target bearing and elevation data. An antenna mounting structure supports the antenna assembly on top of the MK 68 director. Stabilized traversely irrespective of ship pitch and roll, the antenna trains in azimuth with the director and may be tilted in elevation from -26 to +96 degrees.

b. PHYSICAL CONFIGRUATION — [See Figure 4-60(a)].

c. DIMENSIONS — (1) Height: 80½ inches (total including support structure; antenna dish alone is 60 in.); (2) Width: 60 inches (total); (3) Depth: 41 inches (total); (4) Weight: 163 pounds (antenna only).

d. FREQUENCY RANGE — X-band.

e. RF POWER RATING — [Classified].

f. POLARIZATION — Vertical.

g. GAIN — 39 dB.

h. TRANSMISSION LINE TYPE — RG-51/U waveguide.

i. PRIMARY POWER REQUIRED — (1) 440 vac, 60 Hz, three-phase; (2) 115 vac, 60 Hz, one-phase.

j. REFERENCES — (1) Radar Set AN/SPG-53A, NAVSEA OP-2782. (2) Radar Set AN/SPG-53B, NAVSEA OP-2782, Supplement 1, Part 1. (3) Radar Set, AN/SPG-53F, NAVSEA OP-4169. (4) Technical Manual, Gun Director Mark 68 MOD 0 through 7, NAVSEA OP-3480. (5) Technical Manual, Gun Fire Control System Mark 68 MOD 13, NAVSEA OP-4180.

ANTENNA AS-1868/SPG-53B

a. TYPE/USE — The AS-1868 antenna assembly, designed especially for use with the AN/SPG-53B/C radar system, is a dual-purpose parabolic dish very similar in operation and construction to the AS-2649/SPG-51 antenna described earlier. It consists of two co-layered parabolic reflecting surfaces separated by a honeycombed core, a nutating, dielectric, tracking radar feedhorn, a main narrowbeam CWI feedhorn, and an auxiliary broadbeam rear reference CWI missile guidance feedhorn, plus associated waveguide and power dividing subunits.

The front surface of the parabolic dish is the CWI reflector consisting of closely spaced vertical wires imbedded in the surface. It is, therefore, vertically polarized. The rear surface, on the other hand, is horizontally polarized by virtue of the feed orientation, and is the tracking radar reflector. The tracking radar nutating feed rod provides a choice of conical or spiral patterns in a narrow cone. Thus the antenna functions as a single unit to allow simultaneous radiation of tracking and illuminating energy. A radome housing protects the antenna assembly from the sea environment.

b. PHYSICAL CONFIGURATION — [See Figure 4-60(b)].
c. DIMENSIONS — (1) Height: 71 inches; (2) Width: 64 inches (reflector diameter); (3) Depth: 43 inches; (4) Weight: 318 pounds (antenna only).
d. FREQUENCY RANGE — X-band.
e. RF POWER RATING — [Classified].
f. POLARIZATION — (1) Horizontal (tracking radar); (2) Vertical (CWI).
g. GAIN — (1) 38 dB (tracking radar); (2) 41.6 dB (main CWI); (3) 12.5 dB (auxiliary CWI).
h. TRANSMISSION LINE TYPE — Waveguide.
i. PRIMARY POWER REQUIRED — (1) 440 vac, 400 Hz, three-phase. (2) 115 vac, 60 Hz, one-phase.
j. REFERENCE — Radar Set AN/SPG-53B, NAVSEA OP-2782, Supplement 1, Part 1.

4-3.10 AN/SPG-55

Radar equipment AN/SPG-55 is a long range, tactical, shipboard Fire Control System used integrally with the MK 76 Terrier Guided Missile Fire Control System (GMFCS) for acquiring, tracking or jamming a target; transmitting missile guidance commands; and, as a secondary means, to search and detect. The Terrier weapon system provides a naval combat vessles with missile fire power against airborne, surface, anti-ship missile, and shore targets. In general, the weapon system operation can be divided into five phases: detection and location of targets; evaluation of target threat and engageability; target assignment to the missile guidance radar; acquisition and track of the selected target by the radar; and missile selection, loading, and firing to destroy the target. In support of these operations, the AN/SPG-55 radar performs the following functions:

a. Automatically acquires and tracks a designated primary and secondary target.
b. Provides target position and velocity data to the GMFCS computer.
c. Automatically generates and transmits guidance information for use by *Terrier* beam-riding missiles in flight.

d. Automatically generates and transmits continuous wave illuminator (CWI) information for use by *Terrier* homing missiles in flight.
e. Automatically acquires and tracks a beacon signal returned from an outbound, beam-riding *Terrier* missile (tracking in range only).
f. Automatically generates command signals (arm, detonate, and destruct) to *Terrier* beam-riding missiles i flight.
g. Acquires and tracks jamming sources.
h. Detects, acquires, and tracks anti-ship missiles.
i. Provides a secondary search function for area surveillance.

In order to provide these capabilities, the SPG-55B contains a C-band pulse transmitter and receiver for target tracking, a capture beam transmitter, a guidance beam transmitter, and an X-band transmitter. The capture and guidance transmitters are used with beam riding missiles and the X-band transmitter provides continuous wave (CW) illumination for homing missiles. The major above-deck portion of the SPG-55 is the radar director, which is mounted on a barbette and houses the electronic, electromechanical, and mechanical components used for remote and local control of the radar director. The radar director consists principally of upper and lower train assemblies.

Upper Train Assembly—The upper train assembly consists of the pedestal yoke housing, left and right elevation shaft housing assemblies, main antenna assembly, and capture antenna assembly.

The pedestal assembly, which is bolted to the director foundation, contains the servo drives and gearing used to rotate the movable portion of the radar director train. The pedestal assembly also supports the yoke housing, a hollow U-shaped structure which can rotate continuously in train in either direction. The two arms of the yoke housing assembly support the left and right elevation shaft housing assemblies which in turn support the main antenna assembly and contain the units used to position the main and capture antenna in elevation from +85° to -35°.

Lower Train Assembly — The lower train assembly is fastened to the yoke housing assembly and is suspended through the pedestal assembly and director foundation. The lower train assembly consists of the electronic rack assembly, which rotates in train with the radar director, contains the track, capture, and guidance transmitters and the continuous wave (CW) electronic circuits.

The classified nature of much of the AN/SPG-55 equipment characteristics precludes disclosing any further details of the system here.

ANTENNA AN/SPG-55

a. TYPE/USE — The An/SPG-55 antenna system is a very large and complex structure made up of a variety of elements. The main antenna, UD-50 is the most prominent member. Located between, and supported by, the elevation shaft housing (refer to Figure 4-61) it is a drum-like shell with a hemisperically-shaped aluminum rear cover and a conical, honeycomb-molded, fiberglass radome fastened to the front flange of the antenna support structure. Inside the shell are (1) the main parabolic dish reflector, (2) a hyperbolic grid subdish reflector, (3) a CWI feedhorn sub-assembly comprised of X-band primary and secondary horns, a C-band track and guidance comparator four-horn cluster feed projecting out of the center of the main dish, and a large assortment of associated RF equipment. At the rear of the main antenna assembly, behind the parabolic reflector, are three gyrostabilization data in elevation, traverse, and cross-traverse axes.

   Mounted externally to, and rotating in train and elevation with, the main antenna are four small ancillary antennas:

   (1) An ECCM C-band omnidirectional receiving horn, UD 50-14, bracketed to the outer top surface of the support structure near the capture antenna;
   (2) A capture antenna, UD-60, located on the outboard side of the elevation shaft housing and consisting of a Cassegrain-feed, polarized parabolic reflector, and polarization converter, all covered by a radome. The circular waveguide feed rotates at 30 Hz to produce a conically-scanned capture beam;
   (3) An X-band CW cluster horn pillbox antenna, UD-16, composed of a reflector, feed panels, support, waveguide assembly, and radome, mounted on the upper side of the main antenna opposite the ECCM horn, near the Doppler antenna;
   (4) A doppler antenna assembly, UD-65, secured to the outboard side of the elevation shaft housing assembly directly opposite the capture antenna. The doppler antenna has a radome with a back-plate made of honeycomb fiberglass imbedded with metal strips. It also has, as part of its assembly, a comparator, and track microwave components.

b. PHYSICAL CONFIGURATION — [See Figure 4-61].
c. DIMENSIONS — (1) Height: 281 inches (total radar director assembly); (2) Width: 175 inches (total); (3) Depth: 124 inches (total); (4) Weight: 12,970 pounds (total).
d. FREQUENCY RANGE — C-band, X-band.
e. RF POWER RATING — [Classified].
f. POLARIZATION — Vertical.

g. GAIN — [Classified].
h. TRANSMISSION LINE TYPE — Waveguide.
i. PRIMARY POWER REQUIREMENTS — (1) 440 vac, 60 Hz, three-phase; (2) 115 vac, 400 Hz, one-phase.
j. REFERENCES — (1) Radar Set, AN/SPG-55B, NAVSEA OP-4059. (2) Terrier Guided Missile Weapon System MK 76 MOD 7, DLG-6 Class, NAVSEA OP-4087.

4-3.11 AN/SPG-60

Radar Set AN/SPG-60 is a coherent doppler, X-band, monopulse radar used with the shipboard MK 86 Gun Fire Control System to provide long-range, high and low altitude detection, bearing, and elevation data for tracking of airborne targets at speeds up to Mach 3 and range out to 100,000 yards. It also may be used with the Tartar missile system to supply precise target data for missile positioning and continuous wave illumination (CWI) for missile homing.

Gun Fire Control System MK 86 is a digitally controlled system installed aboard ship for control of lightweight, rapid-fire, five-inch/54 caliber guns against surface, shore, and air targets. The GFCS employs a digital computer that uses stored constants in conjunction with real-time target data to compute gun orders that position contolled guns along the line of fire to future target position. Present target position data for the computer is provided from ship's sources as well as by the GFCS.

During operation the radar is actually a slave to the computer, i.e., the entire process of acquiring and tracking a target is under computer control. The computer initiates the acquisition phase of the radar, computes target range and range rate based on data supplied by the radar, and controls the tracking process.

The radar is capable of operating in four modes: acquisition, auto-track, semi-auto track, and manual. In the acquisition mode, designation of airborne target data is provided from search radar or track-while-scan routine of the GFCS, Visual designation from target designator transmitter or designation from other directors can be provided also. In the auto-track mode the radar set is operated under fully automatic control. In the semi-auto track mode, range and bearing are provided by the radar set; however, elevation is provided by the boresight optics through a manual rate controlled by the operator. A manual mode provides azimuth and elevation control. Range is provided automatically by the radar set.

ANTENNA MK-39 AN/SPG-60

a. TYPE/USE — The MK 39 antenna used with the SPG-60 radar is a rotatable, high gain system consisting of a parabolic dish reflector, a four-horn amplitude comparator monopulse track radar feed assembly, a Cassegrain subdish with a primary CWI waveguide feed projecting out from the center of the dish, and a canted auxiliary CWI feed mounted on the front of the four-horn comparator. The resulting tracking antenna pattern is a sharp pencil beam of 1.2°.

   Additionally, there are several subunits of the antenna system, mainly including a pedestal, turntable, and yoke structure. To control the antenna positioning in elevation from -30° to +95°, an elevation drive is mounted on the port side of the yoke. To train the antenna in azimuth, an azimuth drive assembly is mounted at the base of the turntable.

b. PHYSICAL CONFIGURATION — [See Figure 4-62].

c. DIMENSIONS — (1) Height: 137⅝ inches (total); (2) Diameter: 160 inches (total); (3) Weight: 4015 pounds (total).

d. FREQUENCY RANGE — X-band.

e. RF POWER RATING — 5.5 kW peak.

f. POLARIZATION — Horizontal.

g. GAIN — 41.5 dB.

h. VSWR — 2:1 maximum

i. TRANSMISSION LINE TYPE — Waveguide.

j. PRIMARY POWER REQUIRED — (1) 115 vac, 400 Hz, three-phase; (2) 115 vac, 400 Hz, one-phase.

k. REFERENCE — (1) Technical Manual, Radar Set AN/SPG-60, NAVSEA OP-3646. (2) AEGIS EME Handbook, ECAC, ESD-TR-80-104.

4-3.12 AN/SPG-62

The AN/SPG-62 is a surface ship microwave emitter designed for use as the CW illuminator (CWI) for the MK 99 AEGIS Missile Fire Control System. It is slaved to the primary shipboard AN/SPY-1 radar, and is capable of transmitting both a CW reference signal to the missile and CW illumination energy to the target, which is then reflected back to the missile as a homing signal. The SPG-62 equipment is representative of the Navy's latest technology. Because of its classified nature further details of the system philosophy cannot be disclosed here.

ANTENNA AS-3156/SPG

a. TYPE/USE — The AS-3156 is a rotatable, highly directive, dual-frequency microwave radiator used with the AN/SPG-62 system as the CW illuminator for the AN/SPY-1 radar. It consists of a 7½-foot parabolic dish reflector and two feeds: (1) an offset main pencil beam CW feedhorn and (2) a widebeam auxiliary horn. [The classified nature of the AN/SPG-62 allows only very limited technical characteristics to be given here].

b. PHYSICAL CONFIGURATION — [See Figure 4-63].
c. DIMENSIONS — (1) Diameter: 90 inches (dish reflector); (2) Weight: 3500 pounds (with pedestal).
d. FREQUENCY RANGE — [Classified].
e. RF POWER RATING — [Classified].
f. POLARIZATION — Vertical.
g. GAIN — [Classified].
h. VSWR — 1.7:1 maximum.
i. TRANSMISSION LINE TYPE — RG-67/U waveguide.
j. PRIMARY POWER REQUIRED — 115 vac and 440 vac; 60 and 400 Hz; one- and three-phase.
k. REFERENCES — AEGIS EME Handbook, ECAC, ESD-TR-80-104.

4-3.13 AN/SPQ-5A

Radar Set AN/SPQ-5A is a long-range, C-band, missile guidance system capable of acquiring and automatically tracking high speed targets. As a functional part of the MK 73 FCS, the SPQ-5A has integrated acquisition and track radars as well as two additional CW transmitters used for missile guidance. The SPQ-5A is designed to accept target designation data from other shipboard sources (e.g., surveillance radars); however, it is also capable of searching and designating its own targets. There are three primary modes of operation to the track and acquisition function: designate, acquisition, and track; and four auxiliary modes are avaiable also: search, local designate, shore, and surface. A brief description of each mode follows:

(*1*) The designate mode is normally used first in obtaining a target, i.e., the SPQ-5A radar responds to a designation from a remote sensor (normally a ship surveillance radar). Until such a remote designation is made, the SPQ-5A operates in a search mode.

(*2*) In the acquisition mode, the SPQ-5A antenna is positioned more accurately with respect to the target, continuously scanning a rectangular sector of 6° in elevation and 9° in azimuth. Once the target has been properly seized the radar is switched to the track mode.

(*3*) The track mode employs a narrow range gate to lock on the target. As soon as lock-on occurs, automatic tracking is initiated. An additional function of the track mode is missile guidance, through the use of a capture beam and a guidance beam. The capture beam is a wide angle signal for initial control of the missile after launch. After control has been achieved, a narrow beam guidance signal homes the missile to target.

(*4*) The search mode is an emergency operation used when the ship's primary search radar fails, or target data is not otherwise available from a remote source. In this mode the SPQ-5A operates on only two axes, train and elevation, using the track beam to search out and detect its own targets.

(*5*) In the local designate mode the operator places the radar in search, and, after receiving range bearing and elevation target data, uses manual elevation and train to position the antenna.

(*6*) In the shore mode the radar system is used to direct missile fire against a designated shore target.

(*7*) The surface mode is used for the radar to track targets on or near the surface of the water.

The large, multifunction antenna system used with the SPQ-5A is described below:

ANTENNA ASSEMBLY UD304 AN/SPQ-5A

a. TYPE/USE — The SPQ-5A antenna is a complex assembly of mechanical and microwave RF components mounted on deck as a unit with pedestal, yoke frame, and traverse member gimbal, on a cylindrical steel plate director foundation. Three axis movement of the antenna, in train, elevation, and traverse, allows stabilized, hemispherical coverage, 360° in azimuth and from -23° to +118° in elevation.

The main tracking radar antenna is an eight-foot octagonally formed, loaded lens microwave radiator attached to the box-shaped traverse member gimbal where it is swiveled on the U-yoke. The lens, seventeen inches thick and weighing 1500 pounds, is a multicellular array of over 4000 individual cells. Each cell contains a slow-wave, serrated metal, plastic supported waveguide element. Both the horizontally polarized track energy and the vertically polarized acquisition energy are thus focused by the lens with negligible intercoupling. A fiberglass radome covers the loaded lens to protect it from the sea environment.

Inside the traverse member housing, at the rear center, is the track microwave assembly containing the bulk of the microwave components, including the four-horn track radar monopulse feed cluster. The combined effect of the horn cluster comparator and the loaded lens produces a 1.6° circular pencil beam. Additionally, the comparator and loaded lens combination is used to transmit the guidance beam along which the missile rides to the target.

Also inside the traverse member box, located on the front interior deck is the acquisition scanner assembly shown in Figure 4-65(a). This assembly, which provides the scanning capabilities for the acquisition radar, consists of (1) a horn device which receives microwave energy from the scanner via an E-plane bend element and then directs it to a folded reflector; (2) an acquisition scanner composed of a tapered slotted rotor enclosed in a shaped drum (slotted stator) which causes the acquisition beam to scan in a traverse direction; (3) an E-plane bend connecting the horn and scanner; and (4) a folded reflector so designated that all paths of energy from the feedhorn are of equal length. Because of the narrowness of the track beam, it is difficult to designate targets accurately enough to be picked up directly by the track beam. Therefore, the acquisition radar is used to acquire the target before starting the track function. By means of the high speed scanner and low speed reflector an area in space approximately 6° in elevation by 9° in azimuth is scanned.

Mounted externally atop the traverse member, above the lens, is a separate capture antenna, the UD 347. As seen in Figure 4-65(b), this antenna is a small parabolic-section reflector fed by a mechanically nutating, eccentric radiator projecting out from the center. The capture antenna is used to gain control of the missile immediately after launch by scanning a beam in a circular motion about a null axis at the rate of 30 revolutions per second. The missile seeks this null as the line along which the amplitude of the received capture signal remains constant while the beam circles around it.

b. PHYSICAL CONFIGURATION — [See Figures 4-64 and 4-65].
c. DIMENSIONS — (1) Height: 379 inches (with pedestal); (2) Width: 194 inches; (3) Depth: 180 inches; (4) Weight: 37,200 pounds (total).
d. FREQUENCY RANGE — 5400 to 5900 MHz.
e. RF POWER RATING — 250 kW peak.
f. POLARIZATION — (1) Horizontal (track radar); (2) Vertical (acquisition radar).
g. TRANSMISSION LINE TYPE — Waveguide.
h. PRIMARY POWER REQUIRED — (1) 115 vac, 60 Hz, three-phase; (2) 440 vac, 60 Hz, three-phase; (3) 115 vac, 400 Hz, three-phase.
i. REFERENCE — Radar Set AN/SPQ-5A, NAVSEA OP-2389.

4-3.14 AN/SPQ-9A

Radar Set AN/SPQ-9A is part of the shipboard MK 86 Gun Fire Control System as its prime sensor of range and azimuth data on surface and low altitude targets. The MK 86 GFCS, used to control lightweight, rapid fire five-inch/54mm guns, employs the AN/SPG-

60 monopulse radar for air track and the SPQ-9 for surface search capability. The large, distinctively bubble-shaped radome readily identifies the SPQ-9A on naval surface ships.

ANTENNA AS-2367A/SPG-9

a. TYPE/USE — The AS-2367 is a rotatable, highly directive, X-band microwave antenna used with the AN/SPQ-9A surface ship fire control radar system. It consists of a rectangular paraboloid-section reflector with offset feedhorn and waveguide supported by two upper struts. The resulting radiated beam rotates at a constant 60 rpm. The azimuthal pattern is a low-sidelobe, sharp beam of 1.35° which provides high horizontal resolution. In the vertical plane the antenna produces a broader shaped lobe of 3° beamwidth pointed directly at the horizon. With the reflector fixed at a slight upward tilt, this pattern shape permits acquisition of low-flying airborne targets as well as surface targets and shore installations.

   The antenna is mounted on a pedestal which is stabilized for ship pitch and roll. To protect it from the sea environment the entire assembly is covered by a rigid, bubble-shaped radome fabricated of polyester skin laminated over a nylon phenolic honeycomb core.

b. PHYSICAL CONFIGURATION — [See Figure 4-66].

c. DIMENSIONS — (1) Height: 70⅞ inches (radome is 96 inches high); (2) Width: 54½ inches (radome is 120 inches in diameter); (3) Depth: 80½ inches; (4) Weight: 1185 pounds (total with radome).

d. FREQUENCY RANGE — X-band.

e. RF POWER RATING — 1.2 kW peak.

f. POLARIZATION — Vertical.

g. GAIN — 37.5 dB.

h. TRANSMISSION LINE TYPE — RG-68/U waveguide.

i. PRIMARY POWER REQUIRED — 115 vac, 400 Hz, three-phase.

j. REFERENCE — Technical Manual, Radar Set AN/SPQ-9A, NAVSEA OP-3855.

4-3.15 AN/SPW-2

Radar Set AN/SPW-2 is a missile guidance, C-band radar used with the MK 77 shipboard Guided Missile Fire Control (GMFCS). It operates in conjunction with the AN/SPG-49 tracking radar described in Section 4-3.5 above, and, as such, is also obsolescent. It is included here for historical reference since this system might be reactivated in the future along with heavy cruisers currently in moth balls.

After target tracking and acquisition by the SPG-49, followed by engagement evaluation, the MK 77 weapon direction equipment sends launcher assignment data to its computer which, in turn, sends positioning data to the selected launcher and radar. The SPW-2 then begins to radiate the guidance beam. On orders from the weapon direction equipment, the missile is fired into this beam. After the missile enters the guidance beam and is captured, the missile beacon begins transmitting signals which are received by the guidance radar. Throughout the guidance period, the tracking radar provides target range, range rate, and angle data to the computer. The guidance radar also supplies the computer with range and angle data as determined from the beacon return signal. The computer processes these data to produce guidance orders for positioning the radar beam along which the missile is directed.

ANTENNA UD39 AN/SPW-2

a. TYPE/USE — Antenna UD39, a part of antenna group OA-962/SPW-2, consists principally of an aluminum 42-inch parabolic reflector (type AT-635) with a nutating teflon dielectric feedrod projecting out slightly offset from the reflector center. The reflector collimates the radiated energy to provide a narrow, conically-scanned, guidance beam. Additionally, there are two major subassemblies: (1) the elevation and train system and (2) the pedestal, which is bolted to a barbette. The train member is a U-shaped structure able to rotate continuously in azimuth. The elevation member can move from about -20 to +90 degrees vertically. When the antenna is in operation it follows ballistic-corrected guidance orders which rapidly train it to the computed position for missile capture, then continuously corrects this position for change in fire control and ship pitch and roll.
b. PHYSICAL CONFIGURATION — [See Figure 4-67].
c. DIMENSIONS — (1) Height: 95 inches (with pedestal); (2) Diameter: 54⅛ inches; (3) Weight: 3000 pounds.
d. FREQUENCY RANGE — C-band.
e. RF POWER RATING — [Classified].
f. POLARIZATION — Variable (normally vertical).
g. TRANSMISSION LINE TYPE — Waveguide.
h. PRIMARY POWER REQUIRED — (1) 115 vac, 60 Hz, one-phase; (2) 115 vac, 400 Hz, three-phase; (3) 440 vac, 60 Hz, three-phase.
i. REFERENCES — (1) Radar Set AN/SPW-2, NAVSEA OP-2810. (2) CG-10 Class Talos Guided Missile Weapon System, NAVSEA OP-4090.

4-3.16 Anti-Ship Missile Defense Weapons Control Antennas

In the complex battle scenario of today's navy, the greatest threat to surface ships is the anti-ship missile.[33] Capable of being launched from a variety of platforms at sea and from the air, these sophisticated and deadly missiles may approach a ship just barely above the surface, or steeply diving from high altitude. They also fly in from very slow to extremely high speeds.

Because of this threat, the U.S. Navy has been fervently striving for many years to ensure an ability to defend itself against these missiles. The emphasis at first was to improve techniques for detecting the missiles and to employ electronic warfare devices to deceive the missile guidance systems. As the missiles increased in sophistication, however, the anti-ship missile defense (ASMD) measures required a marked increase in capability. The main thrust has been to provide a so-called defense-in-depth, resulting in the development of a series of automatic, diverse, complementary weapons systems to protect naval ships. To date these defense-in-depth systems include:

1. Basic Point Defense Surface Missile System (BPDSMS)
2. NATO Seasparrow Surface Missile System (NSSMS)
3. Target Acquisition System (TAS)
4. Rolling Airframe Missile (RAM)
5. Phalanx Close-In Weapon System (CIWS)
6. Electronic Warfare (EW) Systems (AN/SLQ-32)
7. Super-Rapid Bloom Offboard Chaff (SRBOC)

The BPDSMS, NSSMS, TAS, CIWS, and EW systems all use antennas as an integral unit. These systems will be described below, except for the AN/SLQ-32 EW antennas which will be covered in a separate chapter.

Basic Point Defense Surface Missile System Antennas

As a means of meeting the rapidly growing threat of anti-ship missiles, the Navy began urgent development of the shipboard Basic Point Defense System (BPDSMS) nearly twenty years ago.[34] This system is intended to provide naval ships with the capability of launching high performance, supersonic (Seasparrow III) guided missiles. To detect, acquire, and track airborne or surface targets under all weather conditions, BPDSMS employs a MK 115 Fire Control System with a CW radar and an electronically augmented optical sight mounted atop a seven-foot MK 76 director. The MK 115 FCS supplies rear reference signals and prelaunch data to the Seasparrow missile, illuminates the target with coded RF energy, acquires and tracks the detected target, and provides tactical data (e.g., target position and approach velocity) to the FCS operator. The system is especially designed to be simple, highly reliable, and able to operate either independently or as an integral part of existing ship weapons systems.

The MK 76 director-illuminator functions as an important source of target data. It incorporates the CW radar antennas, computing circuits, and optical sight used in acquiring, tracking, and illuminating the target. Tracking error signals are visually displayed in the optical sight, so that the director operator, by movement of a handlebar mechanism, manually aims the director at the target. By keeping the error signal centered in the sight, the operator achieves radar lock-on, assuring continuous illumination of the target until missile impact. Therefore, even though BPDSMS employs this "man-in-the-loop" operation—resulting in comparatively long reaction times against modern hostile missiles—it can be used very effectively to intercept, at ranges over four nautical miles, a wide variety of airborne threats coming in at high speeds, from very low to high altitudes.

ANTENNAS MK33 and MK19 BPDSMS

a. TYPE/USE — The two microwave antennas associated with the BPDSMS MK115 FCS, one for X-band CW transmit (MK33) and one for receive (MK19), are mounted on top of the MK 76 director as illustrated in Figure 4-68. Consisting of a two-foot parabolic dish with center feedhorn, the MK 33 transmit antenna is swiveled on a cradle assembly to allow movement in elevation. Sealing the feed from the weather is sufficient to preclude the need for a radome.

   The MK 19 receive antenna is a two-foot flat plate, planar array utilizing 52 slotted aluminum waveguide sections segmented into four quadrants. This arrangement results in a phased array to provide electronic scanning. Combining the received energy in phase from all four waveguide quadrants produces a high gain, narrow beam, broadside array with the direction of maximum sensitivity perpendicular to the plane of the antenna. However, by properly shifting the phase between quadrants, the beam is made to conically scan.

b. PHYSICAL CONFIGURATION — [See Figure 4-68].

c. DIMENSIONS — (1) Height: 84 inches (with pedestal); (2) Width: 79 inches (84-inch diameter working circle; (3) Depth: 60 inches (84-inch diameter working circle); (4) Weight: 1800 pounds.

d. FREQUENCY RANGE — X-band.

e. RF POWER RATING — [Classified].

f. POLARIZATION — (1) Horizontal (transmit); (2) Vertical (receive).

g. GAIN — [Classified].

h. TRANSMISSION LINE TYPE — Waveguide.

i. PRIMARY POWER REQUIREMENTS — (1) 115 vac, 60 Hz, one-phase. (2) 115 vac, 400 Hz, one-phase.
j. REFERENCES — (1) Guided Missile Fire Control System, Mark 115 MOD O, NAVSEA OP-3467. (2) Basic Point Defense Surface Missile System, NAVSEA OP-3656.

NATO Seasparrow Surface Missile System Antennas

Since 1975, an upgraded point defense system has been placed in the fleet "to counter enemy missiles whose capabilities were undreamed of when the original basic point defense system was designed."[35] This newly improved system is the NATO Seasparrow Surface Missile System (NSSMS), a short-range rapid reaction self-defense system for use against aircraft, and particularly against air-to-surface and surface-to-air anti-ship missiles. An essential component of NSSMS is its MK91 Guided Missile Fire Control System, employed optionally in either a single or dual MK95 radar configuration. The primary function of the radar is to automatically acquire and track a target designated from an external source, or, internally, by the FCS radar operator. An important additional feature for the radar is to illuminate the target for guidance of the Seasparrow missile. The antennas associated with the radar, one for transmit and one for receive, are mounted on a director assembly which allows complete training 360° in azimuth, and elevation from -35 to +85 degrees.

NSSMS ANTENNAS

a. TYPE/USE — Two separate antennas are used with the NATO Seasparrow system, one for transmit and one for receive. The transmit antenna is a 39-inch parabolic dish reflector with a high power Cutler type center feed having two inputs: one to produce a sharply focused beam with low sidelobes and the other a null-filling horn attached to the side of the Cutler feed to provide wide angle radiation to the Sparrow missile (for the rear reference signal) while in flight. A visor and phenolic honeycomb radome cover the antenna to protect it from the sea environment. Strip heaters are imbedded in the outer radome skin for anti-icing.

   The receive antenna is also 39 inches in diameter and has a Cassegrain twist reflector protected from the weather by a phenolic honeycomb radome. Imbedded in the radome are heater elements for deicing. Also, mounted on the upper rim of the receiver antenna housing is the sidelobe horn antenna.

b. PHYSICAL CONFIGURATION — [See Figure 4-69].

c. DIMENSIONS — (1) Height: 96 inches, radius (total director); (2) Width: 50 inches, radius (total); (3) Depth: 50 inches, radius (total); (4) Weight: 3,315 pounds (total).
d. FREQUENCY RANGE — X-band.
e. RF POWER RATING — 2kW.
f. POLARIZATION — [Classified].
g. GAIN — [Classified].
h. TRANSMISSION LINE TYPE — Waveguide.
i. REFERENCES — (1) NATO Seasparrow Surface Missile System MK57, MODS 0 and 1, NAVSEA OP-4004. (2) Guided Missile Fire Control System MK91 MODS 0 and 1, NAVSEA OP-4005.

Target Acquisition System MK 23

The MK 23 Target Acquisition System (TAS) is a major component of what is termed the Improved Point Defense Surface Missile System (IPDSMS). TAS itself is comprised of five integrated subsystems: (1) a range-gated pulse-doppler radar; (2) a computer unit; (3) a display unit; (4) an IFF system; and (5) a combination radar/IFF antenna to be described later. Designed for use in conjunction with shipboard fire control systems and weapons, TAS locates, acquires, tracks, classifies, and designates high speed, small radar cross-section hostile threats which dive from high altitudes or "pop up" off the horizon just 30 to 60 seconds before intended impact. TAS, therefore, must automatically identify, react, and designate such targets to the NATO Seasparrow Surface Missile System (NSSMS) in a matter of a very few seconds, and it must do so with precision in all sea environments, including clutter and jamming. To be successful in countering, the corresponding point defense weapon systems (e.g., NSSMS, CIWS, RAM, five-inch guns), of course, must also be highly automatic so as to minimize lock-on-time, firing, and guiding to intercept.

In its operation, TAS offers a selection of four modes:

(*a*) *Normal* — this is the routine, automatic, point defense mode with instrumented range of 20 nautical miles.
(*b*) *Medium Range* — offering radar surveillance and air control beyond 90 nautical miles.
(*c*) *Mixed* — combining Normal and Medium Range to enable the ship to perform simultaneous multiple operation, such as point defense and air operation, from a single radar.
(*d*) *EMCON* — allowing selectable sector scanning to minimize hostile detection of the ship.

ANTENNA TAS

a. TYPE/USE — As seen in Figure 4-70, the TAS antenna is a combined rotating sensor with radar and IFF functions mounted back-to-back on a common, stabilized pedestal. Coverage is 360° continuous in azimuth, and from 0 to 75 degrees in elevation. The radar is a 2-D, L-band, fan beam (3.3° horizontal beamwidth) radiator used for search, detection, and acquisition. It is composed of 26 horns with evenly distributed corporate feed. Associated with the radar antenna, located one at each end, are two small backfill horn antennas; one for sidelobe blanking and one for coherent sidelobe cancellation. Physically mounted 180° behind the radar antenna is the AS-2189/UPX IFF antenna. A small IFF backfill radiator is located on top for IFF sidelobe cancellation.

b. PHYSICAL CONFIGURATION — [See Figure 4-70].

c. DIMENSIONS — (1) Height: 129 inches (with pedestal); (2) Width: 231 inches; (3) Depth: 76 inches; (4) Weight: 2000 pounds (total).

d. FREQUENCY RANGE — L-band.

e. RF POWER RATING — 200 kW peak.

f. POLARIZATION — Vertical (radar antenna).

g. GAIN — 21 dB (radar antenna).

h. TRANSMISSION LINE TYPE — Waveguide.

i. PRIMARY POWER REQUIREMENTS — (1) 115 vac, 60 Hz, three-phase; (2) 115 vac, 400 Hz, three-phase; (3) 440 vac, 400 Hz, three-phase.

j. REFERENCES — (1) Target Acquisition System MK 23, NAVSEA OP-4199. (2) Installation Guidance Notebook for Target Acquisition System MK 23, NAVSEA SW230-AO-INS-010. (3) *Surface Warfare Magazine,* May 1980, p. 15.

Phalanx Close-in Weapon System

The final shipboard self-defense weapon employed as part of the Navy's defense-in-depth protection against anti-ship missiles is the Phalanx Close-in Weapon System (CIWS). That is to say, it is CIWS which is relied upon to, without any assistance from other ship systems, automatically engage and destroy any missiles which penetrate the ship's primary defense screen. CIWS will, in a secondary role, accept targets designated from other sources and engage them up to 80-degree elevation relative to the deck. Additionally, CIWS can operate against small surface targets as directed by optical designators.

CIWS is a self-contained weapon system, consisting of a search radar, a track radar, and a six-barrel 20 mm Gatling gun firing 3000 rounds per minute. The system automatically performs search, detect, target declaration, track, evaluation, declares engageability, firing, and target destruction. In its primary autonomous mode, the CIWS surveillance radar searches for targets with elevation between -35 and +90 degrees, through 360° of azimuth. Once a target is detected, its range, range rate, height, and bearing are assessed to determine the threat. If the target is certified as a threat, the system transfers immediately from the search to acquisition function. The gun mount is ordered to slew to the target bearing and elevation, and the track radar begins a localized concentrated search for the target. When the track radar acquires the target, the CIWS computer takes control of the antenna and gun to maintain the track. Radar feedback data is then used to generate gun aiming orders, including the lead angle, and as soon as the target reaches the predetermined engagement range, firing commences. The track radar and computer rapidly calculate any projectile error to make instantaneous corrections until the target is destroyed.

CIWS ANTENNAS

a. TYPE/USE — Two separate Ku-band pulse doppler radar antennas are used with CIWS, one for search and the other for track. The upper antenna (shown sketched within the radome housing in Figure 4-72) is the search antenna. It is a conventional, rotating, pulse radar antenna consisting of a parabolic-section reflector with single feedhorn. The lower antenna is the monopulse track radar. It uses a scannable parabolic reflector, and, for the greater accuracy required in the track mode, four feedhorns transmitting four simultaneous overlapping lobes. By amplitude comparison of the return signals, precise error corrections are developed.

b. PHYSICAL CONFIGURATION — [See Figures 4-71 and 4-72].

c. DIMENSIONS — (1) Height: 185 inches (total weapon group); Width: 100 inches overall; (3) Depth: 124 inches (working circle radius is 104 inches); (4) Weight: 12,150 pounds (total with 980 rounds of ammunition).

d. FREQUENCY RANGE — Ku-band.

e. RF POWER RATING — [Classified].

f. POLARIZATION — [Classified].

g. GAIN — [Classified].

h. TRANSMISSION LINE TYPE — Waveguide.

i. REFERENCE — Close-in Weapon System MK 15, MODS 1 through 6 (Phalanx), NAVSEA OP-4154.

4-3.17 MK 92 Fire Control System: CAS and STIR

The MK 92 is a lightweight, quick-reaction shipboard fire control system that provides simultaneous tracking of up to two air and two surface targets while, at the same time, maintaining air and surface surveillance. The system is completely self-contained to provide full control for assignment, aiming, and firing of ship guns and missiles against surface and highspeed airborne targets. The MK 92 (MOD 2) employs two separate antenna units, one called a Combined Antenna System (CAS) and the other a Separate Track Illuminating Radar (STIR). The radome-enclosed CAS contains a continuously scanning, high data rate search antenna, and an independent track radar antenna. The STIR antenna provides additional long-range tracking and CW illumination for missile guidance. Furthermore, other shipboard search radars supply data to the FCS, with the result that this total complement allows operators multitarget display and rapid defense air surveillance out to 200,000 yards. The MK 92 FCS is representative of the Navy's latest weapons control radar systems; therfore, much of the detailed technical information is classified and cannot be given here.

ANTENNAS UD 401 CAS and UD 417 STIR

a. TYPE/USE — The UD 401 CAS consists of search and track antennas mounted on a stabilized platform enclosed in a radome. The search antenna provides air and surface stabilized surveillance and is utilized in a track-while-scan mode to allow tracking of two surface targets during surveillance. Both moving target indicator (MTI) and pulse amplitude search capabilities are available. The high rotational speed of the search antenna (60 rpm) allows for detection of pop-up targets. Consisting of a parabolic-section reflector fed by a conventional flared waveguide horn, the search antenna is additionally equipped with 10 vertical dipoles for IFF interrogation of targets. As can be seen in Figure 4-73, the search antenna is located at the bottom of the CAS assembly.

Mounted at the top of the assembly column is the CAS track antenna. This antenna is used primarily for 3-D air target acquisition and tracking, CW target illumination, and rear reference signals for missiles. A monopulse radar, the track antenna uses a parabolic dish with Cassegrain feed, and a separate rear reference horn. Rotating unrestricted in the horizontal plane, the antenna is positioned vertically by an elevation drive unit. The CAS assembly is protected from the sea environment by an eight-foot diameter radome made of lightweight, honeycomb reinforced fiberglass, with entrance hatch at the bottom.

The UD 417 STIR director antenna provides a longer range tracking capability than the CAS, and it is also equipped with a CWI for guidance of semi-active homing missiles. The main antenna, operating in the pulsed tracking mode, consists of a large, rotating, parabolic dish reflector with a multi-mode monopulse feedhorn. The antenna functions separately in the CW mode as a Cassegrain type, with the radiation from the CW feedhorn reflected from the sub-reflector toward the main dish, where it is focused into a sharp pencil beam. Additionally, a CW fill-in horn mounted in front of the monopulse assembly (see Figure 4-74) radiates a broad beam of energy.

b. PHYSICAL CONFIGURATION — [See Figures 4-73 and 4-74].

c. DIMENSIONS — (1) Height: 135½ inches (CAS), 138 inches (STIR); (2) Diameter: 96 inches (CAS), 100 inches (STIR) — Note turning radius is 78 inches; (3) Weight: approximately 6300 pounds (topside equipment only).

d. FREQUENCY RANGE — [Classified].

e. RF POWER RATING — [Classified].

f. POLARIZATION — (1) Horizontal or Circular (CAS Search); (2) Vertical (CAS Track); (3) Horizontal: (STIR Track); (4) Vertical: (STIR CW).

g. GAIN — [Classified].

h. TRANSMISSION LINE TYPE — Waveguide.

i. PRIMARY POWER REQUIREMENTS — (1) 115 vac, 60 Hz, one-phase; (2) 115 vac, 400 Hz, three-phase; (3) 440 vac, 60 Hz, three-phase.

j. REFERENCES — (1) Fire Control System MK 92 MOD 2, NAVSEA OP-4300. (2) "MK 92 Fire Control System," F.N. Ryan and J. Berlin, *Naval Engineers Journal*, August 1979, p. 69.

4-3.18 AN/SPY-1

The AN-SPY-1 is a multifunction, electronically steerable, phased array radar featuring fully automatic surveillance, detection, tracking, threat evaluation, and missile communication. As such, it is the primary air and surface sensor for the Navy's AEGIS combat system. AEGIS, designed to be a virtually impenetrable shield against intense air attack, is an integrated shipboard detection, control, and engagement system that, with near instantaneous and completely automatic reaction, directs high performance anti-air weaponry against hostile targets. The SPY-1 radar, utilizing digital computer control, has multiple target capability and high track capacity. With automatic frequency shifting and Moving Target Indicator (MTI), the radar is highly effective against a wide spectrum of target characteristics even in quite adverse operating environments such as heavy clutter and ECM.

To be installed on guided missile cruisers, production of the AN/SPY-1 began in the early 1980s. As might be expected, therefore, with AEGIS representing the very latest in naval combat systems technology, the SPY-1 is highly classified. Consequently, detailed system and equipment characteristics cannot be revealed here. To the extent possible a general overview of its phased array antenna is given below.

ANTENNA AS-3159/SPY-1

a. TYPE/USE — The AN/SPY-1 radar system consists of an antenna group, transmitter group, signal processor group, control group, and various auxiliary equipment. It is the antenna group that transfers, forms, and steers the high power transmit beam. The AS-3159 antenna system is comprised of four individual 12-foot by 12-foot octogonal phased array faces to provide a combined total 360° azimuth and elevation up to zenith, i.e., hemispherical coverage. Each face contains 140 array modules, each module having 32 radiating elements and phase shifters driven by eight identical four-channel boards. The result is over 4000 radiating elements in each of the four arrays. The SPY-1 is a major departure from the radar antennas described before in this chapter in that it is fixed rather than rotatable and steered completely by electronic circuitry.

b. PHYSICAL CONFIGURATION — [See Figure 4-75].

c. DIMENSIONS — (1) Height: 159½ inches (each face with mount); (2) Width: 153 inches (each face); (3) Depth: 66 inches (including RF component) and interconnect clearance); (4) Weight: 13,030 pounds (per face).

d. FREQUENCY RANGE — S-band.

e. RF POWER RATING — [Classified].

f. POLARIZATION — [Classified].

g. GAIN — [Classified].

h. TRANSMISSION LINE TYPE — Waveguide.

i. PRIMARY POWER REQUIREMENTS — (1) 115 vac, 400 Hz, one-phase; (2) 115 vac, 400 Hz, three-phase, (3) 440 vac, 60 Hz, three-phase; (4) 440 vac, 400 Hz, three-phase.

j. REFERENCES — (1) W.T. Patton, "The AN/SPY-1 Radar System," *Microwave Journal,* March 1979, p. 70. (2) Interface Control Drawing, Antenna AS-3159/SPY-1A, NAVSEA 2898958.

REFERENCES

1. Edward L. Beach, "Radar and Submarines in World War II," *Defense Electronics,* October 1979, p. 48.
2. Louis A. Gebhard, *Evolution of Naval Radio - Electronics and Contributions of the Naval Research Laboratory* (Washington, D.C.: Naval Research Laboratory Report 8300, 1979), p. 170.
3. Merrill I. Skolnik, *Introduction to Radar Systems* (New York, New York: McGraw-Hill Book Company, 1962), p. 8.
4. S. G. Marconi, "Radio Telegraphy," *Proceedings of the IRE,* Vol. 10, No. 4, 1922, p. 237.
5. Gebhard, *op. cit.,* p. 169.
6. John H. Dunn, "Radar Evolution and Innovation at NRL," *Signal,* October 1976, p. 8.
7. Merrill I. Skolnik, *Radar Handbook* (New York: McGraw-Hill Book Company, 1970), pp. 1-12.
8. Gebhard, *op. cit.,* p. 176.
9. *Ibid.,* p. 180.
10. *Ibid.,* pp. 180-183.
11. Dunn, *op. cit.,* pp. 9-10.
12. Thomas E. Walsh, "Military Radar Systems," *Microwave Journal,* November, 1978, p. 87.
13. Skolnik, *Intro. to Radar Systems,* p. 12.
14. A. B. Crawford and M. A. Fanuele, "Combat Applications of Radar," *Signal,* October 1976, p. 16.
15. Dunn, *op. cit.,* p. 11.
16. Skolnik, *Intro. to Radar Systems,* p. 1.
17. Walsh, *op. cit.,* p. 92.
18. Gebhard, *op. cit.,* p. 186.
19. *Ibid.,* p. 191.

20. Roy J. Biondi, *Handbook for Shipboard Surveillance Radars* (Washington, D.C.: Naval Sea Systems Command 0967-LP-587-0010, 1976), p. 66.
21. Naval Sea Systems Command, "SPS-58 Series Improved Air Search Upgraded," *Surface Warfare,* June 1980.
22. Gebhard, *op. cit.,* p. 187.
23. Aldo Bartolomei, "SPS-67 New Surface Search Radar," *Surface Warfare,* June 1980, p. 17.
24. John Beach, "New Surface Search Radar to Replace Navy's SPS-10," *Defense Electronics,* October 1979, p. 42.
25. F. J. Wylie, *The Use of Radar at Sea* (London: Hollis and Carter, 1973), p. 4.
26. *Ibid.,* p. 24.
27. Gebhard, *op. cit.,* p. 279.
28. *Ibid.,* p. 195.
29. *Ibid.,* p. 196.
30. *Ibid.,* p. 198.
31. *Ibid.,* p. 198.
32. *Ibid.,* p. 200.
33. ASMD Project Office, Naval Sea Systems Command, "ASMD—Tough Defensive Lineup," *Surface Warfare,* May 1980, p. 3.
34. Larry D. Laird, "BPDSMS—Getting Down to Basics," *Surface Warfare Magazine,* May 1980, p. 5.
35. Improved Basic Point Defense Surface Missile System Project Office, Naval Sea Systems Command, "NATO Seasparrow—Multinational Missile," *Surface Warfare Magazine,* May 1980, p. 6.
36. J. R. Bond, "963 + AEGIS = DDG-47," *Surface Warfare Magazine,* April 1978, p. 11.

FIG. 4-1 First Shipboard Radar Antenna 200 MHz Experimental Equipment

FIG. 4-2 Model XAF Early Navy Shipboard Radar Antenna *USS New York*

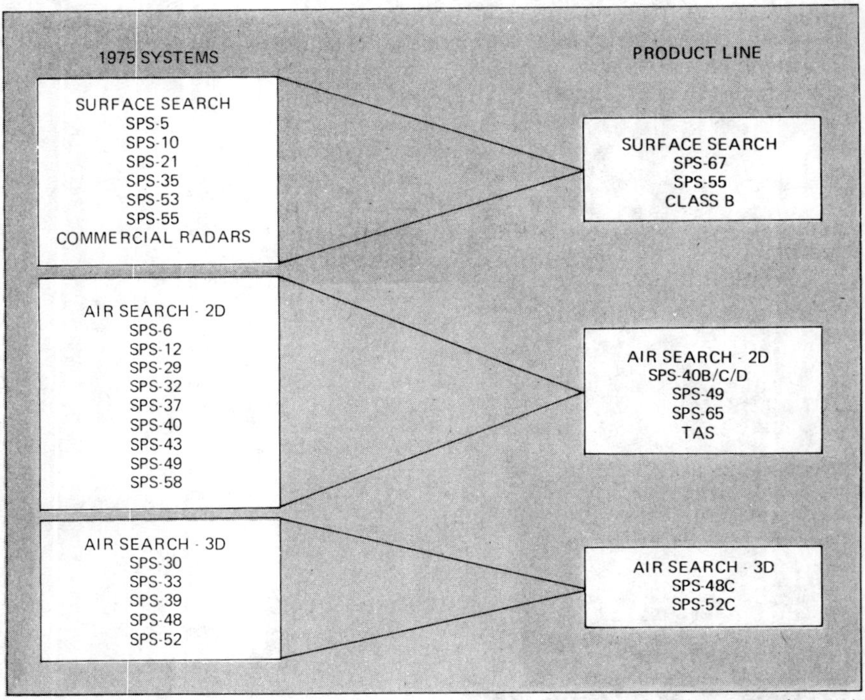

Microwave Journal

'IG. 4-3 Projected Standardized Navy Radars

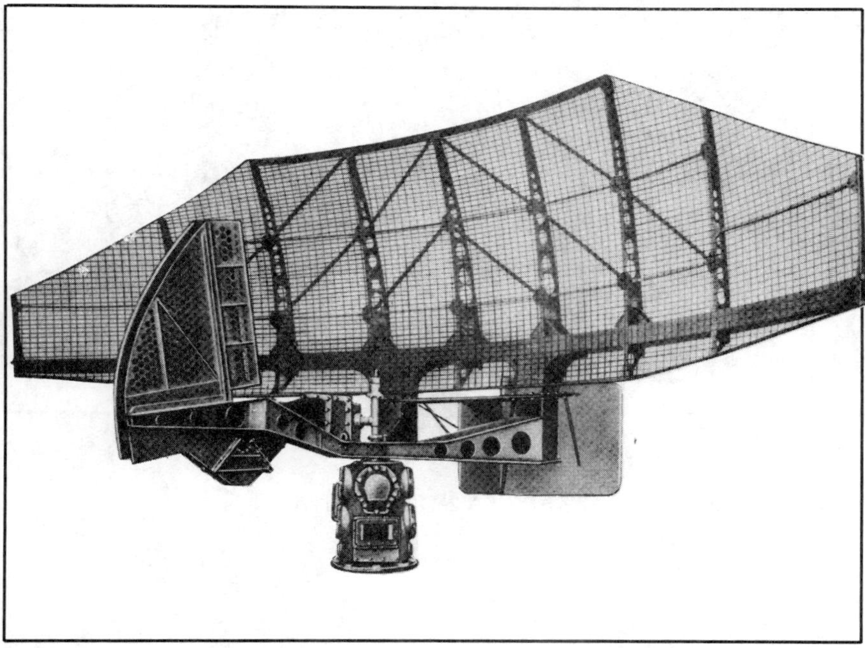

FIG. 4-4 Antenna AS-402/SPS-6 2-D Air Search

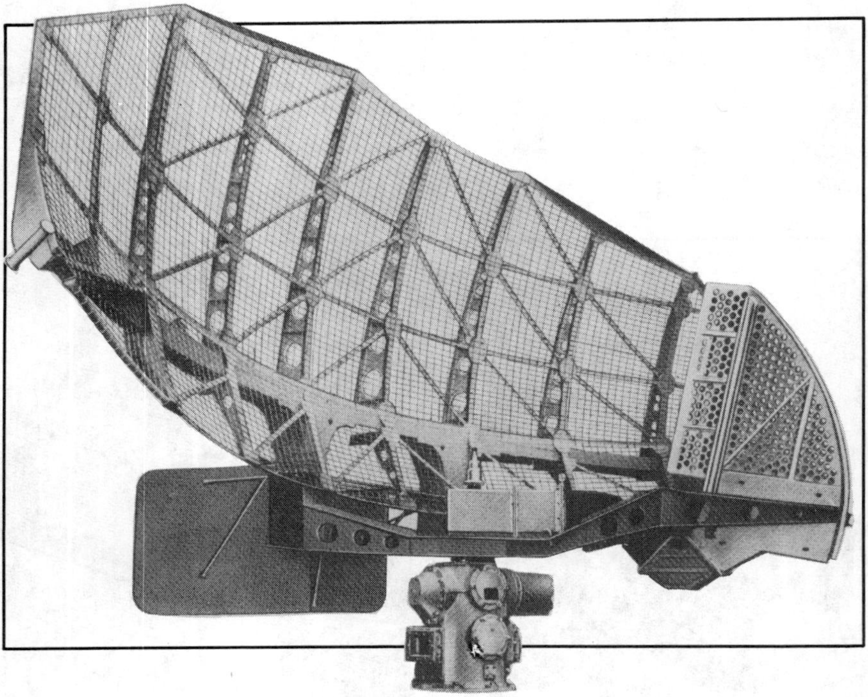

FIG. 4-5 Antenna AS-429/SPS-6A 2-D Air Search

FIG. 4-6 Antenna AS-430A/SPS-6B 2-D Air Search

FIG. 4-7 AS-430B/SPS-6C 2-D Air Search

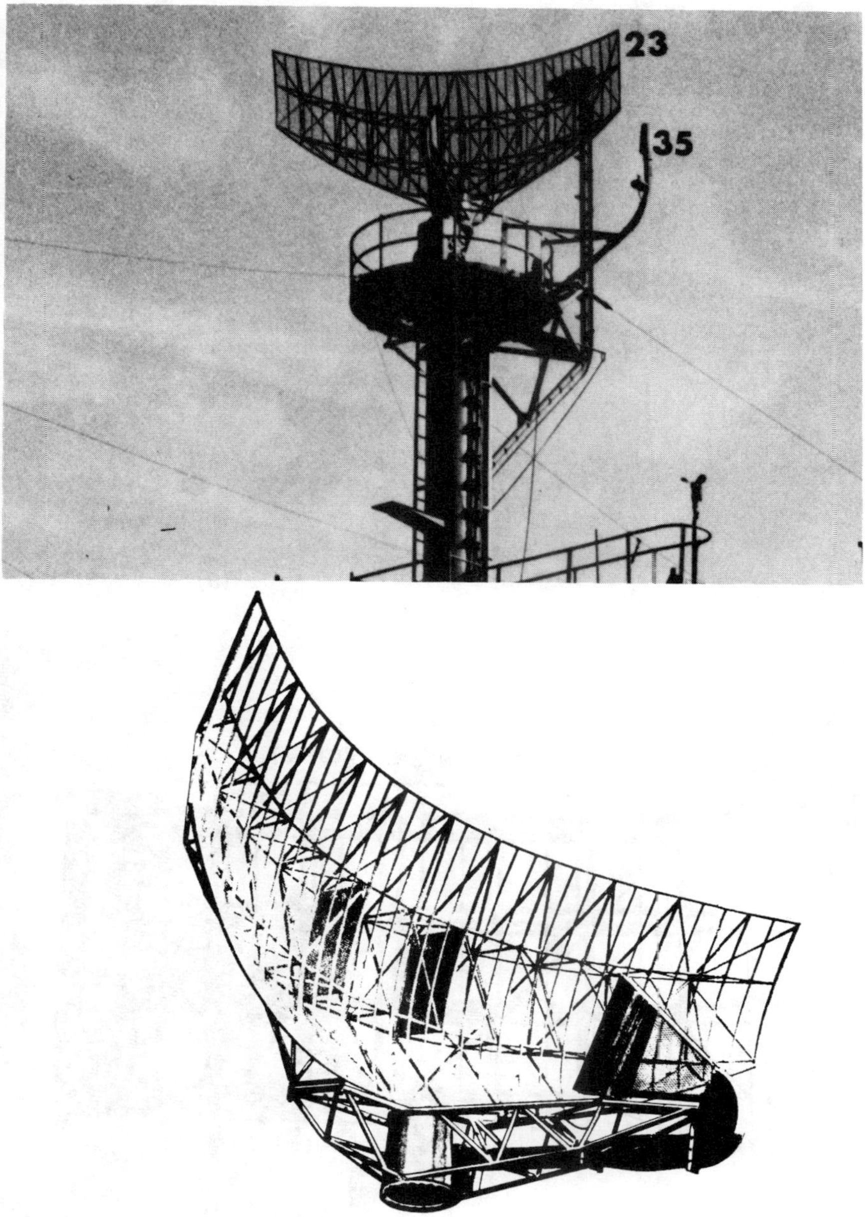

FIG. 4-8 AS-603/SPS-12 2-D Air Search Radar

(a)

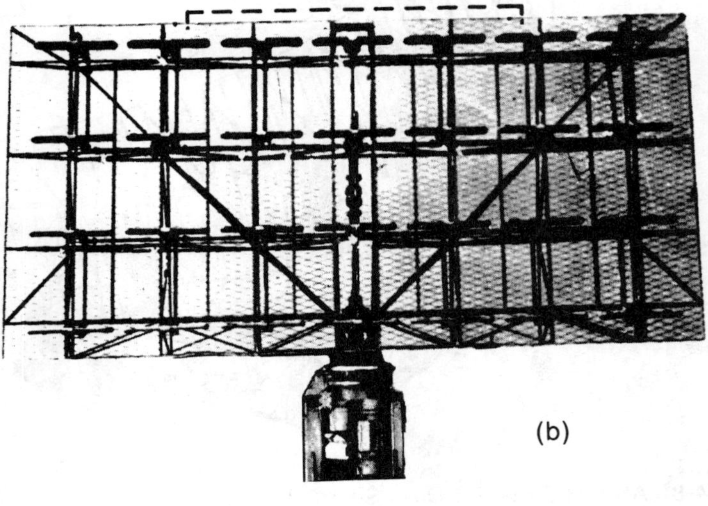

(b)

FIG. 4-9 AS-943/SPS-29 2-D Air Search Radar

FIG. 4-10(a) AS-1092/SP 2-D Air Search Radar Top Of Mast Installation

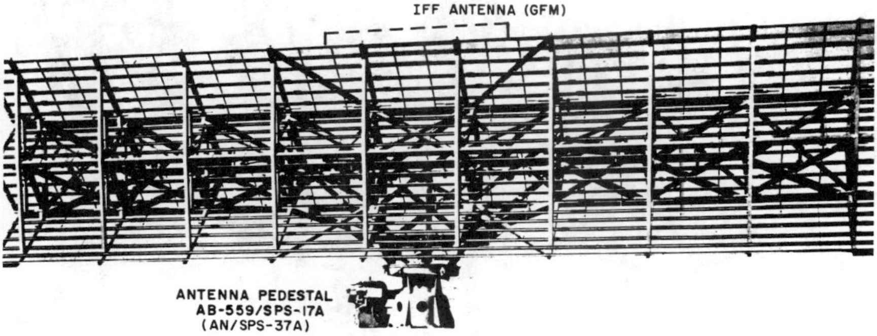

FIG. 4-10(b) AS-1092/SP 2-D Air Search Radar Physical Configuration

FIG. 4-11(a) AS-2782/SPS-40B 2-D Air Search Radar Shipboard Installed

FIG. 4-11(b) AS-2782/SPS-40B 2-D Air Search Radar Physical Configuration

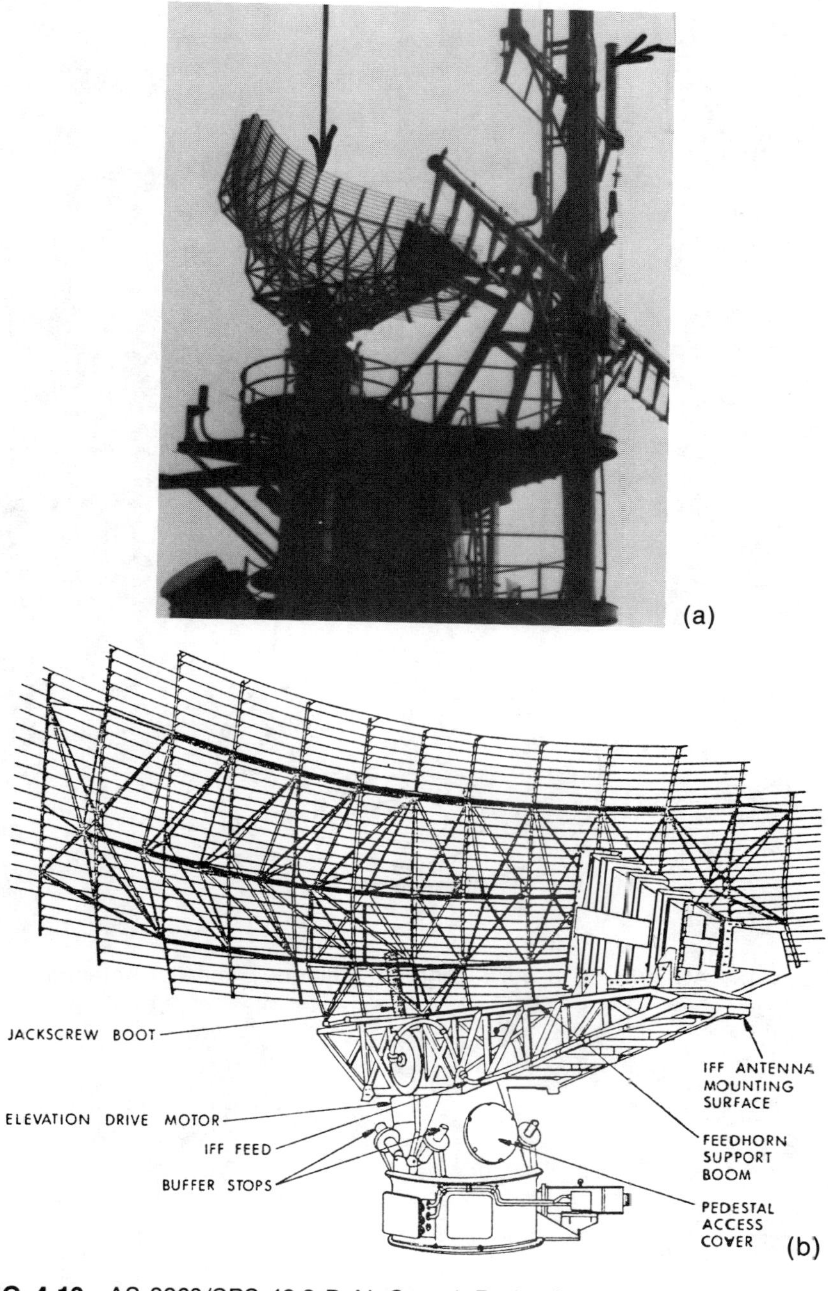

FIG. 4-12 AS-3263/SPS-49 2-D Air Search Radar Antenna

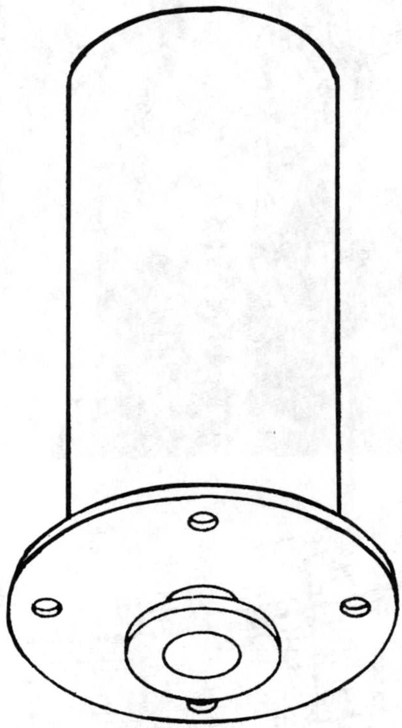

FIG. 4-13 AS-3077, 3078, 3079 Coherent Sidelobe Canceller Antenna

FIG. 4-14 AS-2607/SPS-58 2-D Air Search Radar Antenna

FIG. 4-15 AS-936C/SPS-10B Radar Antenna Used With AN/SPS-65 2-D Air Search Antenna

FIG. 4-16 Model SX 3-D Air Search Antenna

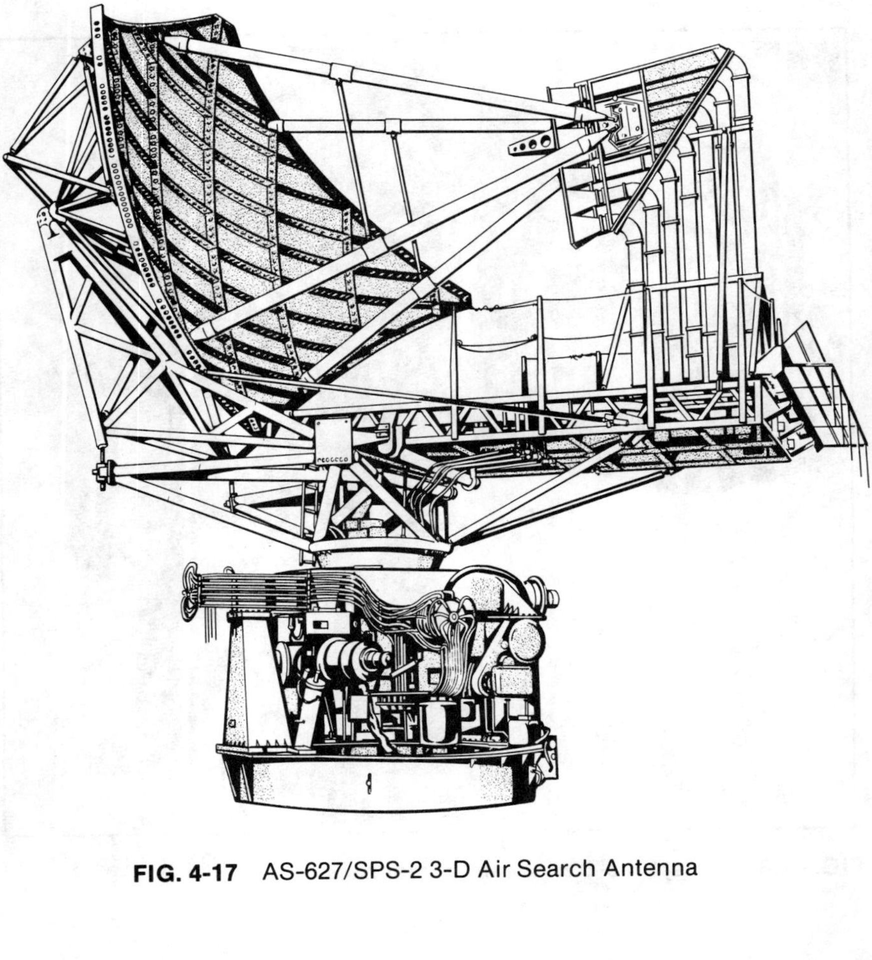

FIG. 4-17 AS-627/SPS-2 3-D Air Search Antenna

FIG. 4-18 AS-434/SPS-8 3-D Air Search Antennas

FIG. 4-19(a) AS-828/SPS-8B 3-D Air Search Radar Antenna

FIG. 4-19(b) AS-1167/SPS-30 3-D Air Search Radar Antenna

FIG. 4-20(a) AS-1167/SPS-30 3-D Air Search Radar Antenna

FIG. 4-20(b) AS-1167/SPS-30 3-D Air Search Radar Antenna

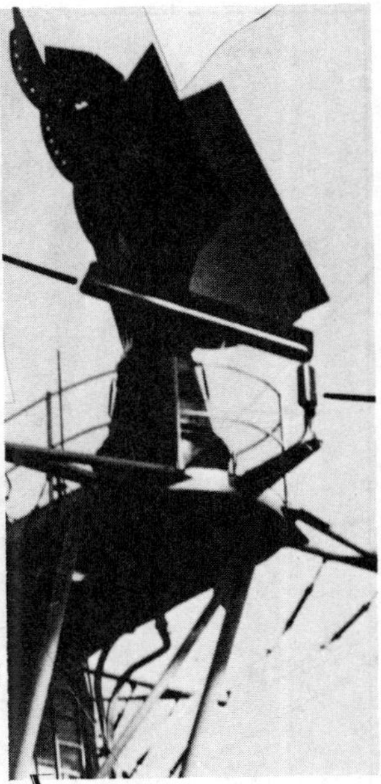

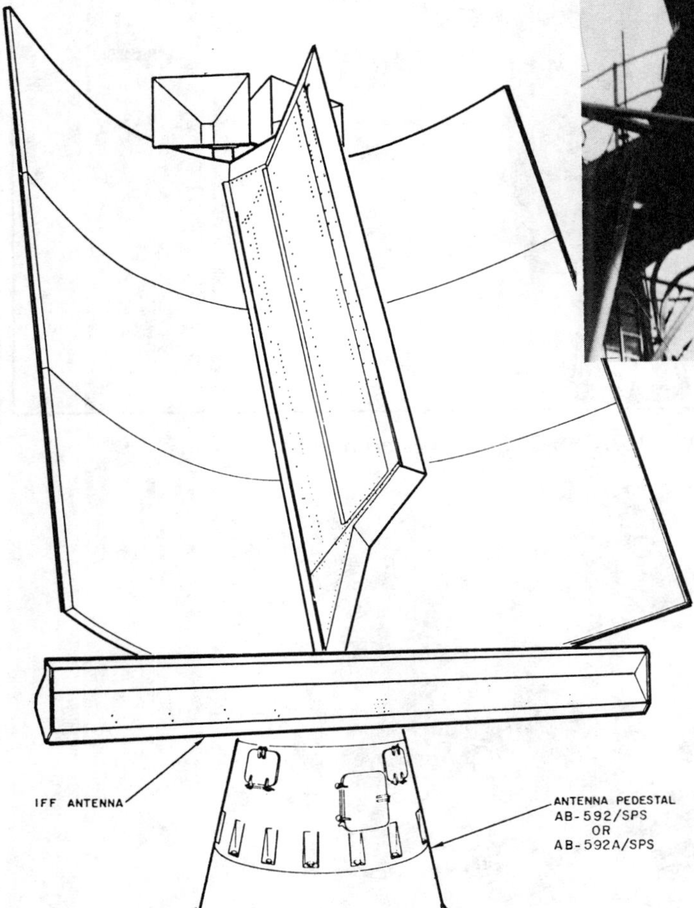

FIG. 4-21 AS-999/SPS-39 and 42 3-D Air Search Radar Antenna (later identical versions called AS-1109/SPS)

FIG. 4-22 AS-1838/SPA-72 3-D Air Search Radar Antenna

FIG. 4-23 AS-1686/SPS-48 3-D Air Search Radar Antenna

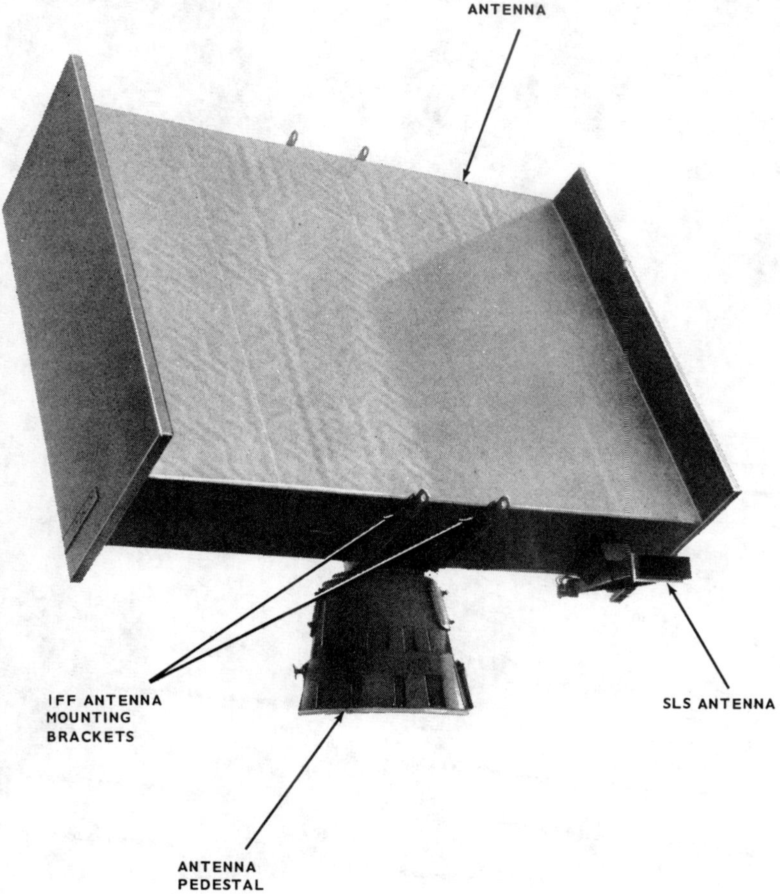

FIG. 4-24 AS-1667/SPA-64 3-D Air Search Radar Antenna

FIG. 4-25 Model SG Surface Search Radar Antenna

FIG. 4-26 AS-511/SPS-5 Surface Search Radar Antenna

FIG. 4-27 AS-651/SPS-5B Surface Search Radar Antenna

FIG. 4-28 AS-1002/SPS-5C Surface Search Radar Antenna

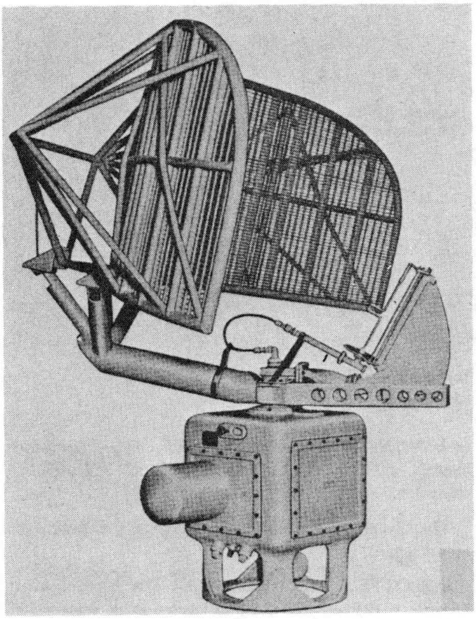

FIG. 4-29 AS-615/SPS-10 Surface Search Radar Antenna

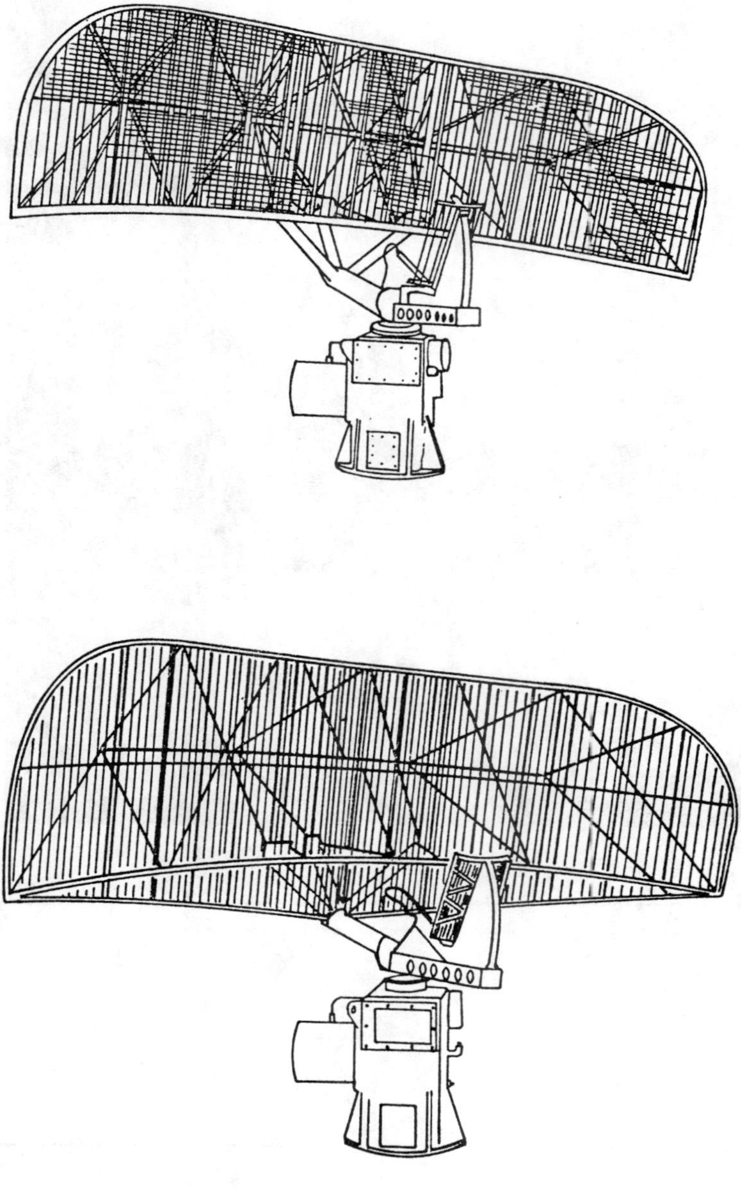

FIG. 4-30 AS-936/SPS-10B Versions Of Surface Search Radar Antenna

Fig. 4-31 AN/SPS-53 Surface Search Radar Antenna (AS-1652, AS-2321, AS-2322, AS-2645)

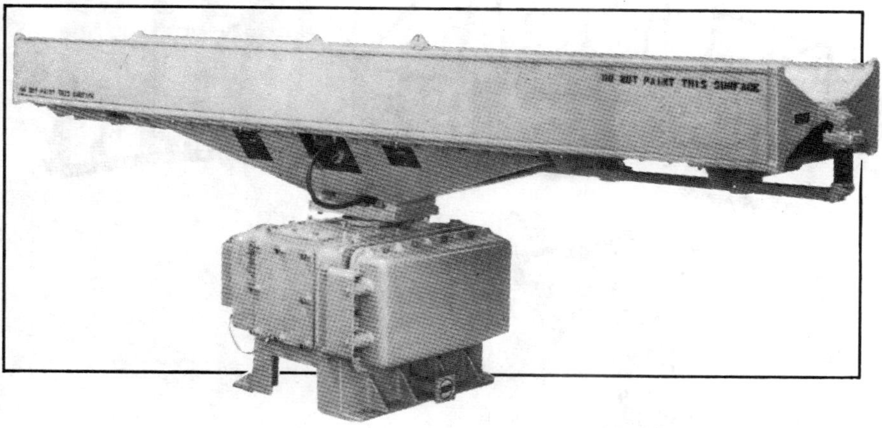

FIG. 4-32 AS-2953/SPS-55 Surface Search Radar Antenna

FIG. 4-33 AS-710B/SPS-21 Surface Navigation Radar Antenna

FIG. 4-34 AS-923/SPS-35 Surface Navigation Radar Antenna

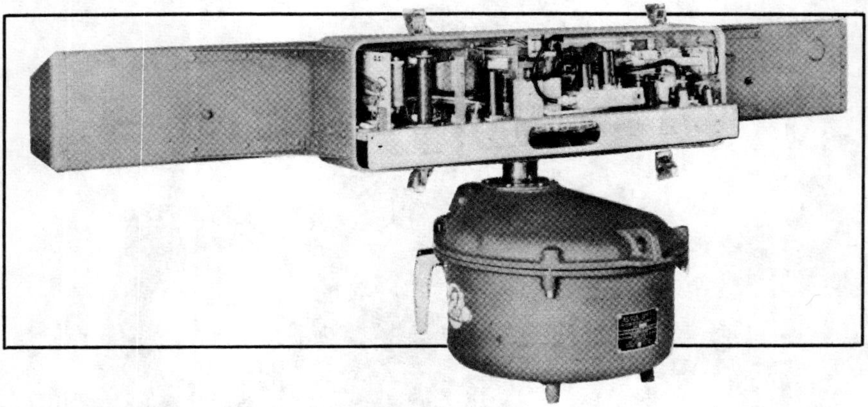

FIG. 4-35 AS-925/SPS-36 Surface Navigation Radar Antenna

FIG. 4-36 AS-1004/SPS-41 Surface Navigation Radar Antenna

FIG. 4-37 AS-1066/SPS-46 Surface Navigation Radar Antenna

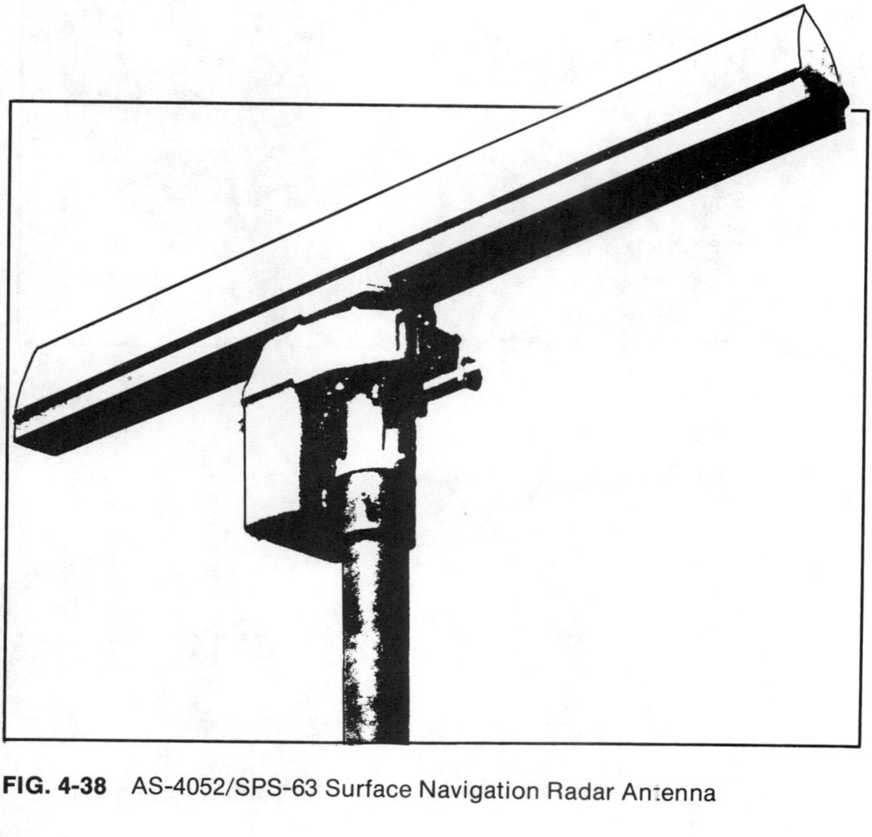

FIG. 4-38 AS-4052/SPS-63 Surface Navigation Radar Antenna

(a)

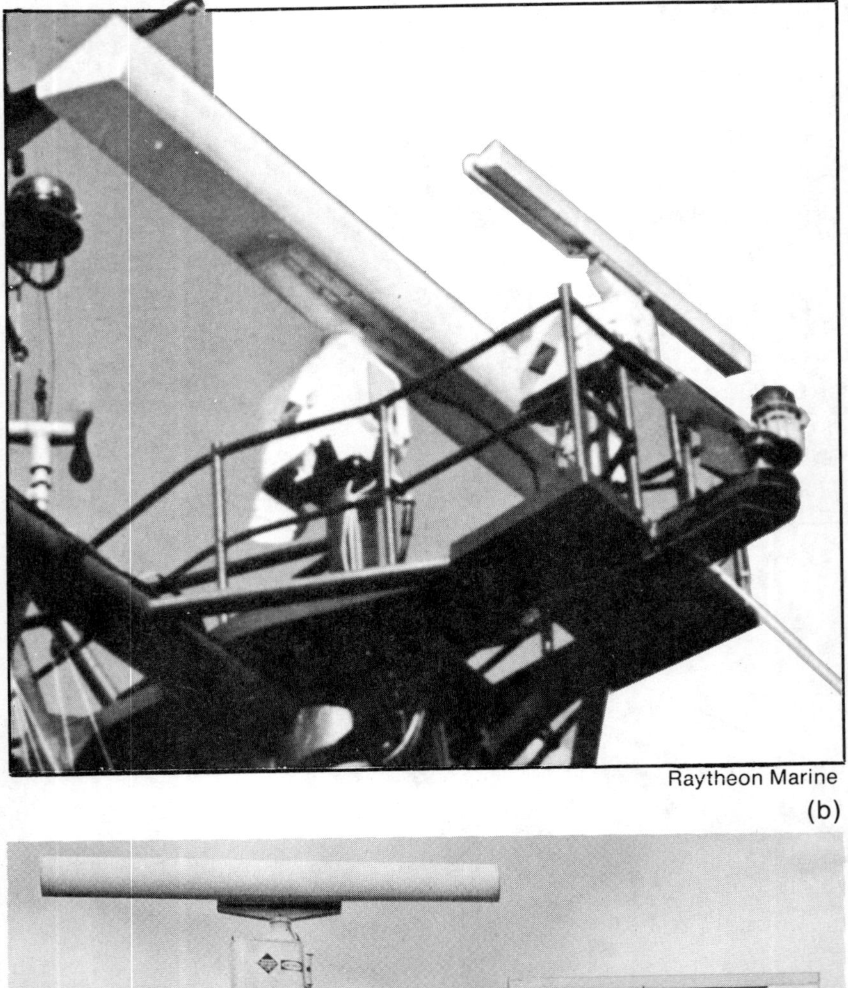

Raytheon Marine

(b)

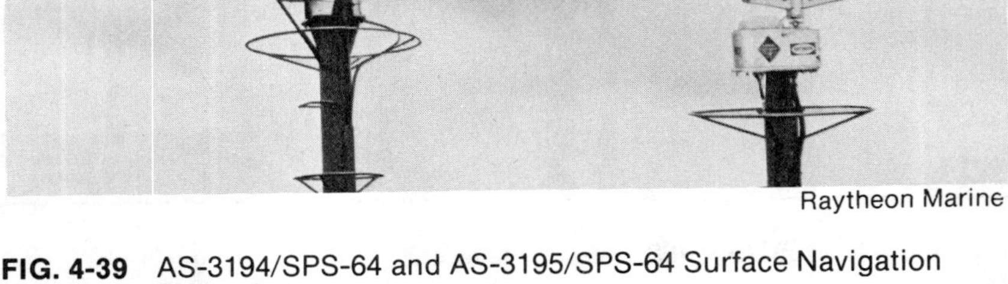

Raytheon Marine

FIG. 4-39 AS-3194/SPS-64 and AS-3195/SPS-64 Surface Navigation Radar Antennas

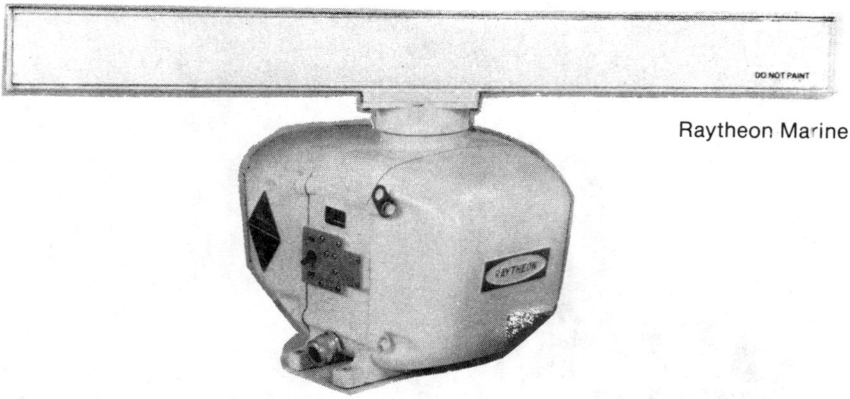

Raytheon Marine

FIG. 4-40(a) AS-4072/SPS-66A Suface Navigation Radar Antenna

Raytheon Marine

FIG. 4-40(b) AN/SPS-66 Dome-Covered Version Navigation Radar Antenna

FIG. 4-41 LN-66 Surface Navigation Radar Antenna

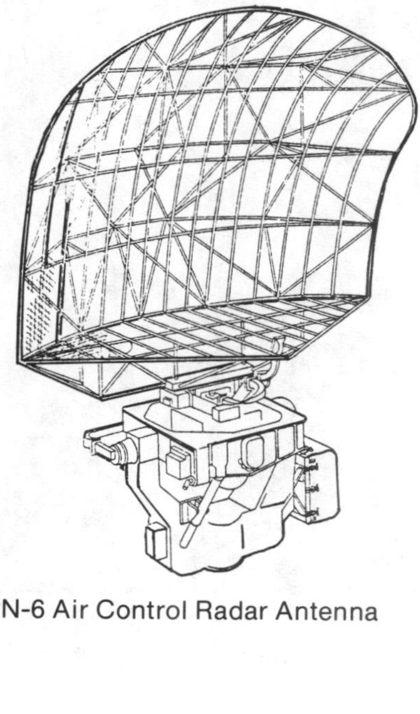

FIG. 4-42 AS-538/SPN-6 Air Control Radar Antenna

FIG. 4-43 AN/SPN-35 Air Control Radar Antenna

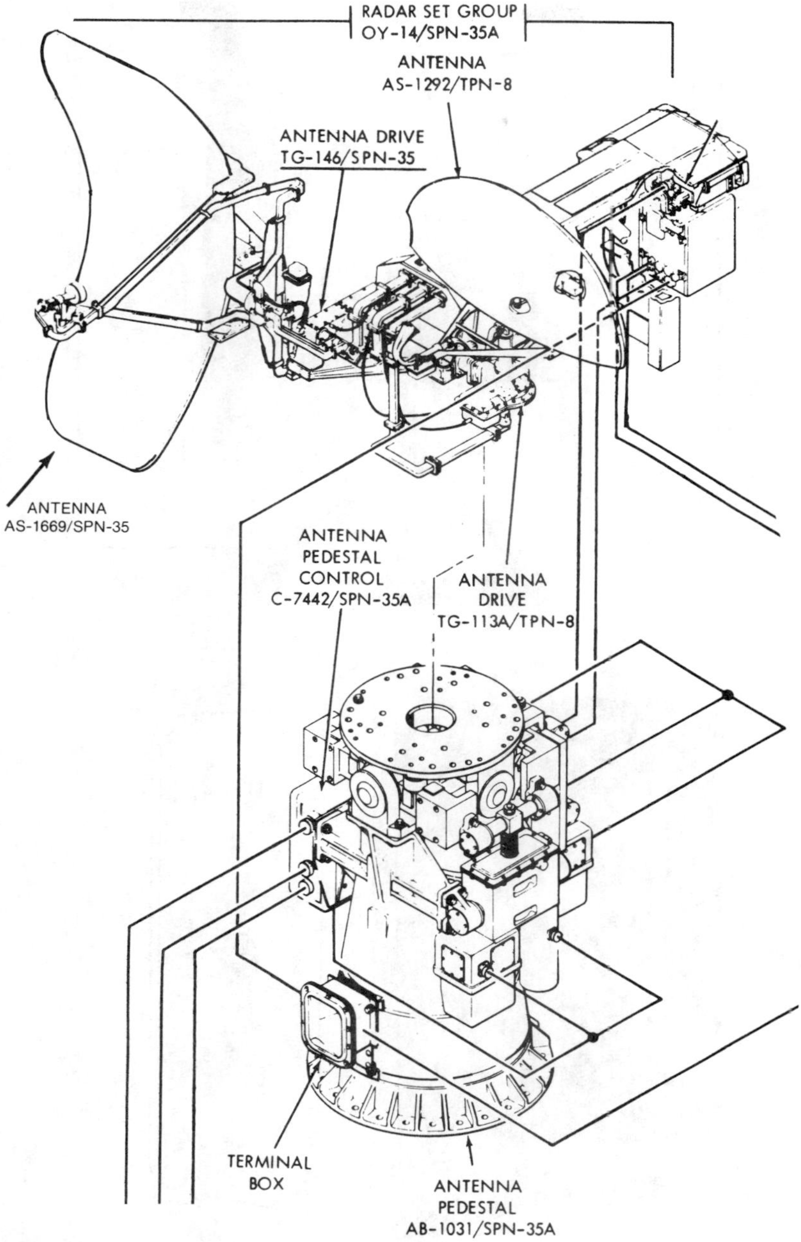

FIG. 4-44 AS-1292 and AS-1669: Air Control Radar Antennas Of AN/SPN-35

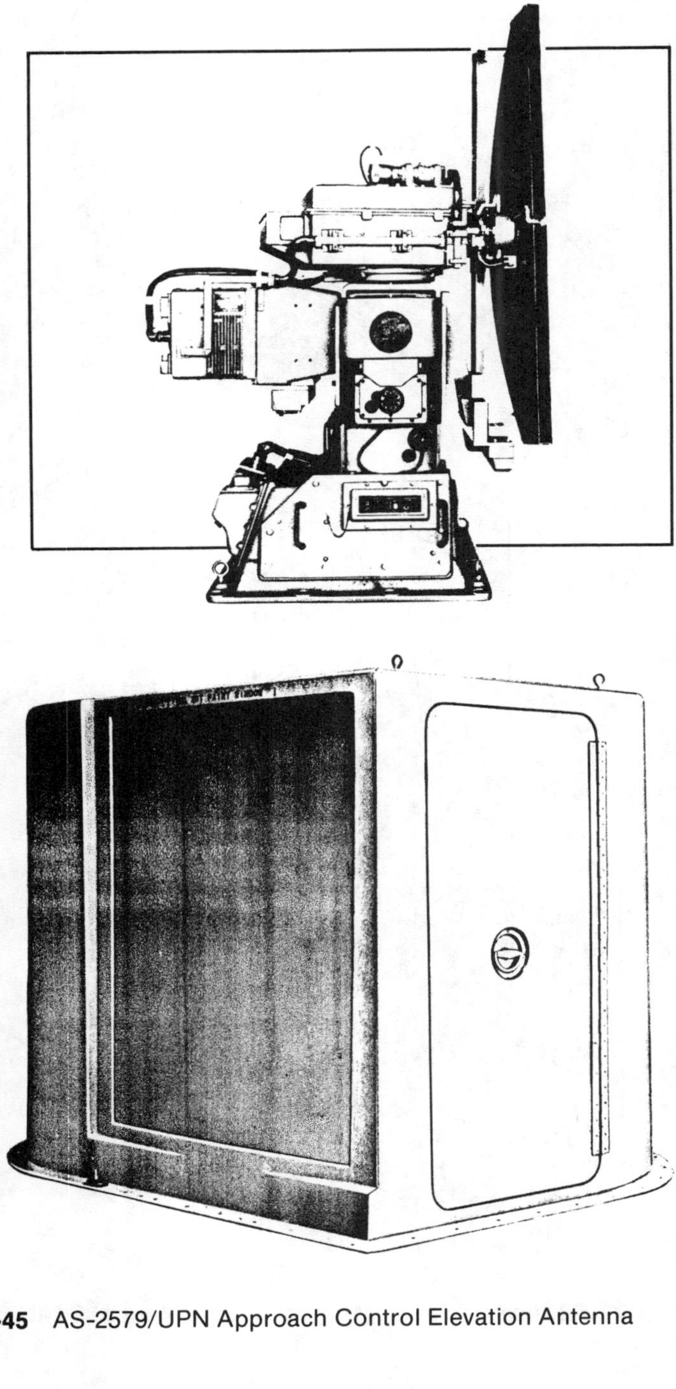

FIG. 4-45 AS-2579/UPN Approach Control Elevation Antenna

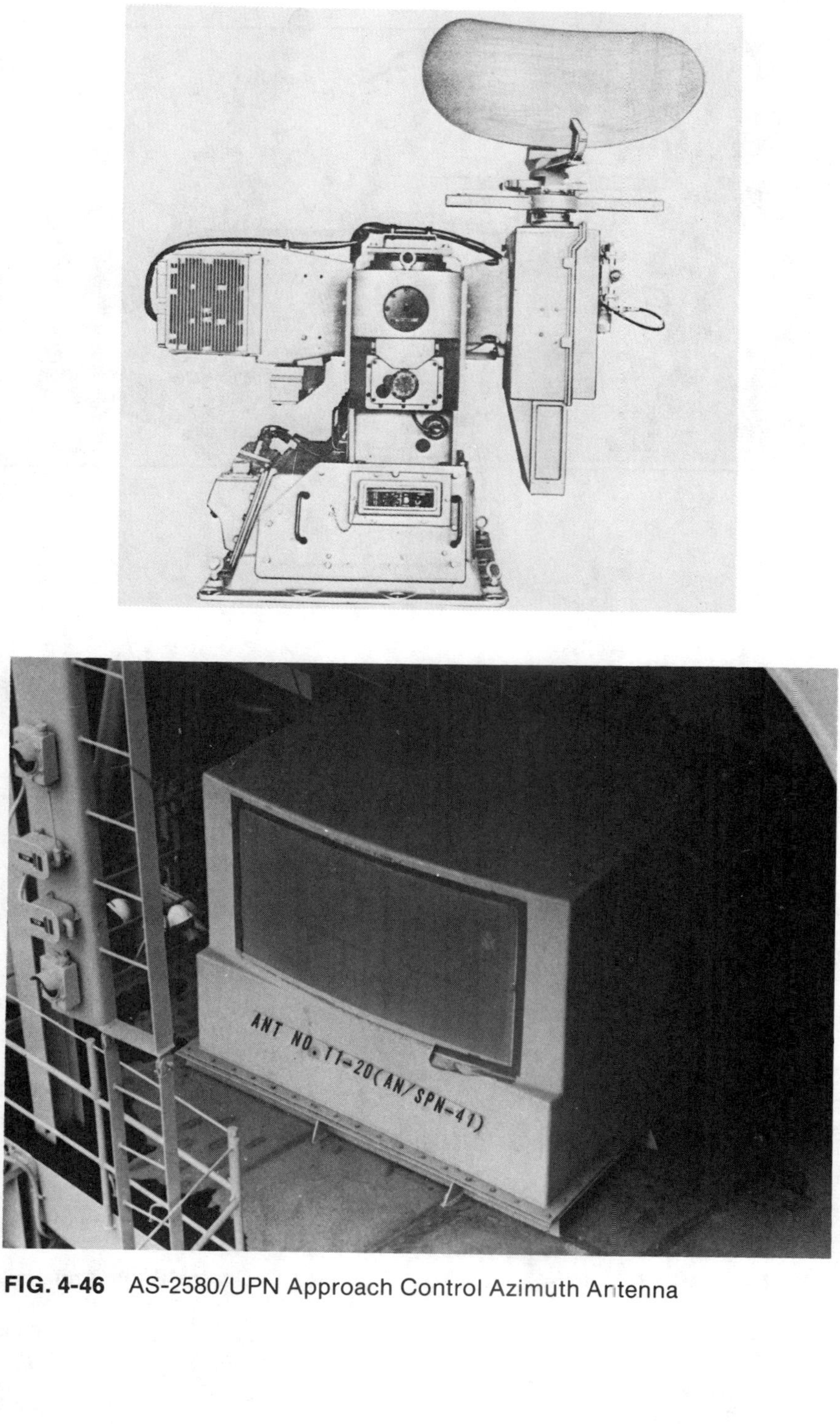

FIG. 4-46 AS-2580/UPN Approach Control Azimuth Antenna

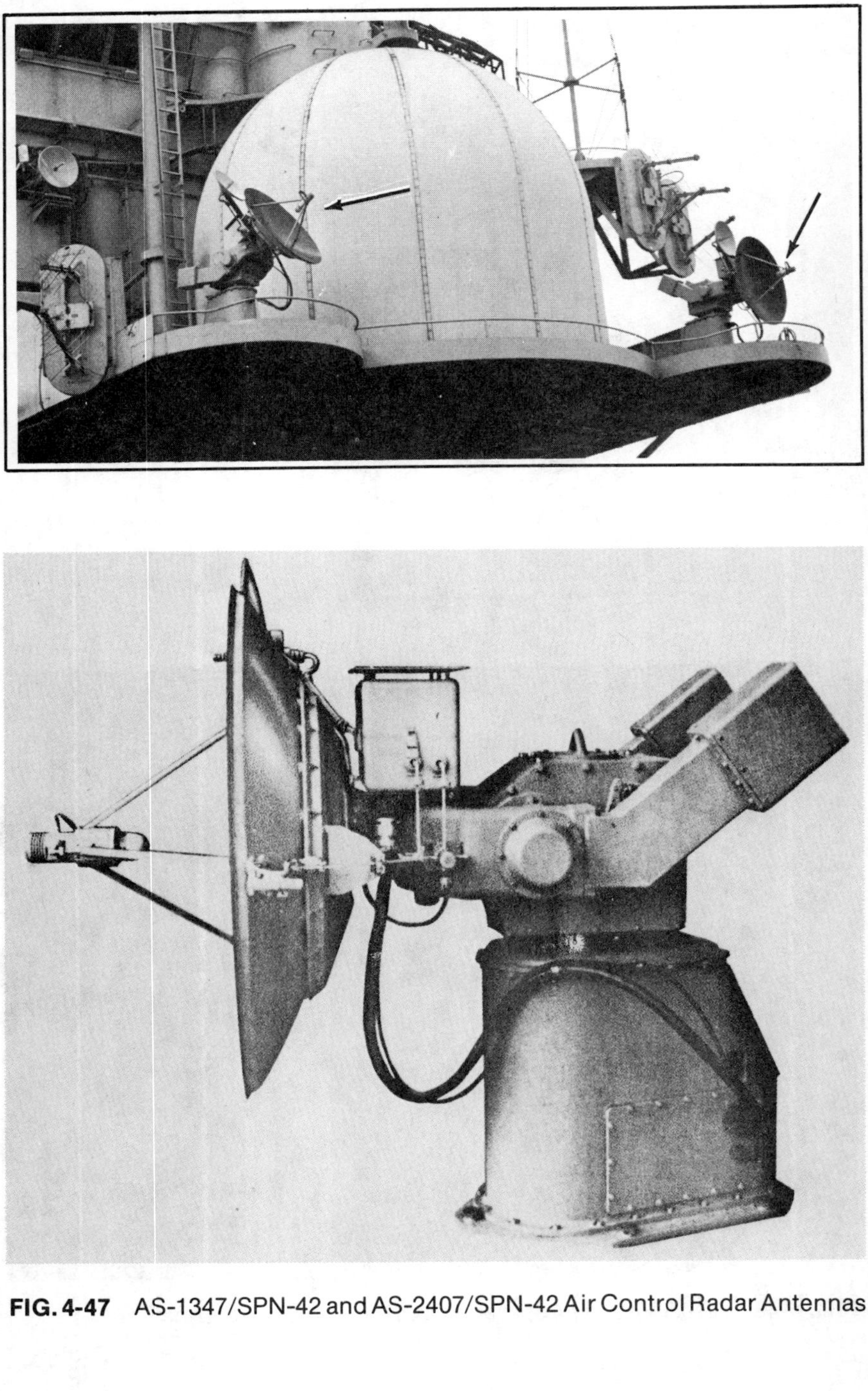

FIG. 4-47 AS-1347/SPN-42 and AS-2407/SPN-42 Air Control Radar Antennas

FIG. 4-48 AN/SPN-42 Air Control Radar Corner Reflector Reference Antenna

FIG. 4-49 AS-2034/SPN-43 Air Control Radar Antenna (with AS-1688 IFF antenna mounted at top of reflector)

FIG. 4-50 AS-2785/SPN-43A Air Control Radar Antenna (with AS-2188/U IFF antenna mounted at top of reflector)

FIG. 4-51 AN/SPN-44 Air Control Radar Antenna

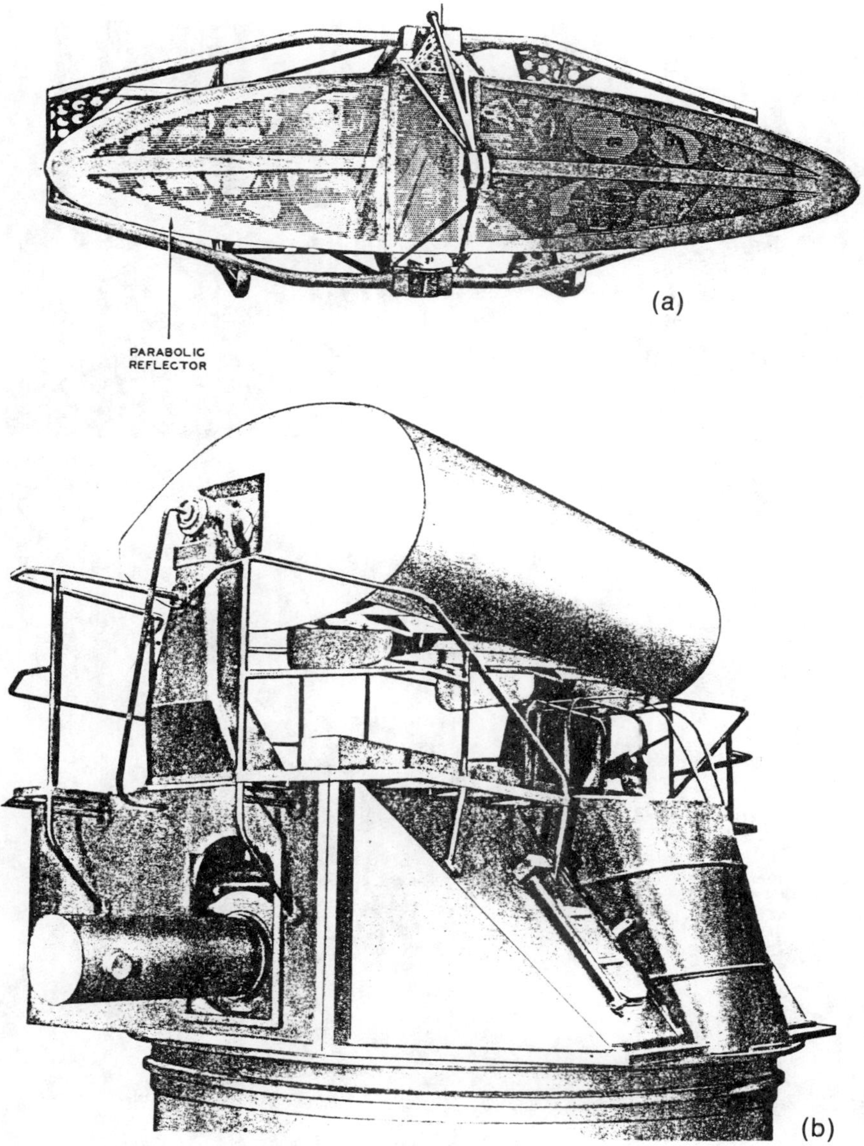

FIG. 4-52 MK 13 Gunfire Control Radar Antenna

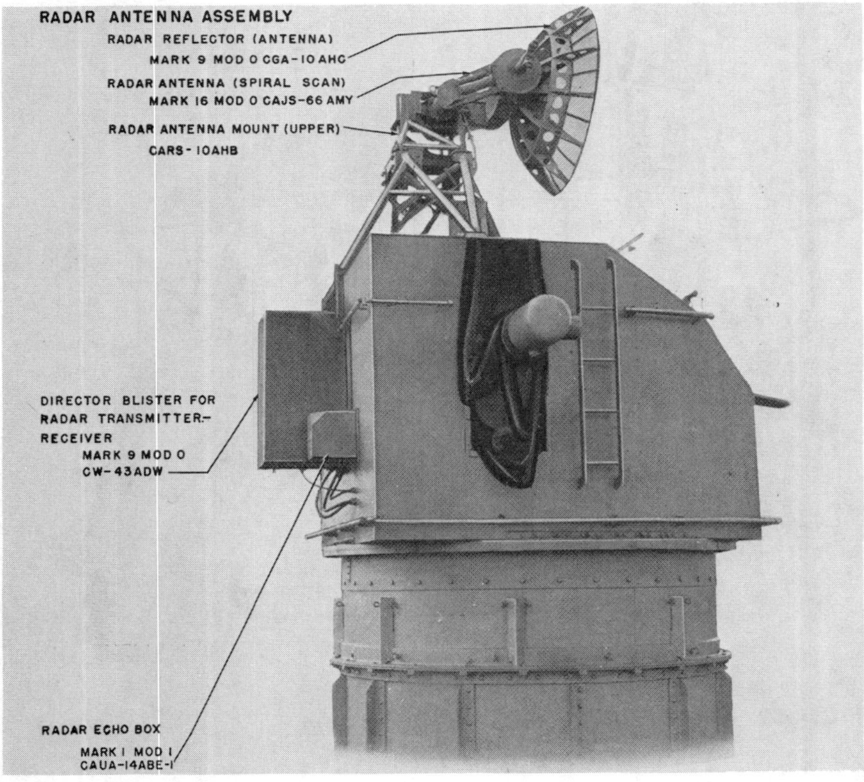

FIG. 4-53 MK-25 (MOD 2) Gunfire Control Radar Antenna

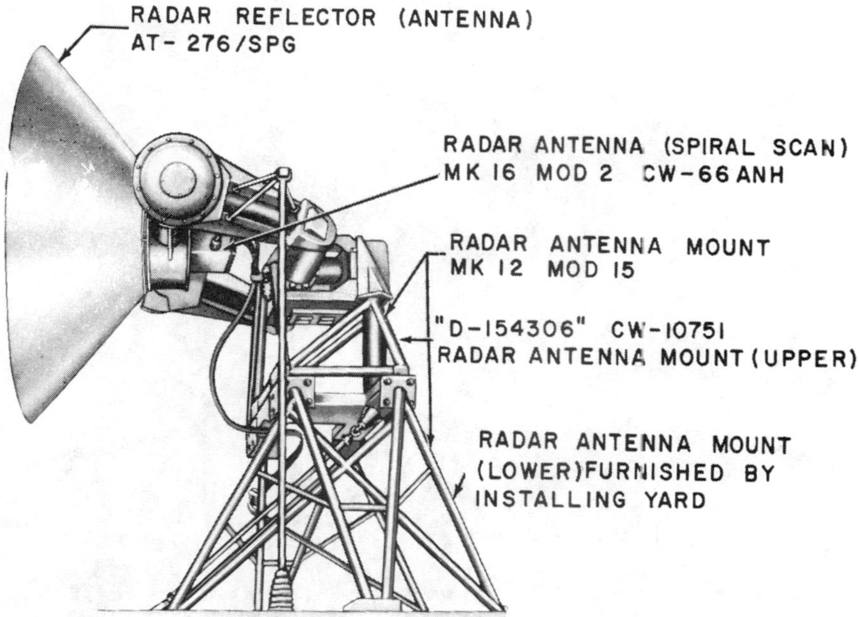

FIG. 4-54 MK 25 (MOD 3) Gunfire Control Radar Antenna

FIG. 4-55 AS-515A/SPG MK 35 Gunfire Control Radar Antenna

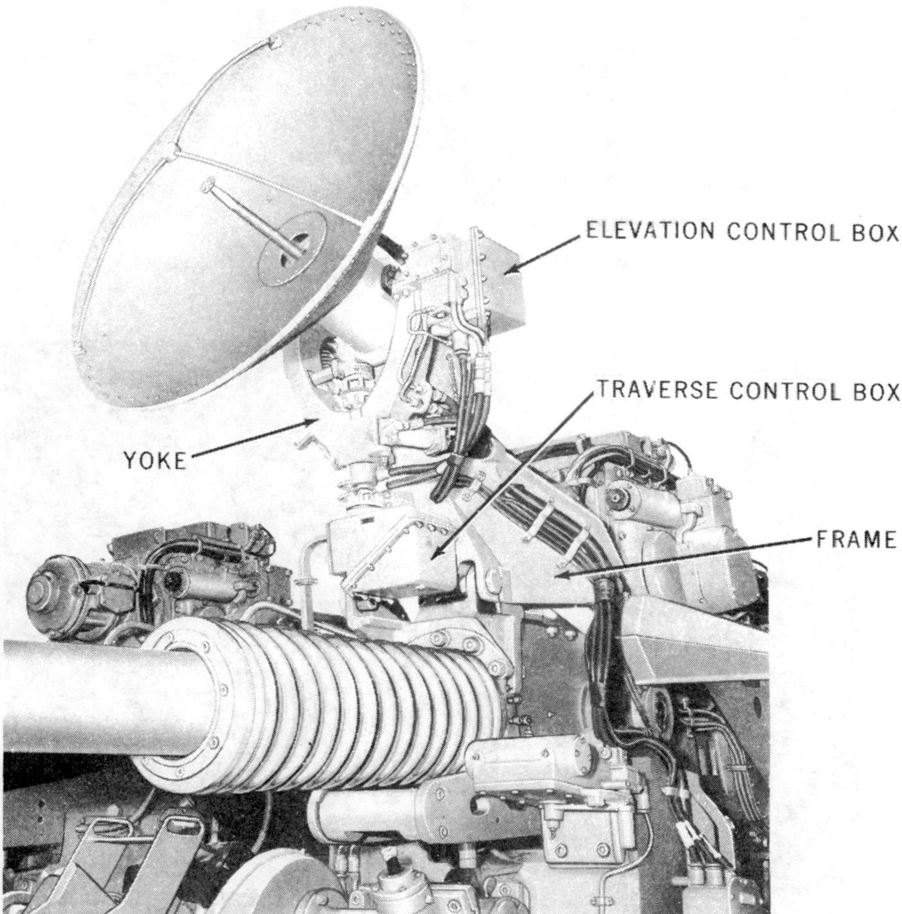

FIG. 4-56 AS-515A/SPG (AN/SPG-34 and AN/SPG-50) Gun Fire Control Radar Antenna

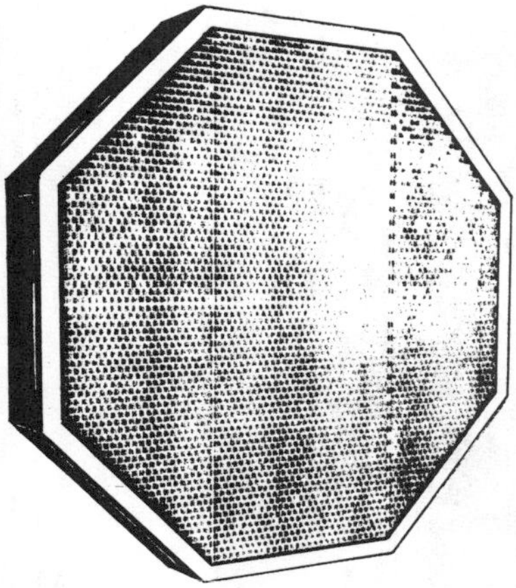

FIG. 4-57(a) UD20 Microwave Lens (Radome Removed)

FIG. 4-57(b) UD 20 AN/SPG-49 Guided Missile Fire Control Radar Antenna

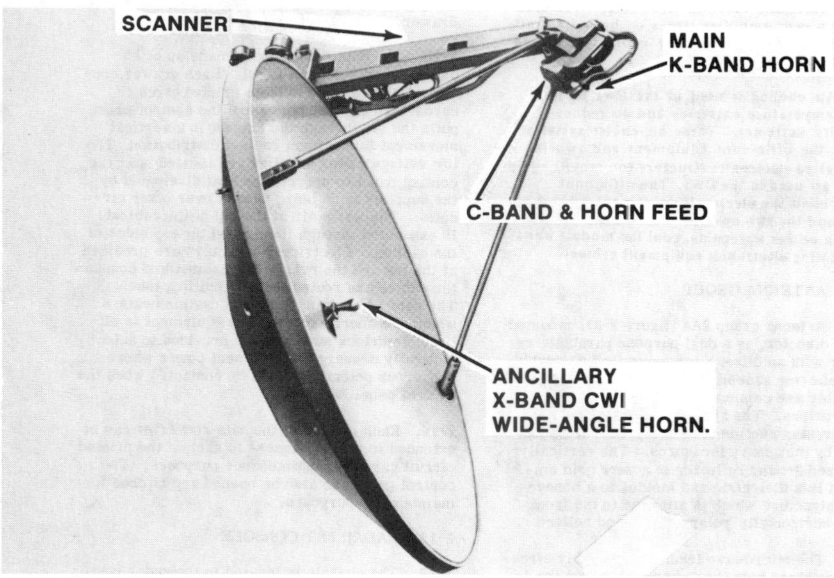

FIG. 4-58 AS-2649/SPG-51C Guided Missile Fire Control Radar Antenna

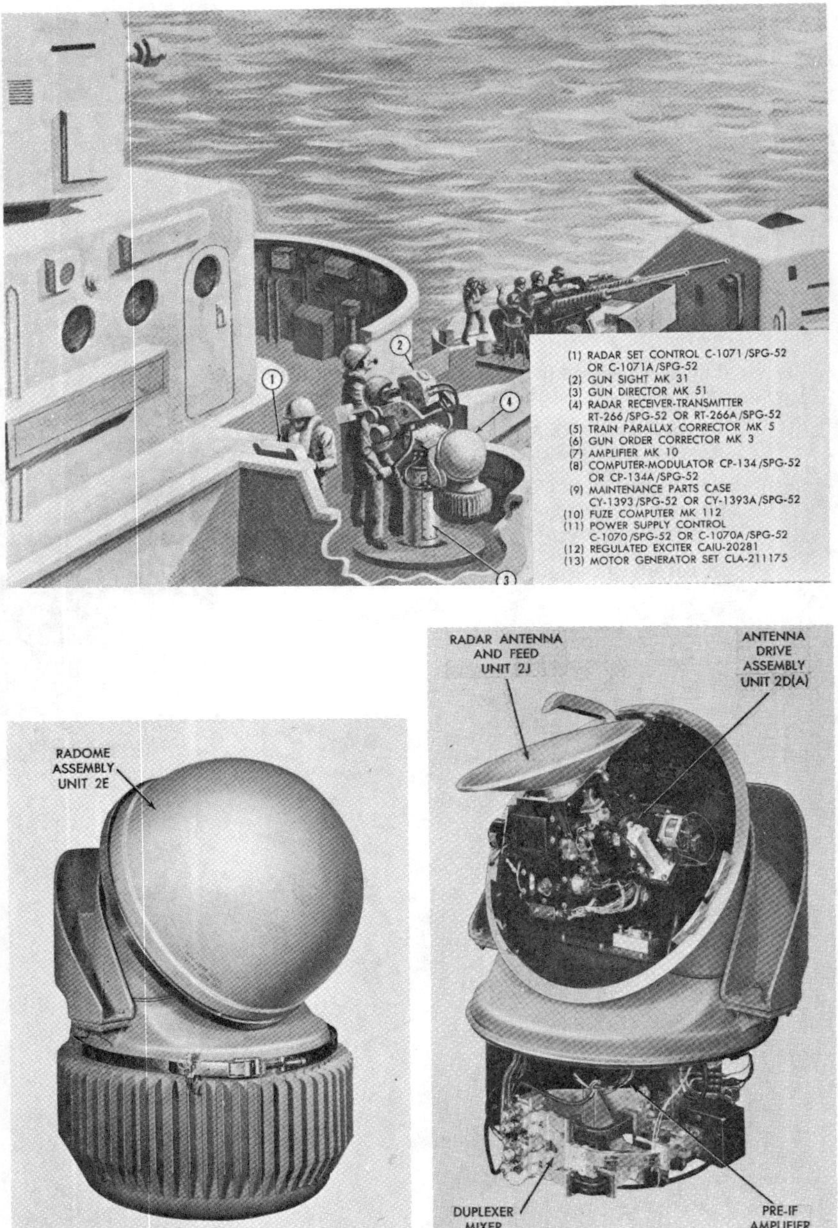

FIG. 4-59 RT-266A/SPG-52 Gunfire Control Radar Antenna

FIG. 4-60 (a) AN/SPG-53 Gun Fire Control Radar Antennas (conventional version)

FIG. 4-60(b) AS-1868 SPG-53 B/C Antenna Modified for CWI.

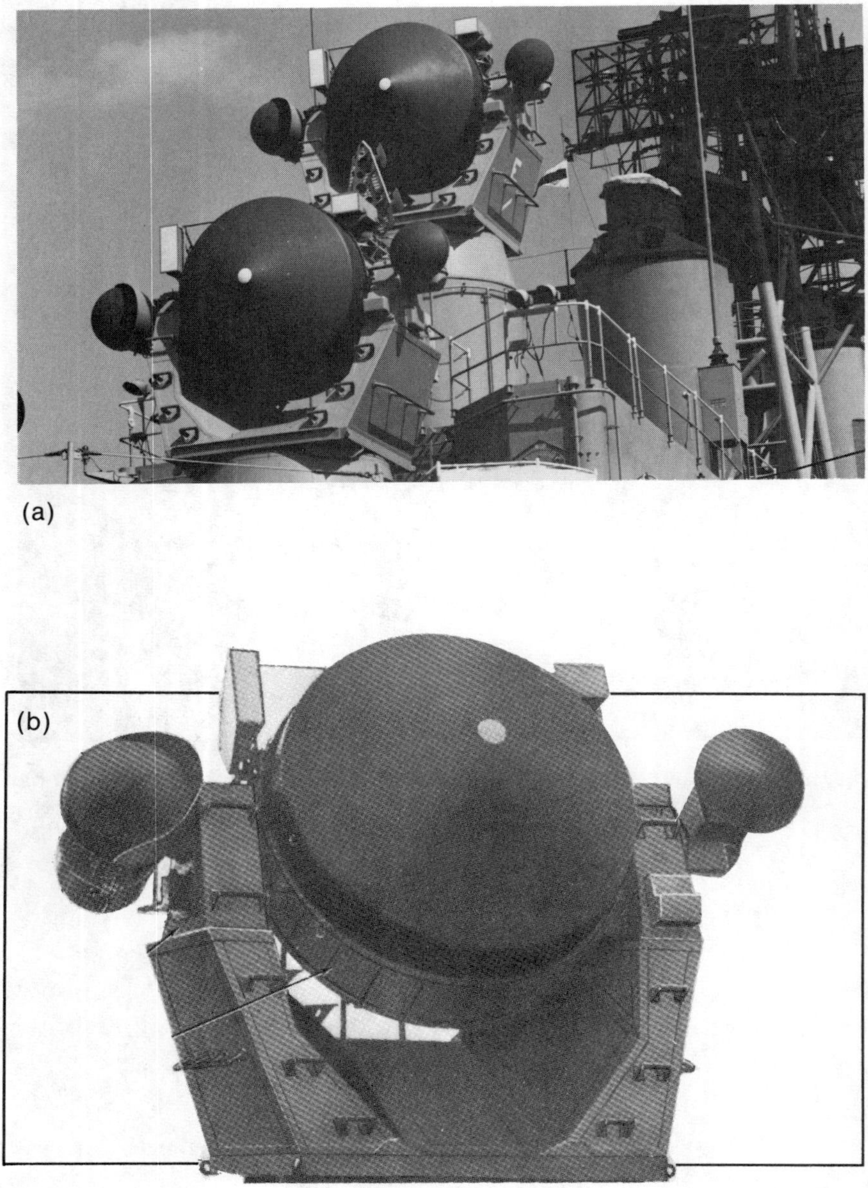

(a)

(b)

FIG. 4-61 AN/SPG-55 Guided Missile Fire Control System Radar Antenna

FIG. 4-62 AN/SPG-60 Gun Fire Control Radar Antenna

FIG. 4-63 AS-3156 AEGIS Guided Missile Fire Control Radar CW Illuminator Antenna

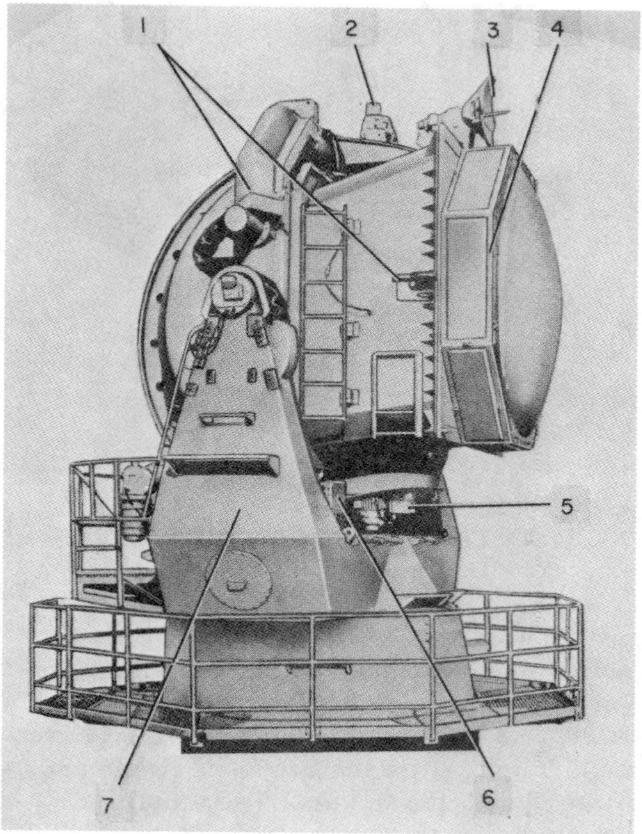

1 Traverse Member and Gimbal UD306
2 Upper Traverse Mechanism UD325
3 Capture Antenna UD347 (Grating Removed)
4 Loaded Lens
5 Lower Traverse Mechanism UD326
6 Lower Traverse Drive Junction Box UD298
7 Yoke Assembly UD305

FIG. 4-64 UD 304 AN/SPQ-5A Guided Missile Fire Control Radar Antenna

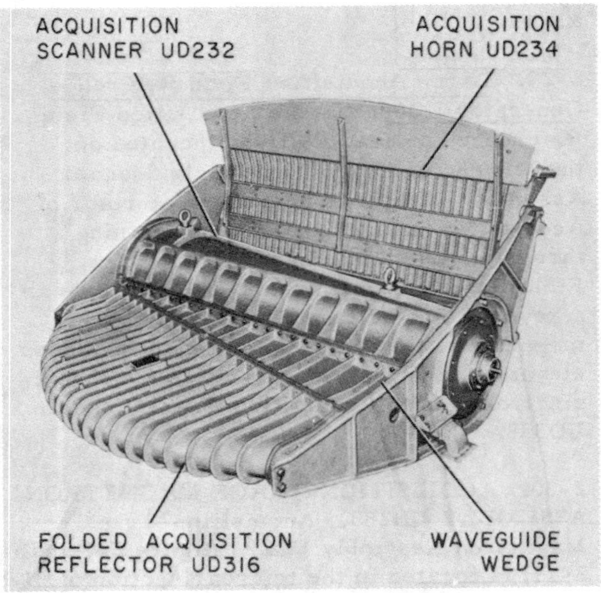

FIG. 4-65 AN/SPQ-5A Fire Control Radar Scanner Assembly and Capture Antenna

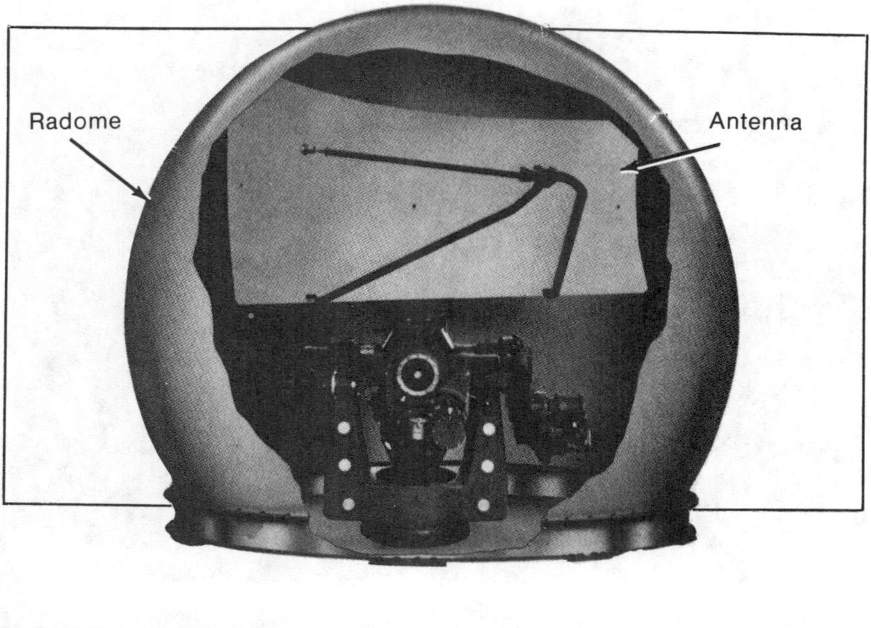

FIG. 4-66 AS-2367A/SPQ-9 Fire Control Radar Antenna

FIG. 4-67 UD 39 AN/SPW-2 Fire Control Guidance Radar Antenna

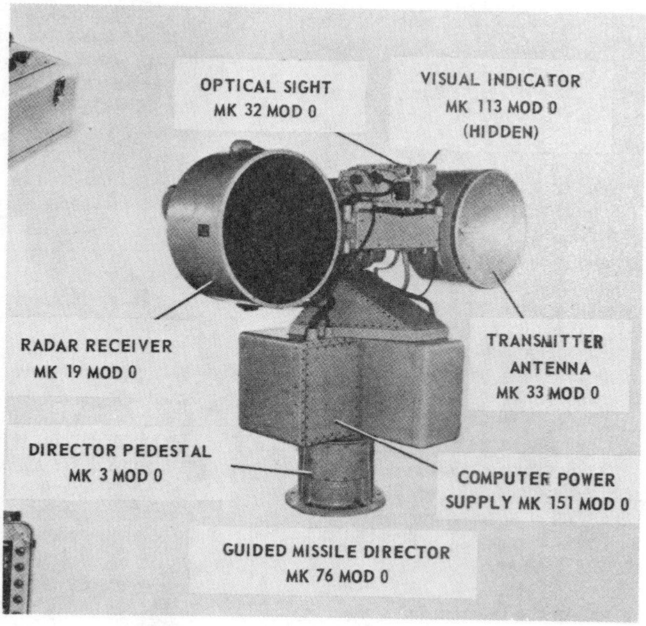

FIG. 4-68 Basic Point Defense System Fire Control Antennas

FIG. 4-69 NATO Sea Sparrow Fire Control Radar Antennas

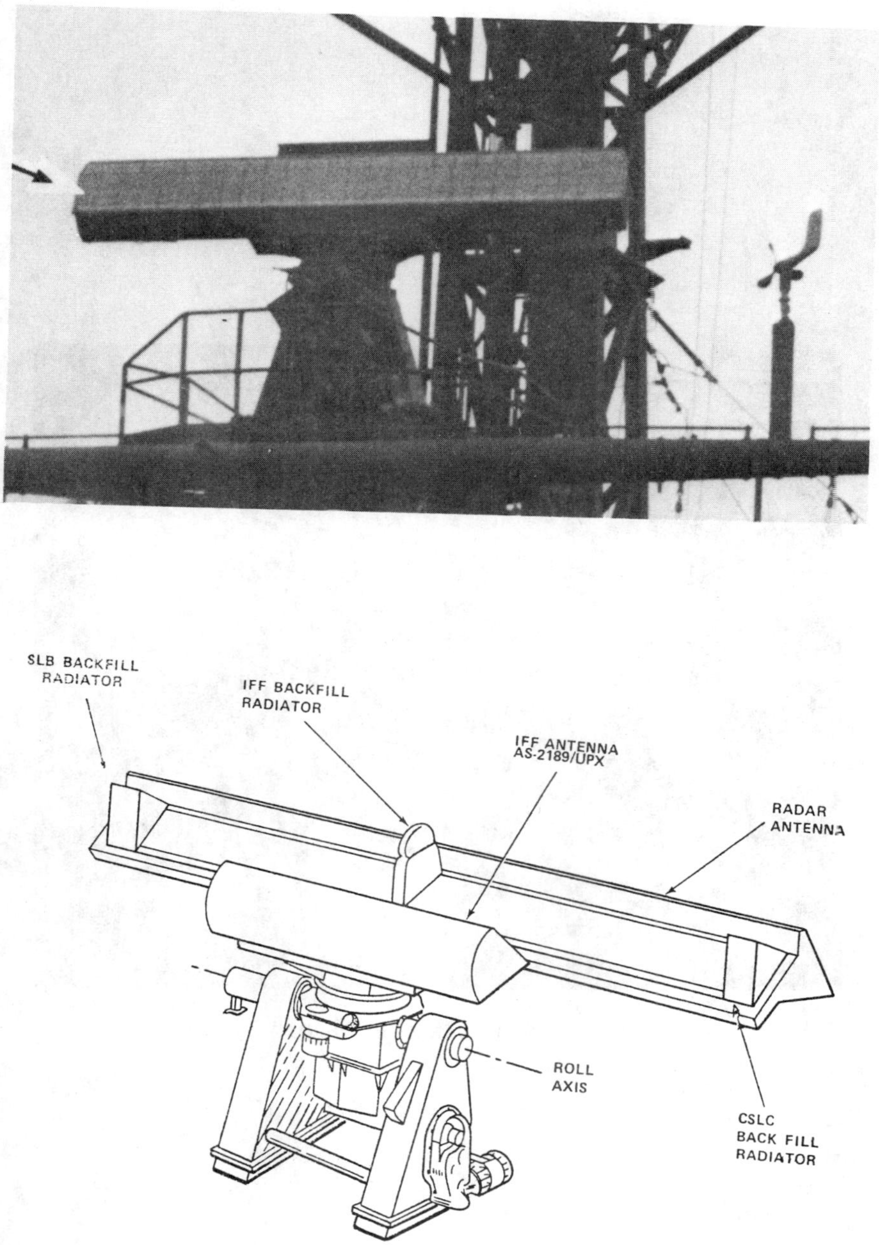

FIG. 4-70 Target Acquisition System (TAS) Antenna

FIG. 4-71 Phalanx Close-in Weapon System

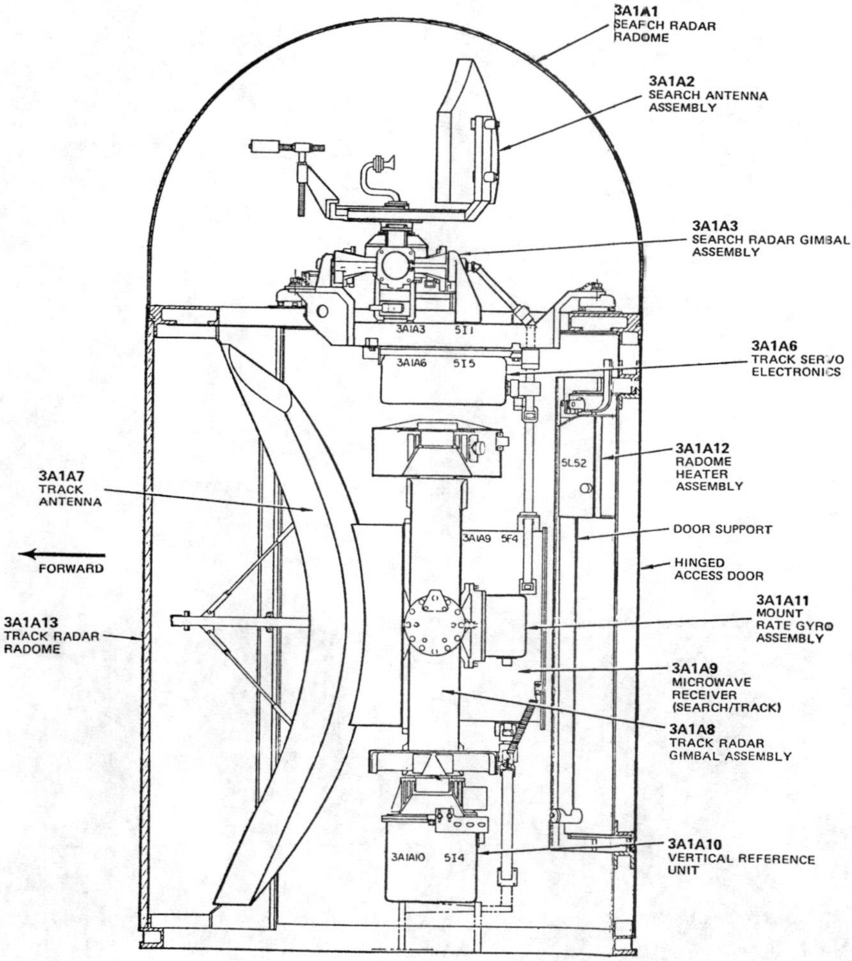

FIG. 4-72: CLOSE-IN WEAPON SYSTEM RADAR ANTENNAS.

FIG. 4-72 Close-in Weapon System Radar Antennas

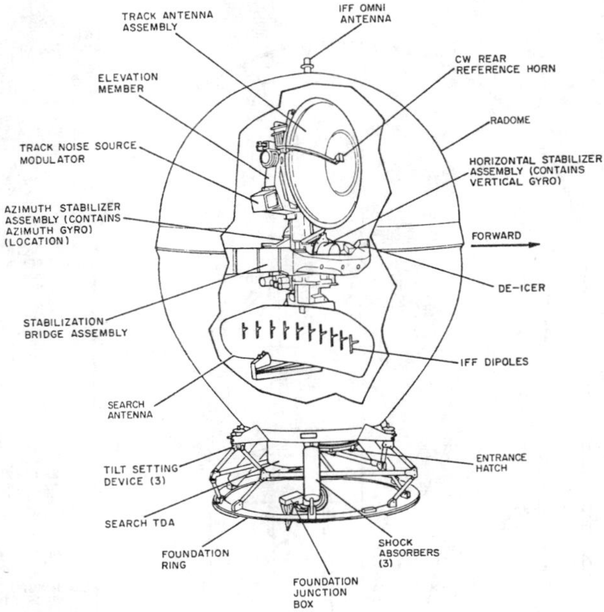

FIG. 4-73 UD 401 Combined Antenna System (CAS)

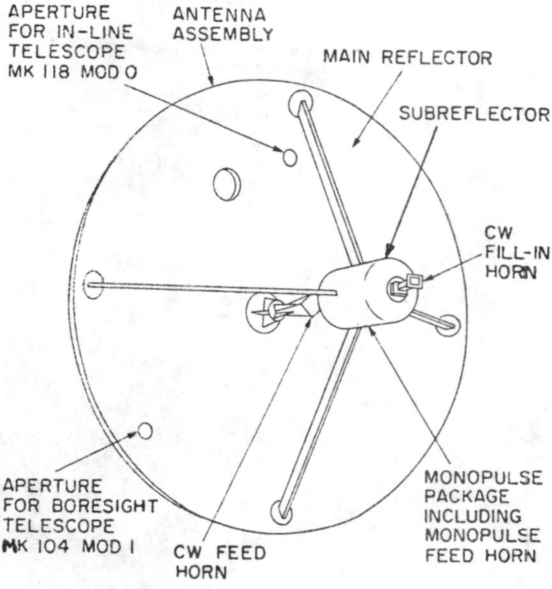

FIG. 4-74 UD 417 Separate Track Illuminating Radar (STIR) Fire Control Antenna

FIG. 4-75. AS-3159/SPY-1 Multifunction Weapons Systems Phased-Array Radar Antenna

Chapter 5

Electronic Warfare Antennas

5-0 INTRODUCTION

It is immediately evident from the material in the preceding chapters concerning shipboard communication, navigation, and radar, that, in just the Twentieth century, the use of electronic systems has revolutionized naval operations. From the smallest sea-going pleasure craft, to the largest commercial vessels plying the oceans, it is virtually impossible to imagine any ship today without at least radio communication and navigation. Use of the electromagnetic spectrum has become an essential and invaluable maritime aid. Moreover, it should be apparent from Chapter Four that electronic systems have become an inseparable, interwoven part of naval warfare as well. From that very reliance upon electronics in combat systems — that vital dependency upon the electromagnetic medium — an entirely new discipline peculiar to our times has arisen. It is known as electronic warfare.

EW, as it is more commonly called, was founded on the concept that any technology so totally committed to electromagnetic radiation must possess an inherent weakness. Therefore, it should be possible to probe for and exploit that weakness. In radio electronics the weak link is simply that RF energy, once radiated, is highly susceptible to detection and interference.[1]

Electronic warfare has been defined in a formal manner as military action involving the use of the electromagnetic spectrum, and (on the other hand) action that retains friendly use of the electromagnetic spectrum. Robert J. Schlesinger has expressed it more fundamentally and informally by noting that, through interaction between two or more electronic systems, any intentional interference *is* electronic warfare. The end purpose is to interfere by electronics with the successful operation of an opponent's weapon system. "The problem concerned with when, where, and how to generate this electronic interference," and what to do "to counter its detrimental effects are of fundamental importance to electronic warfare."[2] Moreover, if the essence of EW is to employ electronic devices against an enemy's electronic capabilities, then the inevitable result is that a highly competitive struggle ensues whereby the opponent counters by making improvements to his combat systems so as to be resistant to interference. The interferer in turn counters the opponent's improvements by devising ever more clever means to penetrate the new resistance and it is this continuing technological duel that reveals the true nature of electronic warfare.

How then did it all begin? It began very early in this century, nearly as soon as wireless communication was adapted to shipboard use, when it was realized that the location of an enemy fleet could be determined by monitoring radio transmissions. Thus, some believe that the very first application of electronic warfare occurred during the Russo-Japanese War, early in 1904, when a quick-thinking Russian wireless operator at Port Arthur, hearing the radio signals being passed between Japanese heavy cruisers and their spotter destroyer near shore, slammed down his own transmit key to effectively jam the Japanese communicators. Without spotting information from the destroyer, the big guns of the cruisers were able to inflict only minor damage to the port.

Yet it was World War I that afforded the first conclusive though primitive strategic uses of an infant EW. It is thought that the very first combat exploitation through electronic means occurred in 1914 at the Battle of Tannenberg. The German Army, intercepting uncoded Russian operational orders and thus knowing their plans in advance, encircled and completely destroyed the Russian forces.[3] Two years later, having broken the German naval code, the British intercepted messages to their advantage in the major naval engagement known as the Battle of Jutland. But perhaps the clearest example of the intentional use of EW techniques as a combat aid was the employment of loop antenna radio direction finders by American destroyers to locate and sink German U-boats. (See Chapter Three.) By this same technique, Allied convoys were directed safely around lurking enemy submarines.[4]

The value of these early rudimentary experiences in the art of electronic warfare was not lost on the U.S. Navy. Laboratory research and development programs were funded for "avoiding enemy detection, and detecting enemy transmissions—and creation of interference for the enemy," i.e., devising elementary jamming methods.[5] Simple jamming interference was generated by sweeping noise transmissions across selected frequency bands, while avoidance of enemy jamming could be accomplished through the use of specialized selective receiver circuitry.

By the time of World War II, the underlying principles of electronic warfare were understood to the extent that hurried measures could be introduced in order to successfully ward off new threats. As an example, to reduce the devastating effects of British antiaircraft guns and fighter planes, the *Luftwaffe* switched to night bombing using radio navigation, in the form of two intersecting narrow beams, for guidance over the target area under cover of darkness. However, in an electronic countermove, the British used deliberate signal interference to alter the radio navigation beams, with the result that the misdirected German bombers dropped their bombs harmlessly into the English Channel.[6] To further confuse the enemy pilots, the British transmitted deceptive messages on correct frequencies, using German-speaking announcers.*

It was radar—at once the latest aid to modern warfare and the newest threat—that ushered in a wholly new era of electronic countermeasures, with far-ranging results. The obvious first strategem to thwart radar detection was to employ brute-force jamming; i.e., to transmit sufficiently high-power noise to obliterate the target signals displayed on the enemy's radar scope. In one of the more cleverly improvised uses of this technique, the Germans succeeded in conditioning British coastal radar operators to increasing levels of noise appearing at a specific time each morning for several successive days. This resulted in the British attributing the interference to natural atmospheric conditions occurring with regularity at dawn. Finally, at an appointed hour, with the help of foul weather, and with jammers activated, the Germans slipped their great warships *Gneisenau* and *Scharnhorst* out of France and up the English Channel, past the unsuspecting British, through the North Sea and into Kiel without incident.

The United States, too, made very effective use of jamming, of both the active and the passive types. During bombing raids over enemy territory, American crews transmitted broadband and spot noise

*The reader is referred to Alfred Price's *Instruments of Darkness* (New York: Charles Scribner's Sons, 1978) for an extraordinary account of England's desperate struggle to successfully apply EW techniques during the air war against Germany.

and dropped reflective chaff for screening. Furthermore, since it was known that both Germany and Japan had produced radar systems, equipment such as special search and monitor receivers and high-level noise emitters were developed for signal interception, location, interference, and deception. This was merely the beginning of a great technological explosion occurring during the Second World War and afterwards, which resulted in a surge of new interest in radar for surveillance and fire control. A concurrent emphasis was placed on developing means to oppose the effectiveness of radar electronic support measures (ESM) and electronic countermeasures (ECM) as important subsidiaries of electronic warfare.

ESM, thought more of as passive EW, is used to intercept hostile signals (e.g., communication, radar, navigation) for extracting useful military information to aid the observing side. Thus the objective of ESM is to search for and collect data, determine the location of the source, and analyze the signal characteristics and equipment capabilities, in order to determine the threat and to invoke effective countermeasures.

ECM, on the other hand, is general interference used to either deny the enemy needed information or cause him to accept false information to the detriment of his operation.[7] Simply put, ECM is employed to gain military advantage in a specific situation.[8] As such, it consists of these principal features:[9]

1. *Jamming:* Blotting out the enemy at his receiver by transmitting various types of noise on his frequency. This represents the least sophisticated form of ECM.
2. *Confusion:* Generating a number of signals appearing to be the true desired signal. For example, causing a number of false radar targets, by using either passive reflectors or active pulse transmissions simulating the target. The enemy is then forced to decide which of the many pulse returns is the true signal.
3. *Deception:* Diverting the enemy from the true signal to a false signal which he has been deceived into believing is the true one.

Over the past 30 years, then, electronic warfare has been refined to an increasingly sophisticated and complex science; an art, as some might insist. Important lessons learned from the Viet Nam and Middle East conflicts have led to continued improvements in the quality and variety of electronically aided weapons,[10] forcing even greater sophistication of modern-day EW technology. Nowhere is this technology more crucial than in naval operations of today, where the major defense must be against anti-ship missiles. The recent Battle of the Falklands provided an alarming illustration of the vulnerability of surface ships even in the 1980s when without adequate EW protection. Undoubtedly that war has stimulated a resurgent worldwide interest in electronic warfare. Prophetically, just a matter of weeks before the Falklands crisis, Richard Hartman wrote, "In today's and tommorow's wars, learning to survive in an electronic environment is just as basic as firing a rifle. . .it will be incumbent to basic survival to be able to counter the enemy's electronic weapons."[11]

It should be clearly understood that shipboard antennas play a critical role in the effectiveness of EW at sea. As will be seen in the following sections, EW antennas are found in a wide variety of shapes and sizes, some of which are quite exotic. Many are so sensitively classified that only a cursory description can be given here.

5-1 ELECTRONIC SUPPORT MEASURES (ESM) ANTENNAS

A great many electromagnetic emission detection and radio direction finding systems have been employed as electronic countermeasures devices by the U.S. Navy since the early 1940s for purposes of search, monitor, and analysis. Initially, narrowband countermeasure receivers adapted from the U.S. and British Air Forces were deployed by the Navy for interception of enemy radar signals.[12] Beginning with the World War II AN/SPR-1 (100-950 MHZ) and Model DAQ high-frequency set, successive ESM equipment, progressing up through microwave frequencies, has seen service over the ensuing decades. Shipboard ESM systems at the present time include the AN/SLD-1, AN/SLR-12 and 22, AN/SRD-19 and the AN/WLR-1, 8, and 11A. Associated with these systems is a wide variety of specialized antennas.

5-1.1 AN/SLD-1A

The AN/SLD-1A is a shipboard microwave EW direction finding set used to locate, and display the bearing of, hostile airborne and surface craft. It is meant to operate chiefly as the auxiliary search component of the AN/SLQ-26 ECM system, but may be used independently as well. Type AS-2854 antennas, installed aboard ship in a port and starboard configuration, are used with the SLD-1A as described below.

ANTENNA AS-2854/SLD-1A (formerly AS-2308/SLD-1)

a. TYPE/USE — The AS-2854 receiving antenna is a four-horn array designed for use with the AN/SLD-1A ESM direction finder. (An earlier version AS-2308 was employed with the older model AN/SLD-1 equipment.) By electronically sequenced switching, rapid scanning through 180 degrees in azimuth is accomplished. To provide 360 degrees complementary coverage, two antennas are installed, one on the port side and one on the starboard. In addition to the four microwave horns, each antenna unit contains a diplexer, TWT amplifier, detector and switching control circuitry.

b. PHYSICAL CONFIGURATION — [See Figures 5-1 and 5-2].

c. DIMENSIONS — (1) Height: 9 inches; Width: 27⅜ inches; (3) Depth: 18¾ inches; (4) Weight: 80 pounds.

d. FREQUENCY RANGE — [Classified].

e. INPUT IMPEDANCE — 50 ohms.

f. RF POWER RATING — N/A (receive only).

g. POLARIZATION — Receives signals of any polarization equally well.

h. TRANSMISSION LINE TYPE — RG-224/U coaxial cable.

i. PRIMARY POWER REQUIRED — 115 vac, 60 or 400 Hz, one-phase.

j. INSTALLATION REQUIREMENTS — The frequency range in which this antenna operates requires line-of-sight radio paths. In order to obtain unobstructed omnidirectional coverage, each antenna must be located in a clear area, as close as possible to the outboard side of the ship. The bottom of the unit is fitted with six ½-inch bolt holes for mounting the antenna.

k. SPECIAL CONSIDERATIONS — Each antenna requires pressurization from a source of dry air maintained at one to two psig. The length of the coaxial cables between the antenna and comparator unit must not exceed 400 feet.

l. REFERENCE — Technical Manual, Direction Finder Set AN/SLD-1A, NAVSEA 0967-477-8010.

5-1.2 AN/SLR-12A

The AN/SLR-12A is a wideband shipboard EW direction-finder receiving set used to intercept and display the bearing of microwave pulse signals. The type AS-2209 antenna used with this system is described below.

ANTENNA AS-2209/SLR-12A (formerly AS-1789/SLR)

a. TYPE/USE — The AS-2209 is a rotating, directional microwave receive antenna designed for use with the AN/SLR-12A ESM direction finding system. (An earlier version AS-1789 was used with the older model AN/SLR-12 equipment.) The antenna assembly consists of a small, solid, parabolic-section reflector fed by a waveguide horn housed within the upper portion of a clamp-on radome. Enclosed within the lower portion, as part of the overall assembly, are the servo motor drive mechanism and various electronic components.

b. PHYSICAL CONFIGURATION — [See Figures 5-3(a) and (b)].
c. DIMENSIONS — (1) Height: 31⅞ inches; (2) Diameter: 19 inches; (3) Weight: 99 pounds.
d. FREQUENCY RANGE — [Classified].
e. INPUT IMPEDANCE/VSWR — 50 ohms; 3:1.
f. RF POWER RATING — N/A (receive only).
g. POLARIZATION — Linear in either plane or circular in either sense with separate RF outputs for right and left hand circular.
h. TRANSMISSION LINE TYPE — RG-244/U coaxial cable.
i. PRIMARY POWER REQUIRED — 115 vac, 60 Hz, one-phase.
j. INSTALLATION REQUIREMENTS — The antenna is platform-mounted on the mast. No obstacles are to be within the near field of the antenna (26 feet) and nothing greater than 6 inches in diameter can be within the line of sight of the antenna. If these precautions are not followed, erroneous data will result. Observe the bench mark and mount the unit with the mounting rails perpendicular to the axis of the ship. The center housing has two 23 x 1½ inch angle brackets for mounting the antenna. Each bracket has two 9/16-inch diameter bolt holes on 19¾ inch centers for attachment to a platform or other suitable structure. The antenna should be mounted sufficiently high to provide omnidirectional horizon coverage. It should not be installed in the main beam of any high-power transmitting equipment. The bench mark on the antenna housing must be aligned with the fore and aft centerline of the ship. A minimum of 13 inches above and 9 inches below the antenna must be provided to permit removal of the radomes for servicing.
k. SPECIAL CONSIDERATION — Since the output to the receiver is a video signal, the distance between the antenna and control unit is not critical. However, this distance should not exceed 600 feet.
l. REFERENCE — Technical Manual, Receiving Set Countermeasures AN/SLR-12(U), NAVSEA 0967-306-5010.

5-1.3 AN/SLR-16

The AN/SLR-16 is a shipboard countermeasures receiving set that automatically searches for and acquires radiated energy in the HF range, then simultaneously selects signals of interest for further analysis while blanking out certain segments of the band known to contain no energy of interest. However, the SLR-16 has no antenna system of its own, relying instead on an interconnection with the antenna systems of the AN/SRD-19, a shipboard EW set described next in Section 5-1.4. For further details of the AN/SLR-16 equipment, the reader is referred to NAVELEX Technical Manual 0967-LP-594-0010.

5-1.4 AN/SRD-19

The AN/SRD-19 is a wideband shipboard EW direction-finder set used to locate, process, and display the bearing of signal sources of interest in the LF/MF, HF, and VHF spectra. The set may be operated as an independent system, but more often is employed integrally with the AN/SLR-16 as part of the Shipboard Signal Exploitation System (SSES) known as *classic outboard*. In normal usage an operator controls the equipment; however, commands received by the SLR-16 have priority override and are thus automatically responded to and immediately processed, interrupting any of the SRD-19 DF functions in progress.

To perform its required functions, the SRD-19 employs a number of antenna subsystems. For the LF/MF-HF range a series of 24 small, deck-mounted antennas are located around the edge of the ship's main deck in three groups of eight. Additionally, for acquisition and reference purposes in this frequency range, four general purpose communication type antennas are used, e.g., a 10-30 MHz inverted cone, a 2-30 MHz 35-foot whip, and two small broadband discages. For the VHF range a three-tier mast-mounted array of Adcock dipoles and crossed loops is used. Characteristics of these SRD-19 antennas are given below.

ANTENNA AS-3202/SRD-19

a. TYPE/USE — The AS-3202 is a small, loop antenna unit mounted near the edge of a ship's main deck as shown in Figure 5-4(b). Designed for use in the LF/MF-HF range of the AN/SRD-19 DF system, twenty-four AS-3202s are switched into service by a predetermined sequence during data acquisition periods. This selected sampling is then compared with continuously applied reference antenna signals (provided by general-purpose communications-type antennas) for analysis and determination of signal source location.
b. PHYSICAL CONFIGURATION — [See Figure 5-4(a) and (b)].
c. DIMENSIONS — (1) Height: 9 inches; (2) Width: 20½ inches; (3) Depth: 6½ inches; (4) Weight: 54½ pounds.
d. FREQUENCY RANGE — [Classified].

e. VSWR — Less than 1.7:1.
f. RF POWER RATING — N/A (receive only).
g. POLARIZATION — [Classified].
h. TRANSMISSION LINE TYPE — RF coaxial cable.
i. PRIMARY POWER REQUIRED — 115 vac, 60 Hz, one-phase.
j. REFERENCE — Technical Manual, Direction Finder Set AN/SRD-19, NAVELEX 0967-LP-615-9060.

ANTENNA AS-3112/SRD-19

a. TYPE/USE — The AS-3112, as part of shipboard antenna group OA-236, is a three-tier omnidirectional array used with the AN/SRD-19 to receive signals in the VHF range segmented into three bands. The antenna incorporates both sequentially switched DF receive and a wideband target acquisition/reference subsystem. The upper and center antenna tiers are Adcock dipole elements, while the lower tier consists of crossed loops. The DF function is achieved by taking advantage of the phase and amplitude relationship of a wavefront impinging upon two spatially separated antennas when compared with an established acquisition or reference signal.
b. PHYSICAL CONFIGURATION — [See Figure 5-5].
c. DIMENSIONS — (1) Height: 85 inches; (2) Width: 50 inches; (3) Weight: 142 pounds.
d. FREQUENCY RANGE — [Classified].
e. INPUT IMPEDANCE/VSWR — 50 ohms/less than 2:0.1.
f. POWER RATING — N/A (receive only).
g. POLARIZATION — Vertical.
h. TRANSMISSION LINE TYPE — RF coaxial cable.
i. PRIMARY POWER REQUIRED — 115 vac, 60 Hz, one-phase.
j. INSTALLATION REQUIREMENTS — The antenna, performing a critical electronic warfare DF function, must be placed as high as possible to ensure a clear field of view. Normally it is placed at or near the top of the mast, just under the ship's TACAN antenna. However, the height of the AS-3112 should not exceed the 100-foot maximum RF cable length between the antenna and RF amplifier located below deck.
k. REFERENCE — Technical Manual, Direction Finder Set AN/SRD-19, NAVELEX 0967-LP-615-9060.

5-1.5 AN/WLR-1

The AN/WLR-1 is a high-resolution, high-sensitivity EW receiving set designed for service aboard naval surface ships, submarines, or at shore stations. A very wideband signal intercept system, it tunes from VHF through SHF microwave (50 MHz to 10.75 GHz) in nine selective bands. Rapid analysis of the received signals is accomlished by simultaneous direction finding, emission analysis, and information display. Either automatic frequency scanning or manual tuning is available.

The shipboard installed WLR-1 employs several antennas to cover the broad frequency range. These include the NT-66131 and NT-66132 VHF and UHF omnidirectional antennas; the AS-571, AS-616, AS-899 and the AS-1174 (or AS-3200) VHF through SHF DF antennas. Late models of the shipboard WLR-1 sometimes use an AS-1023 to combine the UHF and microwave range in one antenna, thus replacing the AS-616 and AS-899. Each of these diverse antennas is described below. A good example of a typical AN/WLR-1 antenna suite is illustrated in Figure 5-6.

ANTENNA NT-66131

a. TYPE/USE — The NT-66131 is a small, omnidirectional UHF cone-type fixed antenna designed for use with shipboard countermeasures receiving equipment such as the AN/WLR-1. The assembly consists of the cone antenna element and a groundplane comprised of seven rods projecting out as spokes from a metal center plate. An impedance matching section is contained in the base and a radome cover protects the antenna element from the sea environment. To afford complete 360 degrees azimuth coverage, two antennas are normally employed per system, one port and one starboard.
b. PHYSICAL CONFIGURATION — [See Figure 5-7].
c. DIMENSIONS — (1) Height: 8¼ inches; (2) Diameter: 24 inches maximum; (3) Weight: 12 pounds.
d. FREQUENCY RANGE — 300 to 3000 MHz.
e. INPUT IMPEDANCE/VSWR — 50 ohms/less than 5.0:1.
f. RF POWER RATING — N/A (receive only).
g. POLARIZATION — Horizontal or vertical.
h. TYPE OF FEED — Unbalanced, coaxial cable.
i. PRIMARY POWER REQUIRED — None.
j. INPUT CONNECTOR — Type N.
k. INSTALLATION REQUIREMENTS — The bottom of the ground plane has a 1½ inch IPS threaded coupling for use in mounting the antenna. The coupling is attached at a 45° angle to properly position the antenna for the reception of both horizontally and vertically polarized signals. Normally, the antenna is mounted on yardarms, masts, or platforms by means of a 1½ inch pipe welded to the ship's structure. The length of the mounting pipe should not be less than 22 inches and the center ground rod of each antenna should be pointed 45° from the ship's bow.
l. SPECIAL CONSIDERATIONS — The antenna assemblies should be

located so as to provide as great an unobstructed view as possible. They should not be installed in close proximity to transmitting antennas.

m. REFERENCE — Instruction Book, Antenna Assembly NT-66131, NAVSEA 900, 656.

ANTENNA NT-66132

a. TYPE/USE — The NT-66132 is a small, omnidirectional VHF stub-type fixed antenna designed for use with shipboard countermeasures receiving equipment such as the AN/WLR-1. The assembly consists of a stub radiator constructed of phenolic-impregnated maple covered with a conductive material mounted on a fan-shaped groundplane. An impedance matching section is contained in the base. To afford complete 360° coverage, two antennas are normally employed per system, one port and one starboard.

b. PHYSICAL CONFIGURATION — [See Figure 5-8].

c. DIMENSIONS — (1) Height: 38 inches; (2) Width: 67⅝ inches; (3) Depth: 38 inches; (4) Weight: 30 pounds.

d. FREQUENCY RANGE — 40 to 300 MHz.

e. INPUT IMPEDANCE/VSWR — 50 ohms/less than 5.0:1.

f. RF POWER RATING — N/A (receive only).

g. POLARIZATION — Horizontal or vertical.

h. TYPE OF FEED — RG-8 or 10/U coaxial cable.

i. PRIMARY POWER REQUIRED — None.

j. INPUT CONNECTOR — Type UG-58/U.

k. INSTALLATION REQUIREMENTS — The bottom of the groundplane is fitted with a 1½ inch IPS threaded coupling for use in mounting the antenna. The coupling is attached at a 45° angle to properly position the antenna to receive both horizontally and vertically polarized signals. The antenna should be mounted vertically and may be attached to yardarms, masts, or platforms by means of a 1½ inch pipe welded to the ship's structure. The length of the mounting pipe should not be less than 19 inches. The center ground rod of each antenna should be pointed 45° from the ship's bow.

l. SPECIAL CONSIDERATIONS — The antenna assemblies should be located so as to provide as great an unobstructed view of their sectors as possible. They should not be installed in close proximity to transmitting antennas.

m. REFERENCE — Instruction Book, Antenna Assembly NT-66132, NAVSEA 0967-896-4010.

ANTENNA AS-571/SLR

a. TYPE/USE — The AS-571 is a high-gain, rotatable, double-sided, directional UHF dipole antenna array designed for use with shipboard countermeasures receiving equipment such as the AN/WLR-1 as an aid in determining the presence, frequency, direction, and characteristics of transmitted signals. The antenna assembly consists principally of an open-rib-constructed, rectangular reflector and an array of five dipoles. One side of the reflector acts as a planar surface, while the other side is fashioned as a corner reflector. Three of the five dipoles are mounted vertically on the plane-reflector side and the other two dipoles are oriented horizontally on the opposite corner-reflector side. In this manner the antenna is receptive to both horizontally and vertically polarized energy; however, only one polarization may be selected at a time.

The entire assembly is supported on a pedestal which houses the drive mechanism and is covered by a large, type CW-456/SLR radome for protection from the sea environment. The antenna may be operated in an automatic mode to rotate continuously, or it may be manually positioned to any desired bearing.

b. PHYSICAL CONFIGURATION — [See Figure 5-9].
c. DIMENSIONS — (1) Height: 62 inches (with radome); (2) Width: $45^{11}/_{32}$ inches (diameter of radome); (3) Weight: 250 pounds.
d. FREQUENCY RANGE — 300 to 600 MHz.
e. INPUT IMPEDANCE/VSWR — 50 ohms. VSWR not greater than 4:1 for horizontal array and not greater than 3:1 for vertical array.
f. RF POWER RATING — N/A (receive only).
g. POLARIZATION — Horizontal or vertical, selectable from control unit.
h. TYPE OF FEED — Unbalanced, coaxial cable.
i. PRIMARY POWER REQUIRED — Servo power and control voltages supplied from control unit; 115 vac, 60 Hz, one-phase required for space heaters.
j. INPUT CONNECTOR — Coaxial Type N.
k. BEAMWIDTH — *Horizontal* (1) Horizontal dipoles: 50° at lowest frequency to 30° at highest frequency. (2) Vertical dipoles: 55° at lowest frequency to 30° at highest frequency. *Vertical* (1) Horizontal dipoles: 90° at lowest frequency to 60° at highest frequency. (2) Vertical dipoles: 60° at lowest frequency to 70° at highest frequency.
l. AZIMUTH SCAN — 360° mechanical.
m. SCAN RATE — 0 to 250 rpm, CW, CCW.
n. INSTALLATION REQUIREMENTS — The antenna should be located so as to provide a line-of-sight view of the horizon with a minimum of obstructions. The pedestal casting has four 13/16-inch holes on 16-inch centers for connecting the coaxial transmission line.
o. SPECIAL CONSIDERATIONS — The AS-571/SLR antenna should not be located closer than ten feet from other masts and other structures having large diameters. It should not be located in the strong fields of transmitting antennas.

p. REFERENCES — (1) Instruction Book, Antenna AS-571-SLR, NAVSEA 9067-889-2010. (2) NAVSEA Drawings RE-F2692035 and RE-C2697627.

ANTENNA AS-616/SLR

a. TYPE/USE - The AS-616 is a high-gain, rotatable, two-band, directional, UHF multifunction antenna array designed for use with shipboard countermeasures receiving equipment such as the AN/WLR-1 ESM set. The antenna array consists of four separate and independent antenna subsystems, two of which provide polarization diversity in the low portion of the UHF band, the two others are for polarization diversity in the high portion.

(1) *Lower UHF Band.* On one side of the antenna are six vertically mounted dipole elements in front of an open-ribbed, aluminum, rectangular planar reflector. The dipoles are all identical and delta-matched to form a vertically polarized array. On the opposite side are three identical horizontally oriented dipoles backed by an open-rib-constructed corner reflector. This array receives horizontally polarized energy.

(2) *Higher UHF Band.* Below the dipole arrays, on one side of the assembly are four vertically mounted waveguide horns, identical in construction and operation, to provide reception of vertically polarized signals. On the opposite side, for horizontal polarization, are four of the same type horns mounted in the horizontal plane.

The entire antenna assembly is supported on a pedestal which houses the drive system to allow operation in either a continuous rotation mode or by manual positioning. A large radome, type CW-456/SLR (identical to that used with the AS-571 antenna), protects the antenna assembly from the sea environment.

b. PHYSICAL CONFIGURATION — [See Figure 5-10].
c. DIMENSIONS — Height: 62 inches (with radome); (2) Diameter: 45 11/32 inches (radome); (3) Weight: 425 pounds (total).
d. FREQUENCY RANGE — 550 to 2600 MHz.
e. INPUT IMPEDANCE/VSWR — 50 ohms. VSWR not greater than 4.5:1 over entire frequency range.
f. RF POWER RATING — N/A (receive only).
g. POLARIZATION — Horizontal or vertical, selectable from control unit.
h. TYPE OF FEED — RG-18/U coaxial cable.
i. PRIMARY POWER REQUIRED — Servo power and control voltages supplied from control unit; 115 vac, 60 Hz, one-phase required for space heater, fan, and servo/synchro reference.
j. INPUT CONNECTORS — Type N receptacle.

k. BEAMWIDTH — (1) *Low Band: Horizontal Dipoles.* Horizontal Plane: 35° at lowest frequency to 20° at highest frequency. Vertical Plane: 85° at lowest frequency to 70° at highest frequency.
(2) *Low Band: Vertical Dipoles.* Horizontal Plane: 40° at lowest frequency to 60° at highest frequency. Vertical Plane: 60° at lowest frequency to 70° at highest frequency.
(3) *High Band: Horizontal Horns.* Horizontal Plane: 35° at lowest frequency to 20° at highest frequency. Vertical Plane: 70° at lowest frequency to 50° at highest frequency.
(4) *High Band: Vertical Horns.* Horizontal Plane: 20° at lowest frequency to 10° at highest frequency. Vertical Plane: 95° at lowest frequency to 65° at highest frequency.

l. AZIMUTH SCAN — 360° mechanical.

m. SCAN RATE — 0 to 250 rpm, CW, CCW.

n. INSTALLATION REQUIREMENTS — The antenna should be located so as to provide an omnidirectional line-of-sight view of the horizon with a minimum of obstructions. The pedestal casting has four 13/16-inch bolt holes on 16-inch centers for mounting purposes. A minimum clearance of 12 inches is required below the pedestal for connecting the coaxial transmission lines.

o. SPECIAL CONSIDERATIONS — The AS-616A/SLR should not be located closer than 10 feet from masts and other structures having large diameters. It should not be installed in strong fields of transmitting antennas.

p. REFERENCES — (1) Instruction Book AS-616/SLR Antenna, NAVSEA 0967-892-8010. (2) NAVSEA Drawing RE-C2697634.

ANTENNA AS-899F/SLR

a. TYPE/USE — The AS-899F is a rotatable, very broadband microwave direction-finding antenna designed for use with such countermeasures receiving equipment as the AN/WLR-1 where it is responsive to signals in bands 6, 7, 8, 9, and 10 (1 GHz to 20 GHz). The antenna elements consist of a single, stationary spiral feed, a multiplexer, and a rotating, shaped reflector. (Earlier models of the AS-899 employed a more complex system of four feedhorns, two mounted coaxially at the top center and two coaxially centered at the bottom, with a dual-surface parabolic reflector oriented at 45° in between. These antennas, which were limited in frequency response to bands 6, 7, 8, and 9 of the WLR-1, were subsequently modified to be equivalent to the extended range AS-899F.) The antenna is capable of being operated in an automatic, continuous rotation mode with speeds variable from 0 to 250 rpm, or manually positioned to any desired bearing. To protect it from the sea environment the antenna is enclosed within a radome.

b. PHYSICAL CONFIGURATION — [See Figure 5-11].

c. DIMENSIONS — (1) Height: 82 inches (with radome); (2) Diameter: 34 inches (radome); (3) Weight: 268 pounds (total).
d. FREQUENCY RANGE — 1 GHz to 20 GHz.
e. INPUT IMPEDANCE/VSWR — 50 ohms. VSWR not greater than 4:1 over entire frequency range.
f. RF POWER RATING — N/A (receive only).
g. POLARIZATION — Right hand circular or any sense of linear polarization.
h. TYPE OF FEED — Unbalanced, three coaxial cables and one waveguide for RF.
i. PRIMARY POWER REQUIRED — Servo power and control voltages supplied from control unit; 115 vac, 60 Hz, one-phase required for space heaters.
j. INPUT CONNECTOR — Coaxial Type N. (Waveguide WR-62 for highest band.)
k. BEAMWIDTH — 40° on lowest band to 4° on highest band.
l. AZIMUTH SCAN — 360° mechanical.
m. SCAN RATE — 300 rpm maximum, CW or CCW.
n. INSTALLATION REQUIREMENTS — The antenna assembly should be mounted as high as possible on the ship's superstructure in order to obtain maximum omnidirectional coverage. The assembly base casting has four 1-inch bolt holes on 19¾-inch centers for mounting purposes. A clearance of 12 inches is required beneath the base for cover removal.
o. SPECIAL CONSIDERATIONS — In order to reduce the possibility of receiver overload damage, the AS-899F/SLR should not be located in the strong fields of transmitting antennas.
p. REFERENCES — Technical Manual for Antenna Assembly AS-899E/SLR and AS-899F/SLR, NAVELEX 0967-550-3010.

ANTENNA AS-1023/SLR

a. TYPE/USE — The AS-1023 is a rotatable, broadband, omnidirectional UHF direction finding antenna designed for use with electronic shipboard countermeasures equipment such as the AN/WLR-1 and AN/WLR-8 to determine the presence, frequency, direction, and characteristics of transmitted RF energy. When employed with the WLR-1 or WLR-8 the single AS-1023 with its dual-band capability replaces two antennas: the AS-571 and AS-616. In order to receive both left and right hand senses of circular polarization, as well as any sense of linear polarization (i.e., omnipolarization) the AS-1023 consists of an arrangement of paired, crossed, sleeve-stub dipole elements and shaped, back-to-back reflectors. [See Figure 5-12(a)]. The dipoles, placed at right angles to each other and paired on the same horizontal axis, are fed in phase quadrature so as to be bicircularly polarized. The dipoles, moreover, are fixed; only the reflectors rotate. Each of the dipoles has two outputs which may be used independently or simultaneously by two separate receivers. The AS-1023 is capable of fully automatic operation, with continuous rotation selectable from 0 to 300 rpm, or manual positioning to any desired horizontal bearing. To protect it from the sea environment, the antenna assembly is enclosed within a fiberglass radome.

b. PHYSICAL CONFIGURATION — [See Figure 5-12].
c. DIMENSIONS — (1) Height: 60-5/16 inches (with radome); (2) Diameter: 43 inches (radome); (3) Weight: 320 pounds.
d. FREQUENCY RANGE — (1) Band 1: 300 to 600 MHz. (2) Band 2: 550 to 1100 MHz.
e. INPUT IMPEDANCE/VSWR — 50 ohms. VSWR not greater than 3:1 over entire frequency range.
f. RF POWER RATING — N/A (receive only).
g. POLARIZATION — Left or right-hand circular; any sense of linear polarization.
h. TYPE OF FEED — RG-214/U coaxial cable.
i. PRIMARY POWER REQUIRED — (1) 115 vac, 60 Hz, two-phase (servo motor). (2) 115 vac, 60 Hz, one-phase (heater).
j. GAIN — (1) *Band 1:* 5 to 10 dB. (2) *Band 2:* 7 to 13 dB.
k. INPUT CONNECTOR — Type N receptacle.
l. BEAMWIDTH: (1) *Low Band: Horizontal Dipoles.* Horizontal Plane: 65° to 40°; Vertical Plane: 75° to 45°.
   (2) *Low Band: Vertical Dipoles.* Horizontal Plane: 65° to 35°; Vertical Plane: 65° to 40°.
   (3) *High Band: Horizontal Dipoles.* Horizontal Plane: 55° to 25°; Vertical Plane: 70° to 35°.
   (4) *High Band: Vertical Dipoles.* Horizontal Plane: 45° to 30°; Vertical Plane: 70° to 35°.
m. AZIMUTH SCAN — 360° mechanical.
n. SCAN RATE — 0 to 300 rpm CW, or manually trained.
o. INSTALLATION REQUIREMENTS — The antenna assembly should be mounted as high as possible on a mast platform in such a position as to give a clear view of the horizon with as few obstructions as possible. The pedestal base has four 13/16-inch bolt holes on 16-inch centers for mounting purposes. The pedestal bench marks must be aligned with the ship's centerline. A minimum clearance of 14 inches is required beneath the pedestal for servicing.
p. SPECIAL CONSIDERATIONS — The AS-1023/SLR should be located out of the field of high power radar transmitters. It should not be located closer than 10 feet from masts and other structures having large diameters.
q. REFERENCES — (1) Technical Manual, AS-1023/SLR Antenna Assembly, NAVSEA 0967-441-1010. (2) NAVSEA Drawing RE-B2698009.

ANTENNA AS-1174/SLR

a. TYPE/USE — The AS-1174 is a shipboard omnidirectional ESM receiving antenna used with naval EW systems such as the AN/WLR-1 for the UHF through lower SHF frequencies (bands 6, 7, and 8). The antenna element consists of two strips of aluminum tape wound spirally on a fiberglass cone and covered by a foam-filled protective radome. A triplexing filter is utilized to supply the RF outputs for each of the three ECM bands. Fréquency characteristics of the antenna are determined by the apex angle of the cone. Bandwidth is determined by the rate of spiral about the cone.

b. PHYSICAL CONFIGURATION — [See Figure 5-13].

c. DIMENSIONS — Antenna: (1) Height: 41½ inches; (2) Diameter: 16½ inches; (3) Weight: 11¼ pounds. Triplexing Filter: (1) Height: 7$^{15}/_{16}$ inches; (2) Width; 10½ inches; (3) Depth: 5¼ inches; (4) Weight: 5 pounds.

d. FREQUENCY RANGE — 300 to 7350 MHz. Triplexing Filer: Band 6, 300 to 2450 MHz; Band 7,2450 to 4375 MHz; Band 8,4375 to 7350 MHz.

e. INPUT IMPEDANCE/VSWR — 50 ohms. VSWR not greater than 1.8:1 over entire frequency range.

f. RF POWER RATING — N/A (receive only).

g. POLARIZATION — Circular.

h. TYPE OF FEED — Unbalanced, coaxial cable from antenna to triplexer. Band 6, coaxial cables. Bands 7 and 8, waveguide.

i. PRIMARY POWER REQUIRED — None.

j. INPUT CONNECTOR — Type N. Waveguide adapters are supplied with triplexer.

i. INSTALLATION REQUIREMENTS — The base of the antenna has eight 10-32 size holes equally spaced on a 12-inch diameter bolt circle for mounting the antenna to a pipe support. The triplexer has five 7/32 inch holes for mounting to a flat surface. The antenna should be mounted in as clear an area as possible to give unobstructed omnidirectional coverage, with the centerline of the antenna perpendicular to the normal trim water line of the ship.

l. SPECIAL CONSIDERATIONS — The site selected for mounting the triplexer should be near the antenna, and the coaxial cable between the antenna and filter as short as possible to keep down the high frequency losses.

m. REFERENCES — (1) Technical Manual, Antenna As-1174/SLR, NAVSEA 0967-316-4010. (2) NAVSEA Drawing RE-F2692171.

ANTENNA AS-3200/SLR

a. TYPE/USE — As the AS-1174 type antennas now in the fleet deteriorate due to age or damage they are currently being replaced by the newer model AS-3200/SLR cone antennas. Nearly identical in outward appearance, the AS-3200 differs from the AS-1174 only in that it uses four spirally wound copper bands and is larger in size, being 45 inches high, with a base diameter of 30.3 inches, and weighing 34 pounds.

5-1.6 AN/WLR-8(V)4

Representative of the very latest technology in naval ESM is the AN/WLR-8(V)4, a very wideband shipboard EW system designed to replace the AN/WLR-1 equipment. To be used for rapid interception, analysis, processing, and display of signal information, the WLR-8(V)4 has the additional feature of completely automatic threat/early warning capability in designated priority bands.

At the time of this writing the Navy has only one WLR-8(V)4 installed aboard ship, undergoing extensive system evaluation. Current plans are to use the WLR-1 antenna suite with the WLR-8(V)4; e.g., the AS-899 and AS-1023 combination with AS-1174 as described above. A new lightweight antenna complement expressly designed for the WLR-8(V)4 will be procured in the future.

5-1.7 AN/WLR-11A

The AN/WLR-11A is a wideband microwave countermeasure receiving system used aboard ship for instantaneous frequency measurement and analysis of selected radar signals in the 7 to 18 GHz range. As an inherently wide-open device, the full RF band of the set is monitored on an instantaneous basis rather than the conventional scanning method for radar emission detection. Both aural and visual indicators are supplied to the operator.

The WLR-11A can be operated in conjunction with the WLR-1 to provide acquisition of bands 9 and 10, or it can be interconnected with the SLR-12 for added frequency selection capability. In either case, an omnidirectional AS-2918 antenna with integral preamplifier (AM-6799) is employed with the system, and is mast-mounted as clear as possible to allow maximum unblocked reception.

ANTENNA AS-2918A/S

a. TYPE/USE — The AS-2918A/S is a fixed, broadband, omnidirectional and omnipolarized SHF receive antenna designed for use with the ESM systems such as the AN/WLR-11A. Protected by a fiberglass, cylindrically shaped housing, the antenna is composed of three subassemblies: a biconical horn, a polarizer, and a diplexer.

BICONICAL HORN: The biconical horn consists of two metallic cones set apex to apex, but separated from each other by a small air gap. An RF connector and a section of coaxial transmission line, installed in the lower cone, complete the circuit and allow an RF field to be set up between the cones. The region between the two cones, then, comprises the radiation aperture of the antenna. Due to the configuration and symmetry of the two cones, the biconical horn possesses a peripheral view angle of 360 degrees. In elevation the radiation pattern has a beamwidth of 25 degrees or greater. This device is inherently broadband and operates across the frequency range of 7 to 18 GHz.

POLARIZER: The polarizer is fabricated in the shape of a hollow cylinder. In the antenna assembly, it is concentric with and encloses the biconical horn, and is composed of alternate layers of grid-wire fabric and core material. The function of this structure is to transform in successive steps the vertical-linear field polarization from the biconical horn into slant 45 degrees polarization. However, the antenna is used for receive-mode only; hence, all incident RF signal field polarizations (vertical-linear, horizontal-linear, right-hand circular and left-hand circular) are transformed into vertical linear polarization prior to their passing into the biconical horn. Since the polarizer is nearly transparent to RF, an RF signal of the proper polarization (slant 45 degrees) passes through unattenuated.

DIPLEXER: The diplexer consists of a high bandpass filter and a low bandpass filter connected together at a common junction. This device has one common input, one low-pass output, and one high-pass output. In the antenna assembly, the common port is coupled to the RF connector at the base of the biconical horn. Thus, RF signal information from the biconical horn passes into the common port, then is routed to the appropriate output circuit, based on the signal RF frequency, That is, RF signals within the frequency range 7 to 18 GHz are separated into two bands: 7 to 11 GHz (loss-pass) and 11 to 18 GHz (high-pass). These signals are then presented to the diplexer output terminals for further processing.

b. PHYSICAL CONFIGURATION — [See Figure 5-14].
c. DIMENSIONS — (1) Height: 24 inches (with preamp); (2) Width: 10½ inches (with preamp); (3) Depth: 7⅝ inches (with preamp - diameter of antenna alone is 6.3 inches); (4) Weight: 32 pounds (total with preamp).
d. FREQUENCY RANGE — 7 to 11 GHz and 11 to 18 GHz.
e. INPUT IMPEDANCE/VSWR — 50 ohms/3:1 nominal.
f. RF POWER RATING — N/A (receive only).
g. POLARIZATION — Linear, slant 45 degrees.
h. TRANSMISSION LINE TYPE — RF coaxial cable.
i. PRIMARY POWER REQUIRED — 115 vac, 60 Hz, one-phase, and +28 vdc are supplied from RF unit.
j. INPUT CONNECTOR — Type N.
k. INSTALLATION REQUIREMENTS — The enclosure is fitted with two mounting brackets, each with three ¼-inch bolt holes on 6-inch centers. The unit should be mounted vertically as high as possible, in a

clear area providing maximum omnidirectional coverage. It should not be located in high power radar or communications transmitting fields.

l. SPECIAL CONSIDERATIONS — The enclosure has front access for adjustment and maintenance. The unit should be located in an area where it can be adequately maintained.

m. REFERENCES — (1) Technical Manual, Countermeasures Receiving Set AN/WLR-11A, NAVELEX 0967-LP-544-9010. (2) Technical Manual, Omnidirectional Antenna AS-2918A/S, NAVELEX 0967-LP-550-9010.

5-2 ELECTRONIC COUNTERMEASURES (ECM) ANTENNAS

To this point only passive EW systems, i.e., ESM antennas, have been addressed. Active electronic warfare in the form of ECM plays a major role as well in the combat system effectiveness of modern naval forces. The requirement for ECM emerged dramatically during World War II with the urgent need to jam the then somewhat limited frequency bands of enemy radars. The Navy built the 500-600 MHz TDY-1 spot jammer for use against German shore defense radar, and, jointly with the Radio Research Lab, under Project Elephant went on to develop the S-9000 integrated microwave receiver and jamming system. This equipment, operating at 600 watts over a 2 to 4 GHz band, is considered the forerunner of all modern-day shipboard ECM systems.[13] From that point on, in an effort to deny or disrupt electronic threats at will, there has been a steady upgrading of techniques used to generate and transmit interfering electromagnetic energy. Methods of electronic countermeasures may include various type of jamming, or might involve electronic means of deception. As such, sophisticated ECM equipment now is used to automatically (1) search for signals of interest, (2) stop and hold onto an intercepted signal, (3) analyze characteristics of the intercepted signal to decide whether it is a threat and what mode of action to take, (4) remain locked on and commence jamming, (5) periodically sense ("look-through") to determine whether the intercepted emissions have ceased, and (6) if ceased, begin searching for new signals.

Further ECM improvements in the 1980s must increase the rapidity of operations and capability to jam on several separate frequencies simultaneously, as well as increase reliability and the level of system confidence. At the same time more complex means of confusing and deceiving the adversary in terms of range and angle to the intended victim must be adapted. Instantaneous reaction time and precise accuracy are essential, for example, to successfully counter a bevy of extremely high speed sea-skimming anti-ship missiles closing in from several directions together. As might be surmised, such electronic wizardry makes use of rather impressive antenna systems. Detailed information on these antennas is most often closely guarded for long periods of time. However, that which may be divulged in naval shipboard ECM antennas of an unclassified nature is given below.

5-2.1 AN/SLA-12

The AN/SLA-12 is a large, complex EW antenna group used by the Navy to provide both passive DF and the active emitter functions for such shipboard ECM systems as the AN/ULQ-6 and the variants AN/SLQ-22, 23, and 24. The SLA-12 consists of three major units, two of which are the control indicator subsystems located below deck. The third, and principal, unit is the AS-1751A/SLA-12, a group of trainable antennas installed topside. Complementing the AS-1751A is a separate system of two fixed antennas, the AS-1750A/SL, mounted one above and one below the trainable AS-1751A. Two complete SLA-12 systems are installed per ship, one port and one starboard (usually outboard) in a manner typified by Figures 5-15(b) and 5-16(b).

ANTENNA AS-1750A/SL

a. TYPE/USE — The AS-1750 is a shipboard system of two, fixed, broadbeam omnipolarized microwave ECM antennas, one used for receiving and one for transmitting. In either case each antenna consists simply of a conical horn protected from the sea environment by a radome cover. In the shipboard configuration the receive AS-1750 is placed above the AS-1751/SLA-12 while the transmit AS-1750 is mounted beneath. Each system of two AS-1750s covers 180° in azimuth; therfore, to provide a total 360° of complementary coverage, two separate systems are employed, one on the port side and one on the starboard.
b. PHYSICAL CONFIGURATION — [See Figure 5-15].
c. DIMENSIONS — (1) Height: 12½ inches; (2) Width: 21 inches; (3) Depth: 17¼ inches; (4) Weight: 16¼ pounds.
d. FREQUENCY RANGE — [Classified].
e. INPUT IMPEDANCE/VSWR — 50 ohms. VSWR not greater than 2:1 over frequency range.
f. RF POWER RATING — 25 kilowatt peak, 1000 watt average.
g. POLARIZATION: Circular, either sense.

h. TYPE OF FEED — Waveguide.
i. PRIMARY POWER REQUIRED — 115 vac, 60 Hz and 400 Hz, one-phase.
j. INPUT CONNECTOR — Waveguide.
k. BEAMWIDTH — (1) Horizontal 155°; (2) Vertical 17°.
l. INSTALLATION REQUIREMENTS — The mounting flange has four 23/32-inch bolt holes on 9-inch vertical and 14-inch horizontal centers for use in mounting the antenna. The AS-1750/SL antennas are mounted above and below the AS-1751A/SLA-12 trainable antenna, as far outboard as possible to obtain maximum unobstructed coverage. Generally, the three antennas are mounted on a common vertical structure. The area in the vicinity of the antennas should be clear of reflecting surfaces. In some instances microwave absorbent material has been installed on the mounting structure to reduce reflections.
m. SPECIAL CONSIDERATIONS — The antenna and its associated waveguide require pressurization, not to exceed 2 psig, from a source of dry air. The length of waveguide from the antenna to the transmitting and receiving equipment must not exceed 25 feet.
n. REFERENCES — (1) Technical Manual, Antenna Group AN/SLA-12A(U) and Antenna AS-1750A/U, NAVSEA 0967-031-7000. (2) NAVSEA Drawing RE-D2697956.

ANTENNA AS-1751A/SLA-12

a. TYPE/USE — The AS/1751 is a shipboard system of two, trainable, microwave ECM antennas separated vertically by a mid-section pedestal. Connected together by a common gear shaft assembly, the two antennas rotate simultaneously in parallel horizontal planes. The upper antenna consists of a conical ECM receive horn in the center, with separate skewed DF horns located on either side of the center horn. The lower antenna is an ECM transmit horn. Depending on how the particular SLA-12 system is equipped, the transmit antenna may be either a high gain, narrow beam horn, or low gain, wide beam.

   Even though the AS-1751 antennas are capable of being rotated fully through 360°, normal ship environment restricts coverage to 180° so that two complete AS-1751 systems are installed, one port and one starboard. The antennas are sheltered from the sea environment by radome covers.
b. PHYSICAL CONFIGURATION — [See Figure 5-16].
c. DIMENSIONS — (1) Height: 46½ inches; (2) Width: 33 inches; (3) Depth: 33 inches; (4) Weight: 193 pounds.
d. FREQUENCY RANGE — [Classified].
e. INPUT IMPEDANCE/VSWR — 50 ohms. VSWR of any horn does not exceed 2:1 over frequency range.

f. RF POWER RATING — (1) Transmit Paths: 25 kW, 1 kW average. (2) Receive Paths: 1 kW peak, 10 watts average.
g. POLARIZATION — Either right or left circular may be switched from either control indicator. Antenna polarizers are interlocked making cross polarization impossible.
h. TYPE OF FEED — Waveguide.
i. PRIMARY POWER REQUIRED — 115 vac, 60 Hz and 400 Hz, one-phase.
j. INPUT CONNECTOR — Waveguide.
k. BEAMWIDTH — (1) Wide beam horn; not less than 17° at half power point; (2) Narrow beam horn: not less than 8° at half power point.
l. INSTALLATION REQUIREMENTS — The pedestal has two mounting bosses, each with four 11/32 inch bolt holes on two-inch vertical centers and ten-inch horizontal centers for use in mounting the assembly. The AS-1751A/SLA-12 is usually mounted between the AS-1750A/SL antennas on a common vertical structure, and the assembly should be installed as far outboard as possible to obtain maximum unobstructed coverage, with the area in the vicinity of the antennas clear of reflecting surfaces. The swing circle for this antenna is 16⅞ inches.
m. SPECIAL CONSIDERATIONS — The antenna assembly and its associated waveguides require pressurization not to exceed 2 psig from a source of dry air. The lengths of all waveguides used with this assembly should not exceed 25 feet.
n. REFERENCES — (1) Technical Manual, Antenna Group AN/SLA-12A(U) and Antenna AS-1750A/SL, NAVSEA 0967-031-7000. (2) NAVSEA Drawing RE-D2697959.

5-2.2 AN/SLA-15

The AN/SLA-15 is an EW antenna system designed for use as an operator controlled tracker for such naval shipboard ECM equipment as the AN/ULQ-6. Of the several component parts making up the SLA-15 system, the major topside unit is the AS-2431/SLA-15 trainable antenna subassembly. Mounted above and below the AS-2431 are two of the AS-2430 type fixed auxiliary antennas. For complete coverage, two AN/SLA-15 systems with associated AS-2430s are installed per ship, one port and one starboard, in a manner typified by Figures 5-17(b) and 5-18(b).

ANTENNA AS-2430/SL

a. TYPE/USE — The AS-2430 is a shipboard system of two, fixed, broadbeam, circularly polarized microwave ECM antennas, one used for receiving and one for transmitting. Each antenna consists of a sectoral horn assembly and polarization twister covered by a protective shell. Heater bands are incorporated to prevent icing. In the shipboard configuration the receive AS-2430 is placed above the AS-2431/SLA-15 and the transmit AS-2430 is mounted beneath. Each system of the AS-2430s covers 180° in azimuth; therefore, to provide a total 360° of complementary coverage, two separate systems are employed, one on the port side and one on the starboard.

b. PHYSICAL CONFIGURATION — [See Figure 5-17].

c. DIMENSIONS — (1) Height: 15 9/16 inches; (2) Width: 33 inches; (3) Depth: 14¾ inches; (4) Weight: 38 pounds.

d. FREQUENCY RANGE — [Classified].

e. INPUT IMPEDANCE/VSWR — 50 ohms. VSWR not greater than 1.15:1 over frequency range.

f. RF POWER RATING — 25 kW peak, 1000 watts average.

g. POLARIZATION — Circular, either sense.

h. TYPE OF FEED — Waveguide.

i. PRIMARY POWER REQUIRED — 115 vac, 60 Hz, one-phase for heaters.

j. INPUT CONNECTOR — Waveguide.

k. BEAMWIDTH — (1) Horizontal 155°; (2) Vertical 17°.

l. INSTALLATION REQUIREMENTS — The mounting flange has four 13/32-inch bolt holes on 9-inch vertical and 14-inch horizontal centers for use in mounting the antenna. The AS-2430/SL antennas are mounted above and below the AS-2431/SLA-15 trainable antenna, as far outboard as possible to obtain maximum unobstructed coverage. Generally, the three antennas are mounted on a common vertical structure. The area in the vicinity of the antennas should be clear of reflecting surfaces.

m. SPECIAL CONSIDERATIONS — The antenna and its associated waveguide require pressurization, not to exceed 2 psig, from a source of dry air. The length of the waveguide from the antennas to the transmitting and receiving equipment must not exceed 25 feet. The minimum allowable distance between the horizontal centerlines of a pair of fixed antennas is 10 feet.

n. REFERENCES — (1) Technical Manual, Antenna Group AN/SLA-15 and AS-2430/SL(U), NAVSEA 0967-312-5010. (2) NAVSEA Drawing RE-D2697296

ANTENNA AS-2431/SLA-15

a. TYPE/USE — The AS-2431 is a shipboard system of two, trainable, microwave ECM antennas separated vertically by a mid-section pedestal. Connected together by a common gear shaft assembly, the two antennas rotate simultaneously in parallel horizontal planes at zero degrees elevation. The upper antenna consists of a conical ECM receive horn in the center, with separate skewed DF horns located on either side of the center horn. The lower antenna is an ECM transmit horn. Depending on how the particular SLA-15 system is equipped, the transmit antenna may be either a high gain, narrowbeam horn, or low gain, widebeam. The design is such that conversion from one type to the other is easily made. To provide sheltering from the sea environment, the antennas are covered.

b. PHYSICAL CONFIGURATION — [See Figure 5-18].
c. DIMENSIONS — (1) Height: 42 inches; (2) Width: 36 inches; (3) Depth: 36 inches; (4) Weight: 245 pounds.
d. FREQUENCY RANGE — [Classified].
e. INPUT IMPEDANCE/VSWR — 50 ohms. VSWR of any horn does not exceed 2:1 over frequency range.
f. RF POWER RATING — (1) Transmit Paths: 25 kW peak, 1 kW average. (2) Receive Paths: 1 kW peak, 10 watts average.
g. POLARIZATION — Either right or left circular; may be switched from either control indicator. Antenna polarizers are interlocked, making cross polarization impossible.
h. TYPE OF FEED — Waveguide.
i. PRIMARY POWER REQUIRED — Control voltages supplied from associated control indicator through junction box. Antenna space heater voltage is supplied from junction box.
j. INPUT CONNECTOR — Waveguide.
k. BEAMWIDTH — (1) Wide beam horn: not less than 18° at half power points; (2) Narrow beam horn: not less than 9° at half-power points.
l. INSTALLATION REQUIREMENTS — The pedestal has an upper and lower mounting boss. Each boss has four ⅜-inch bolt holes on 1½-inch vertical centers and 13¾-inch horizontal centers for use in mounting the assembly. The AS-2431/SLA-15 is usually mounted between the AS-2430/SL antennas on a common vertical structure. The assembly should be installed as far outboard as possible to obtain maximum unobstructed coverage. The area in the vicinity of the antennas should be clear of reflecting surfaces. The swing circle radius for this antenna is 18 inches.
m. SPECIAL CONSIDERATIONS — The antenna assembly and its associated waveguides require pressurization not to exceed 2 psig from a source of dry air. The lengths of all waveguides used with this assembly should not exceed 25 feet.
n. REFERENCES — (1) Technical Manual, Antenna Group AN/SLA-15(U), NAVSEA 0967-312-5010. (2) NAVSEA Drawing RE-E2697293

5-2.3 AN/SLA-18

Antenna group AN/SLA-18 is a supplemental antenna system used with shipboard ECM transmitting and receiving equipment. Basically the SLA-18 consists of two AS-2430/SL antennas (described immediately above in Section 5-2.2) with the necessary control and distribution ancillary equipment. For further details refer to Technical Manual NAVSEA 0967-319-9010.

5-2.4 AN/SLQ-17

The AN/SLQ-17 is a new technology naval shipboard ECM equipment designed specifically for aircraft carrier application as a replacement for the AN/ULQ-6 family. An automated, independent EW system, the SLQ-17 is employed to enhance ship defense against missiles and other weapons of similar RF characteristics. It is capable of active ECM and passive ESM simultaneously for detecting, tracking, identification, and recording of emission threats even in the presence of a dense RF environment. In addition, directional information on the threat is provided to coordinate defensive actions. Due to its highly classified nature, little information of the SLQ-17 system can be revealed here except for a few general details of the large antennas used.

ANTENNA AS-2963/SLQ-17A(V)2

a. TYPE/USE — The AS-2963 is a multibeam, electronically steerable, phased array microwave antenna employed with the AN/SLQ-17 shipboard ECM system. The antenna is roll-stabilized and consists of several flared horns and circular slotted waveguide DF arrays to provide simultaneous ESM receive and ECM transmission. The entire assembly, with ancillary phase shifters, drivers, and electronic circuitry, is housed in a large protective aluminum box structure having radome windows for microwave reception and radiation. To allow full 360° complementary coverage, two antennas are installed per ship, one on the starboard side and one on the port. A pictorial cutaway (artist's simulation) of the antenna array is shown in Figure 5-19.
b. PHYSICAL CONFIGURATION — [See Figures 5-19 and 5-20].
c. DIMENSIONS — (1) Height: 68 inches; (2) Width: 71¼ inches; (3) Depth: 50 inches; (4) 1200 pounds.
d. FREQUENCY RANGE — [Classified].
e. RF POWER RATING — [Classified].
f. POLARIZATION — [Classified].
g. TYPE OF FEED — Waveguide.
h. PRIMARY POWER REQUIRED — (1) 440 vac, 60 Hz, three-phase. (2) 115 vac, 400 Hz, three-phase.
i. BEAMWIDTH — [Classified].
j. INSTALLATION REQUIREMENTS — [See NAVELEX Installation Manual FS79-12-64].
k. REFERENCES — (1) Operational Logistic Support Summary for Countermeasures Set AN/SLQ-17A(V)2, NAVELEX P4110.223. (2)

Installation Manual for Countermeasures Set AN/SLQ/17A(V)2, NAVELEX FS79-12-64.

5-2.5 AN/SLQ-22, AN/SLQ-23, and AN/SLQ-24

These shipboard ECM systems are all high gain versions of the AN/ULQ-6 and use the same antenna assemblies (i.e., SLA-12 or SLA-15 described above).

5-2.6 AN/SLQ-26

The AN/SLQ-26 is an ECM system designed for naval ships for use as part of the Anti-Ship Missile Defense (ASMD) improvement program. It provides detection, real time analysis, and automatic generation of response against missile threat emissions. The SLQ-26 utilizes the full complement of ECM antennas currently employed by the AN/ULQ-6 family of equipment; i.e., the AS-2854/SLD-1A, the AN/SLA-12 or AN/SLA-15, and the AS-1750A/SL or AS-2430/SL suites described in detail in the preceding sections of this chapter.

5-2.7 AN/SLQ-32

The AN/SLQ-32, representing a new generation of shipboard electronic warfare equipment, is the Navy's first multibeam, electronically steerable ECM system. Known originally, during the early stages of development, as the "design to price" EW System (DTPEWS) because its paramount specification was cost, the SLQ-32 is to become the standard EW system for most Navy ships, with over 300 installations currently planned. As a computer-controlled set, it has been designed especially for rapid detection, analysis, and identification of highly sophisticated, state of the art missiles and radar threats that use RF emissions over a very broad frequency range. By providing early warning of the presence of hostile emitters, the SLQ-32 complements other surveillance sensors and weapon systems for deep defense of the ship. Moreover, it provides for either automatic or manual launching of chaff decoys (an integral part of the SLQ-32 system) as an aid in altering the courses of enemy missiles, thus causing them to miss the intended victim.

To afford varying levels of EW capability for different classes of naval ships, three separate, modularized versions of the SLQ-32 are available. The *(V)1* and *(V)2* variants are passive, non-jamming ESM systems used for surveillance and alerting of threats. The *(V)1* is intended for such vessels as smaller-class auxiliaries and amphibious, and Knox-class frigates, while the *(V)2*, with an expanded ESM capability, will be installed on guided-missile destroyers and frigates, and the DD-963 class. The *(V)3* system incorporates active, directional jamming at high effective power to provide an EW shield for cruisers, large auxiliaries, and large amphibious ships. (Aircraft carriers are to have the new AN/WLR-8 and AN/SLQ-17 combination rather than AN/SLQ-32s.) It is important to note that the modular, building-block design allows either the *(V)1* or *(V)2* suite to be upgraded to a *(V)3* when desired. In its full capacity the SLQ-32 detects the presence of emitters in three frequency regions designated as Bands 1, 2, and 3. Dedicated antennas and processing units for each of the bands are used to receive and analyze RF signal parameters to determine relative angle, frequency, and amplitude, thereby classifying and identifying the level of threat so that the operator may appraise the tactical situation and respond appropriately.

The SLQ-32(V)1 model is employed to detect RF signals at all azimuths on all Band 3 frequencies used by enemy anti-ship missiles and associated weapons-control radars. As such, the system's receivers are wide open in angle and frequency to alert the ship of hostile threats and to provide control for the launching of chaff. The *(V)1* antenna suite consists of two, large, box assemblies, one port and one starboard, each containing two Band 3 direction-finding lens arrays covering 90° apiece (i.e., 180° total per assembly). In addition, a Band 3 semi-omnidirectional 180° antenna is centered in each of the two box assemblies. This semi-omnidirectional antenna element senses the dominant pulse in a spectrum of intercepted energy, feeding it to an instantaneous frequency-measuring receiver to ascertain the frequency of the received signal. Identification is completed by comparing the observed signal characteristics with parameters stored in a library of possible enemy and friendly emitters contained within the computer memory. A pictorial illustration of the *(V)1* assembly is shown in Figure 5-21(a).

The SLQ-32(V)2 has, with the addition of Band 1 and 2 receiving subsystems, an increased ESM capability. As seen in Figure 5-21(b), this is accomplished through the addition of two more DF lens arrays and another semi-omnidirectional element per assembly for Band 2; and four, small, Band 1 spiral antennas on the mast (e.g., at the end of a yardarm), two each port and starboard; for a combined 360 degrees coverage.

The SLQ-32(V)3 has the same Band 1, 2, and 3 ESM capabilities of the *(V)2*, but, as an ECM system, has added two Band 3 transmit antennas located in the lower portion of each box assembly. This is represented pictorially in Figure 5-21(c). The ECM mode of the *(V)3* may be operated semi-automatically, with the operator initiating jamming on a given emitter, or automatically, under computer control, where the system itself takes action as soon as a threat is identified. In this case the computer selects the most appropriate jamming technique to counter a particular threat, employing either simple repeater methods or barrage jamming, with the capability of engaging several hostile emitters simultaneously.

All three variants of the SLQ-32 use multiple-beam arrays (except for the Band 1 mast-mounted spiral antennas) for ESM receiving. These dielectric lens-fed arrays are considered the heart of the SLQ-32 system. The array of elements is fed by multiple beam, parallel-plate lenses constructed in stripline form using printed circuit technology. The resultant fan pattern is a set of individual, contiguous, high-gain lobes, all existing simultaneously, with each beam possessing the full gain of the array aperture. For any single array, more than one octave in frequency is covered.

ANTENNA AS-3316/SLQ-32 (BAND 1)

a. TYPE/USE — The AS-3316 is a small, mast-mounted planar-spiral ECM antenna designed for use in Band 1 of the AN/SLQ-32 shipboard EW system. Installed in identical pairs, normally port and starboard on the end of a yardarm, each antenna has a 90° quadrant pattern so that a pair covers 180° and the total combination yields 360° azimuthal coverage.
b. PHYSICAL CONFIGURATION — [See Figure 5-22].
c. DIMENSIONS — (1) Height: 18 inches; (2) Width: 25 inches; (3) Depth: 13 inches; (4) Weight: 15 pounds.
d. FREQUENCY RANGE — [Classified].
e. IMPUT IMPEDANCE/VSWR — 50 ohms with VSWR 2.6:1 maximum.
f. RF POWER RATING — Not applicable, receive only.
g. POLARIZATION — Right-hand circular.
h. TYPE OF FEED — Coaxial cable.
i. INPUT CONNECTOR — Type N.
j. INSTALLATION REQUIREMENTS — The Band 1 antennas are assembled in pairs on brackets that bolt to mounting surfaces welded on the outboard port and starboard ends of the yardarm. Mounting hardware includes three ¼-20 UNC cap screws, flat washers, and lock washers through three 0.286-0.295 diameter mounting holes as shown

in NAVELEX EE150-CV-MMO-080, Volume VIII, Chapter 8, Installation Manual, page 8-36.

k. REFERENCE — Technical Manual, Countermeasures Set, AN/SLQ-32(V)2, NAVELEX EE 125-BV-MMO-010.

ANTENNA AN/SLQ-32 (BAND 2 DF ARRAY)

a. TYPE/USE — The SLQ-32 Band 2 direction finding array is a fixed microwave ESM antenna used as an integral part of the AN/SLQ-32 shipboard EW system. Installed in pairs on the box-assembly sidefaces to cover 180 degrees, the antenna receives free-space RF emissions within its designated frequency range. Each Band 2 DF antenna consists of a linear array of 38 vertical (E-plane) sectoral horn elements connected to a constant beamwidth lens through phase-matched coaxial cables. The constant beamwidth lens focuses energy from the 38 horn inputs into nine identical RF modules for receiver processing. Figure 5-23 is an illustration of the antenna array. For protection from the sea environment the array is covered by a rectangular radome housing.

b. PHYSICAL CONFIGURATION — [See Figures 5-21(b) and (c) and Figure 5-23].

c. DIMENSIONS — (1) Overall dimensions of the total box enclosure, of which the Band 2 DF arrays are an integral part, are: 42 inches high x 94 inches wide x 82 inches deep for the *(V)2* system; and 70 inches high x 116 inches wide x 90 inches deep for the *(V)3* system. (2) Weight: 122 pounds each (total *(V)2* antenna assembly is 1200 pounds each; total *(V)3* antenna assembly is 2642 pounds each).

d. FREQUENCY RANGE — [Classified].

e. INPUT IMPEDANCE/VSWR — 50 ohms with VSWR 3.0:1 maximum.

f. RF POWER RATING — Not applicable, receive only.

g. POLARIZATION — Right-hand circular.

h. TYPE OF FEED — Waveguide.

i. INSTALLATION REQUIREMENTS — Mounted integrally with overall *(V)2* or *(V)3* assemblies.

j. REFERENCES — (1) Technical Manual, Countermeasures Set, AN/SLQ-32(V)2, NAVELEX EE 125-BV-MMO-010. (2) Technical Manual, Countermeasures Set, AN/SLQ-32(V)3, NAVELEX EE 150-CV-MMO-010.

ANTENNA AS-3317/SLQ-32 (BAND 2 SEMI-OMNI)

a. TYPE /USE — The AS-3317 is a fixed, directional, microwave ESM antenna used as an integral part of the AN/SLQ-32 shipboard EW system for determining frequency information received in Band 2. It does not provide angular information. One each AS-3317 covering 180° is installed per side of the ship to result in a full 360° field of view.

A curved radome housing protects the antenna from the sea environment.

b. PHYSICAL CONFIGURATION — [See Figure 5-21(b) and (c)].

c. DIMENSIONS — (1) Overall dimensions of the total box enclosure, of which the AS-3317 Band 2 semi-omni antenna is an integral part are: 42 inches high x 94 inches wide x 82 inches deep for the *(V)2* system, and 70 inches high x 116 inches wide x 90 inches deep for the *(V)3* system. (2) Weight: Total *(V)2* antenna assembly is 1200 pounds each; total *(V)3* antenna assembly is 2642 pounds each.

d. FREQUENCY RANGE — [Classified].

e. INPUT IMPEDANCE/VSWR — 50 ohms with VSWR 2.5:1 maximum.

f. RF POWER RATING — Not applicable, receive only.

g. POLARIZATION — Right-hand circular.

h. TYPE OF FEED — Waveguide.

i. INSTALLATION REQUIREMENTS — Mounted integrally with overall *(V)2* and *(V)3* assemblies.

j. REFERENCES — (1) Technical Manual, Countermeasures Set, AN/SLQ-32(V)2, NAVELEX EE 125-BV-MMO-010. (2) Technical Manual, Countermeasures Set, AN/SLQ-32(V)3, NAVELEX EE 150-CV-MMO-010.

ANTENNA AN/SLQ-32 (BAND 3 DF ARRAY)

a. TYPE /USE — The SLQ-32 Band 3 direction finding array is a fixed, microwave ESM antenna used as an integral part of the AN/SLQ-32 shipboard EW system. Installed in pairs on the box-assembly side faces to cover 180 degrees, the antenna receives free-space RF emissions within its designated frequency range. Each Band 3 DF antenna consists of a linear array of 66 sectoral horn elements connected to a constant beamwidth lens through phase-matched coaxial cables. The constant beamwidth lens focuses energy from the 66 horn inputs into 17 identical RF modules for receiver processing. Figure 5-24 is an illustration of the antenna array. For protection from the sea environment the array is covered by a rectangular radome housing.

b. PHYSICAL CONFIGURATION — [See Figure 5-21 and 5-24].

c. DIMENSIONS — (1) Overall dimensions of the total box enclosure, of which the Band 3 DF arrays are an integral part, are: 38½ inches high x 94 inches wide x 55 inches deep for the *(V)1* system; 42 x 94 x 82 inches for the *(V)2* system; and 70 x 116 x 90 inches for the *(V)3*. (2) Weight: 82 pounds (total *(V)1* antenna assembly is 780 pounds each; *(V)2* is 1200 pounds each; and *(V)3* is 2642 pounds each).

d. FREQUENCY RANGE — [Classified].

e. INPUT IMPEDANCE/VSWR — 50 ohms with VSWR 3.0:1 maximum.

f. RF POWER RATING — Not applicable, receive only.

g. POLARIZATION — Right-hand circular.

h. TYPE OF FEED — Waveguide.
i. INSTALLATION REQUIREMENTS — Mounted integrally with overall *(V)2* or *(V)3* assemblies.
j. REFERENCES — (1) Technical Manual, Countermeasures Set, AN/SLQ-32(V)2, NAVELEX EE 125-BV-MMO-010. (2) Technical Manual, Countermeasures Set, AN/SLQ-32(V)3, NAVELEX EE 150-CV-MMO-010.

ANTENNA AS-3318/SLQ-32 (BAND 3 SEMI-OMNI)

a. TYPE/USE — The AS-3318 is functionally identical to the AS-3317 Band 2 semi-omni described above. For technical characteristics and references, refer to the AS-3317.

ANTENNA AN/SLQ-32(V)3 (BAND 3 TRANSMIT)

a. TYPE/USE — The SLQ-32(V)3 Band 3 transmit array is a fixed microwave ECM emitter used as an integral part of the AN/SLQ-32(V)3 shipboard EW system. Installed in pairs on the box-assembly side-faces (each antenna array positioned with its axis aligned 45° off the ship's broadside for 90° quadrant field of view) the total coverage is 180° per side of ship. Operating in Band 3 only, each antenna consists of a beam forming lens-feed and a linear phased array of 35 sectoral horn radiating elements. Commands to the transmitter provide selections of any one of the 32 possible beam directions per array (64 active beams per 180° azimuth coverage).

   Figure 5-25 is a pictorial representation of the horn array complete with its companion transmitter unit. The array is installed in the bottom position of the SLQ-32(V)3 box enclosure (see Figure 5-26), and is covered by a rectangular radome to protect it from the sea environment. To compensate for ship's motion while maintaining the RF radiation beam at an elevation of 6° above horizontal, the SLQ-32(V)3 antenna assembly is electro-hydraulically roll-stabilized.
b. PHYSICAL CONFIGURATION — [See Figure 5-21(c), 5-25, and 5-26].
c. DIMENSIONS — (1) Overall dimensions of the total SLQ-32(V)3 box enclosure of which the Band 3 transmit arrays are an integral part,

are: 70 inches high x 116 inches wide x 90 inches deep. (2) Weight: 250 pounds each (including transmitter).

d. FREQUENCY RANGE — [Classified].
e. GAIN — [Classified].
f. INPUT IMPEDANCE/VSWR - 50 ohms with VSWR 2.0:1.
g. RF POWER RATING — [Classified].
h. POLARIZATION — Horizontal.
i. TYPE OF FEED — Waveguide.
j. INSTALLATION REQUIREMENTS — Mounted integrally with overall *(V)3* antenna assemblies.
k. REFERENCE — Technical Manual, Countermeasures Set AN/SLQ-32(V)3, NAVELEX EE-150-CV-MMO-010.

5-2.8 AN/SLQ-34

At the time of this writing the Navy has under development a new-generation EW equipment, the AN/SLQ-34(V)2 system, for early warning and threat countermeasures. The system is highly classified and detailed characteristics cannot be divulged here. It is expected that two small antennas will be used, one for receive and one for transmit, weighing about 30 pounds each, and having dimensions on the order of 20 inches high, 15 inches wide, and 15 inches deep. The antennas most likely will be mast-mounted and have a waveguide length limitation of about 150 feet.

5-2.9 AN/ULQ-6

The AN/ULQ-6 is a naval shipboard family of ECM equipment operating as a decoy repeater and radar deception jammer at microwave frequencies to provide protection against hostile missiles and weapons radars. Over the past twenty years several models of the ULQ-6 have seen widespread service in the fleet. Modern versions employ the AS-1750/1751 or AS-2430/2431 suites of antennas described in Sections 5-2.1 and 5-2.2 above. The ULQ-6 system is now being replaced by the new AN/SLQ-17 equipment.*

---

* References for the AN/ULQ-6 — (1) Technical Manual AN/ULQ-6, NAVSEA 0967-056-7010; (2) Technical Manual AN/ULQ-6A, NAVSEA 0967-071-6012. (3) Technical Manual AN/ULQ-6B, NAVSEA 0967-050-9010; (4) Technical Manual AN/ULQ-6C, NAVSEA 0967-311-3010.

REFERENCES

1. Robert J. Schlesinger, *Principles of Electronic Warfare* (Los Altos, California: Peninsula Publishing, 1979), p. 2.
2. *Ibid.,* p. 1.
3. Rodney F. Simons, "History and Principles of Electronic Warfare," *Frequency Technology,* November 1969, p. 28.
4. Louis A. Gebhard, *Evolution of Naval Radio-Electronics and Contributions of the Naval Research Laboratory* (Washington, D.C.: Naval Research Laboratory Report 8300, 1979), p. 300.
5. *Ibid.*
6. Lee A. Meadows, "Electronic Countermeasures Technology," *Signal,* March 1978, p. 34.
7. Simons, *op. cit.*
8. Leroy B. Van Brunt, *Applied ECM* (Dunn Loring, Virginia: EW Engineering, Inc., 1978), p. 2.
9. Simons, *op. cit.*
10. Meadows, *op. cit.*
11. Richard Hartman, "More EW Realism Needed," *Defense Electronics,* March 1982, p. 18.
12. Richard D. Curtis, "An Overview of Surface Navy ESM/ECM Development," *Journal of Electronic Defense,* March 1982, p. 27.
13. *Ibid.,* p. 30.

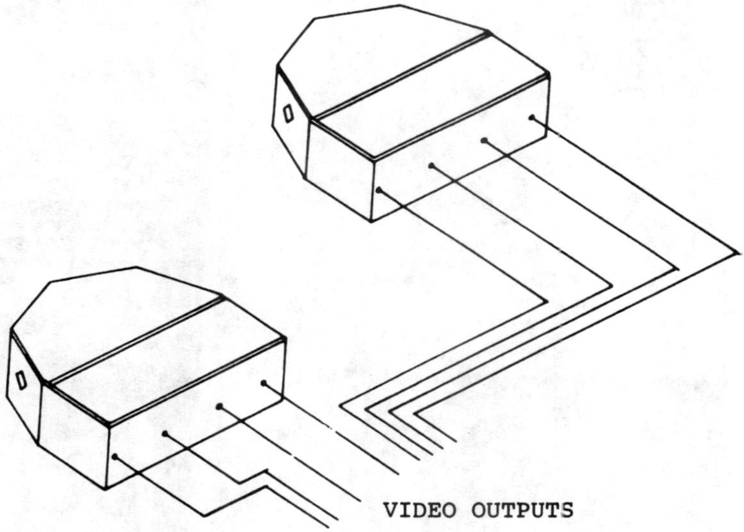

FIG. 5-1 AS 2854/SLD-1A ESM Direction Finder Antennas

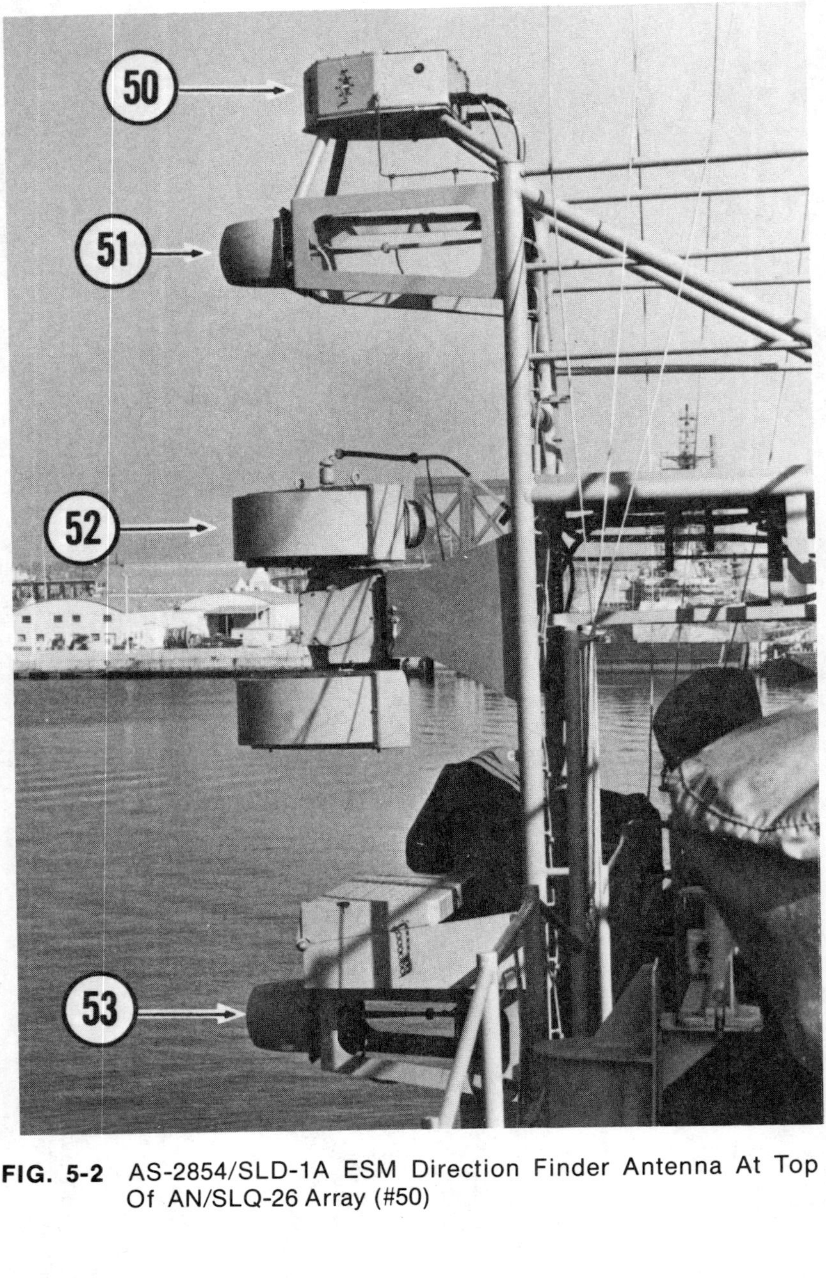

FIG. 5-2 AS-2854/SLD-1A ESM Direction Finder Antenna At Top Of AN/SLQ-26 Array (#50)

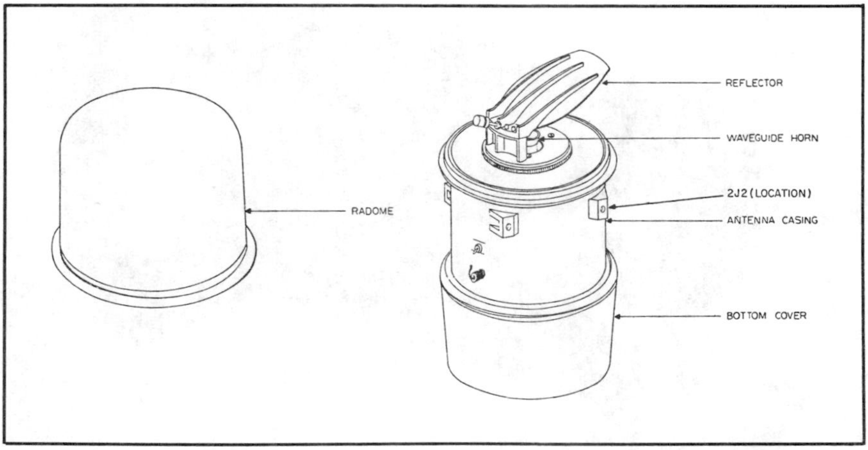

FIG. 5-3(a) AS-2209/SLR-12A ESM Direction Finder Antenna

FIG. 5-3(b) AS-2209/SLR-12A ESM Direction Finder Antenna (installed)

(a)

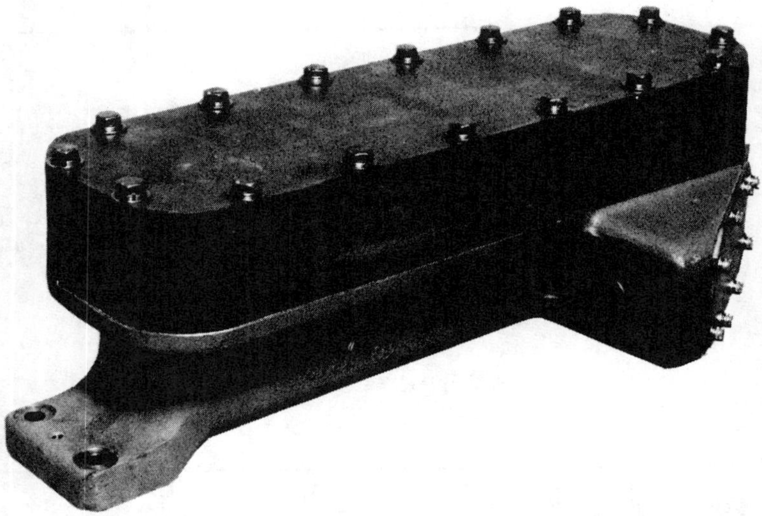

(b)

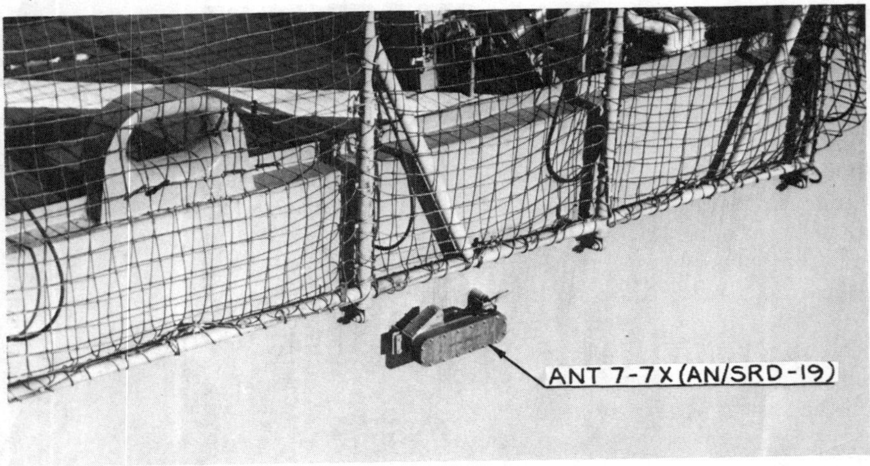

FIG. 5-4 AS-3202/SRD-19 Deck Edge ESM Direction Finder Antenna

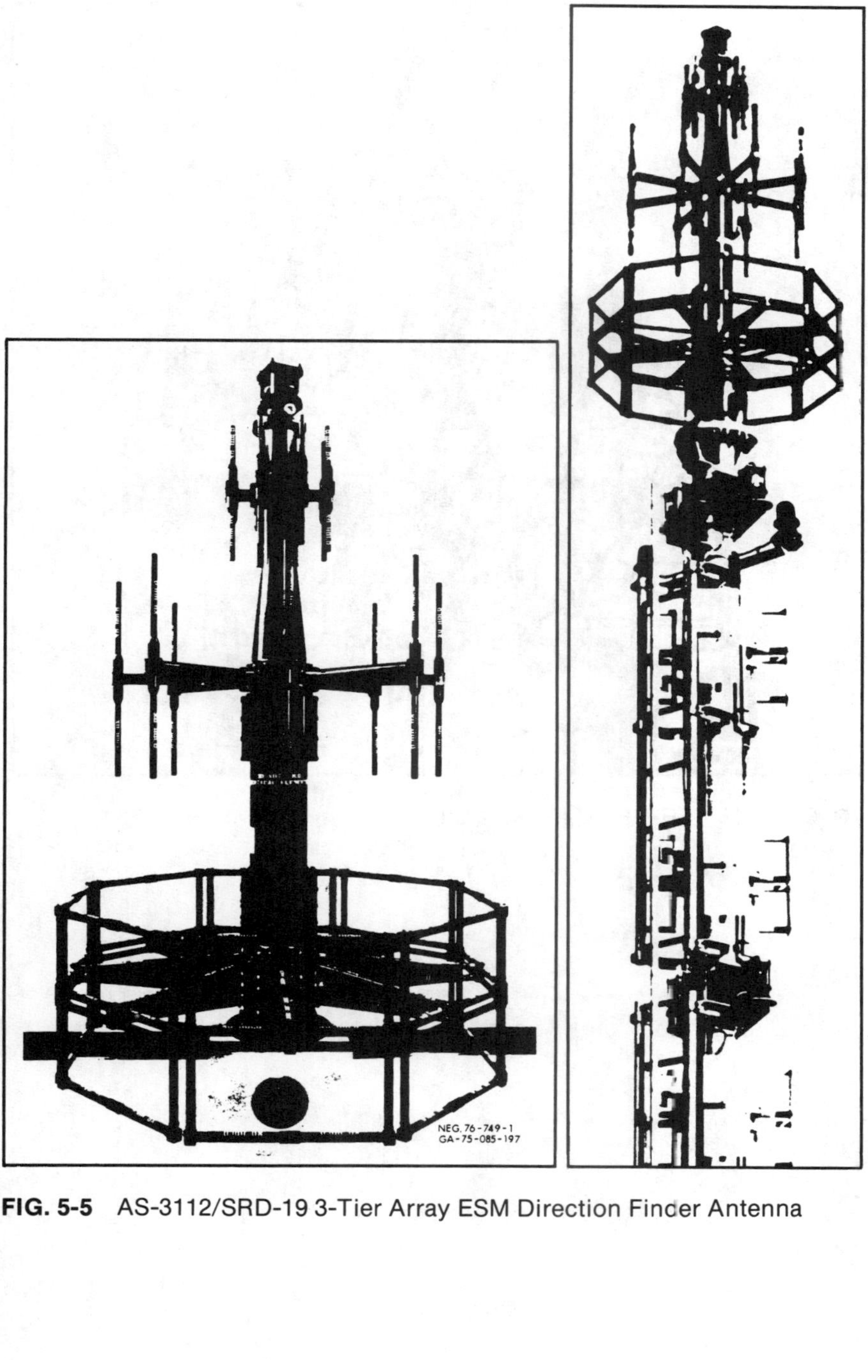

FIG. 5-5 AS-3112/SRD-19 3-Tier Array ESM Direction Finder Antenna

FIG. 5-6 AN/WLR-1 ESM Antenna Suite

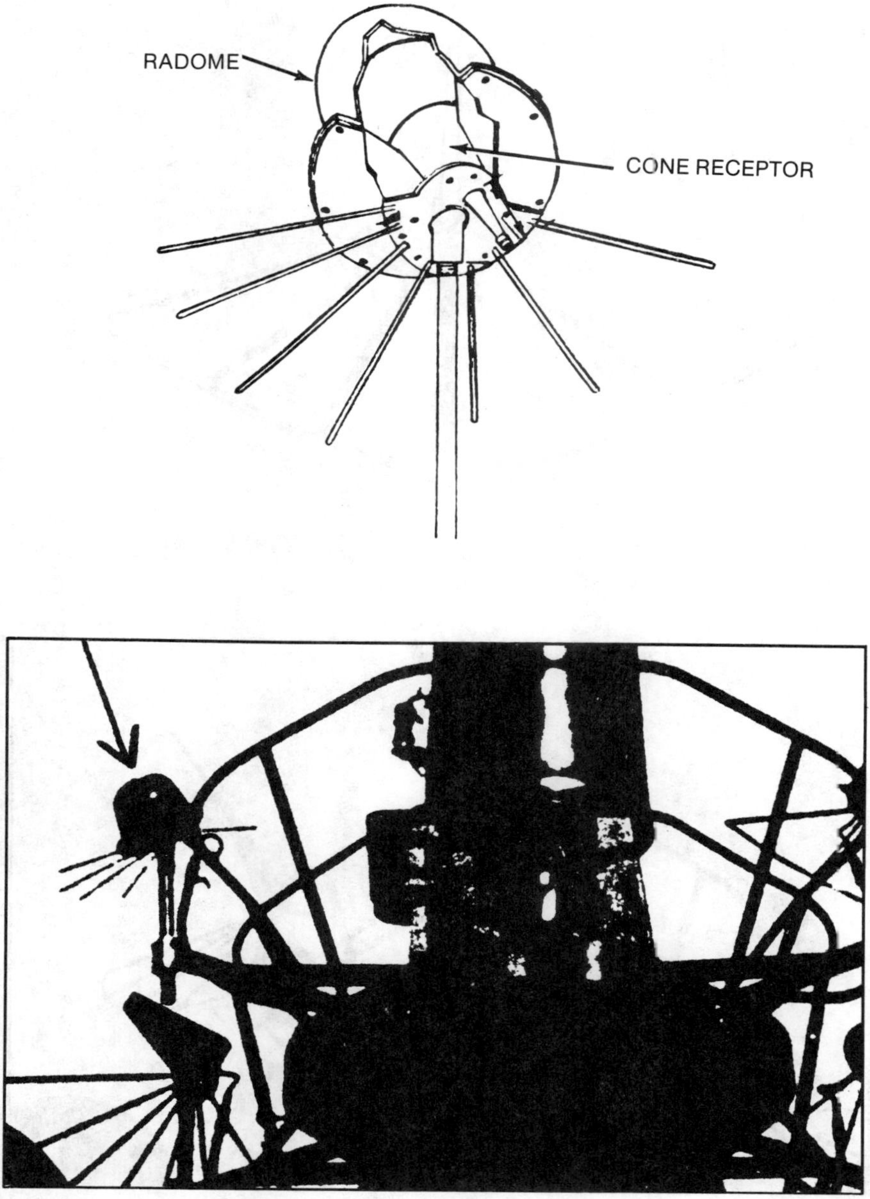

FIG. 5-7 NT-66131 UHF ESM Receive Antenna

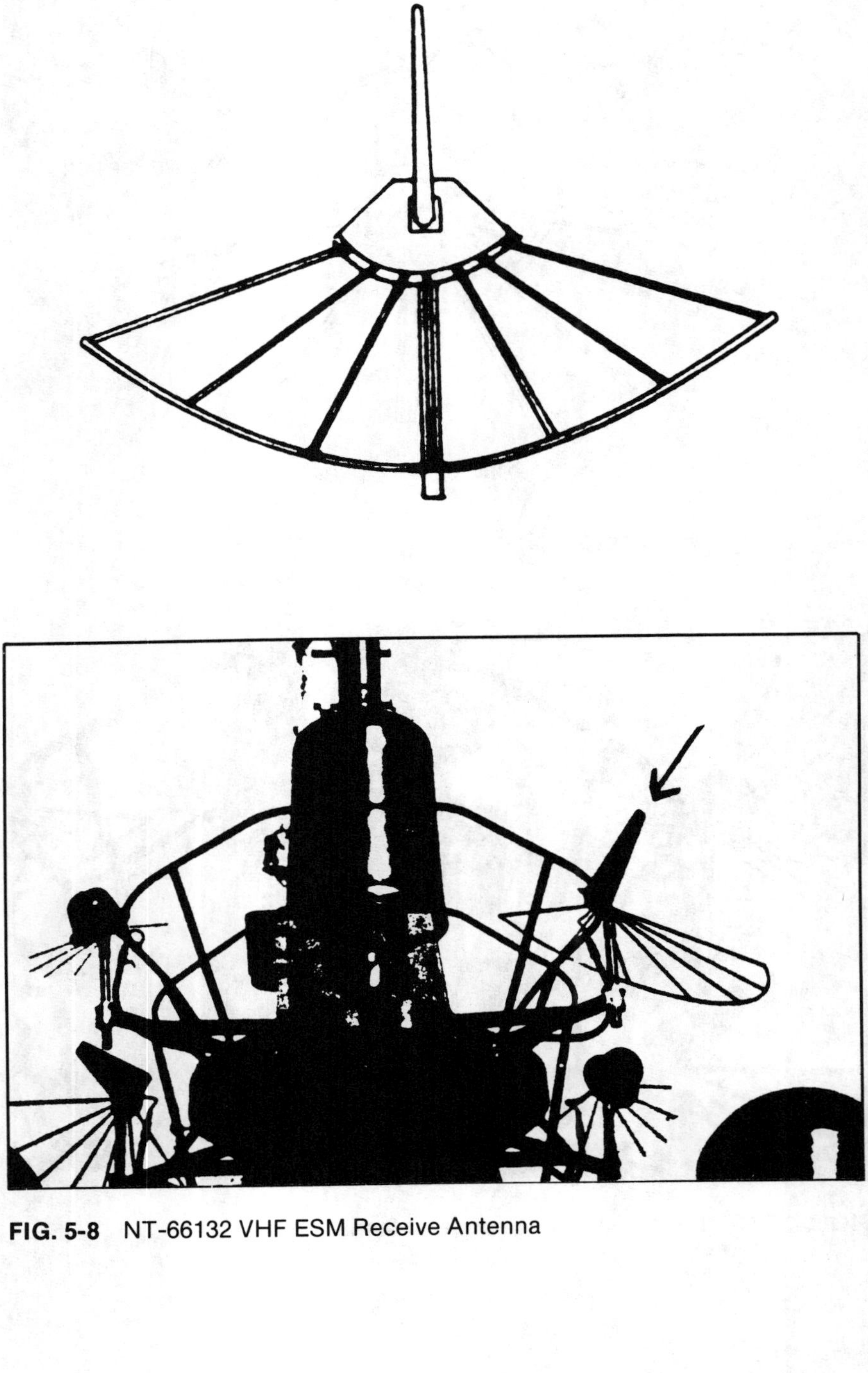

FIG. 5-8 NT-66132 VHF ESM Receive Antenna

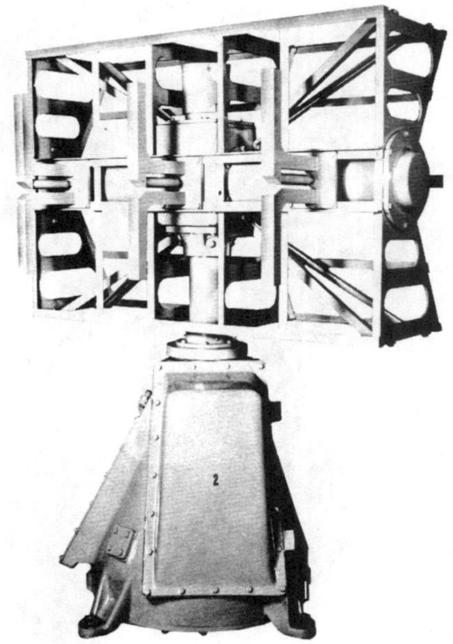

FIG. 5-9 AS-571/SLR UHF ESM Receive Antenna

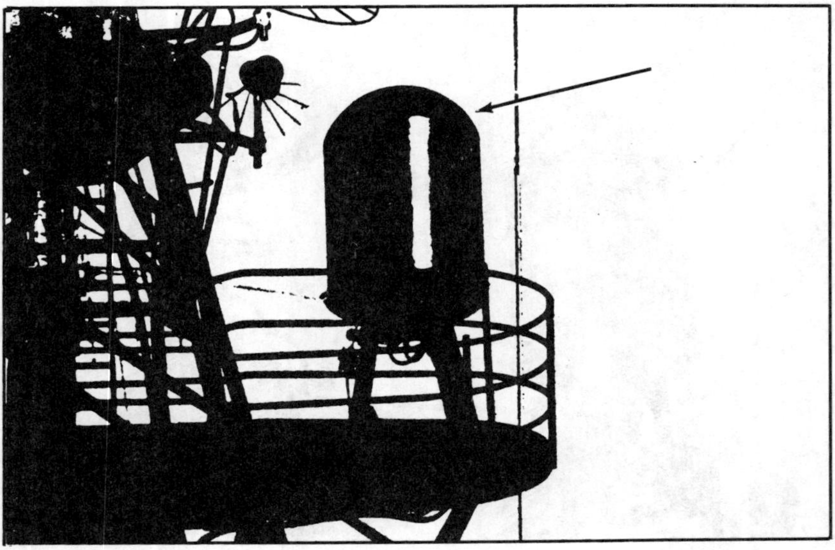

FIG. 5-10 AS-616/SLR UHF Receive Antenna

FIG. 5-11 AS-899/SLR SHF ESM Receive Antenna

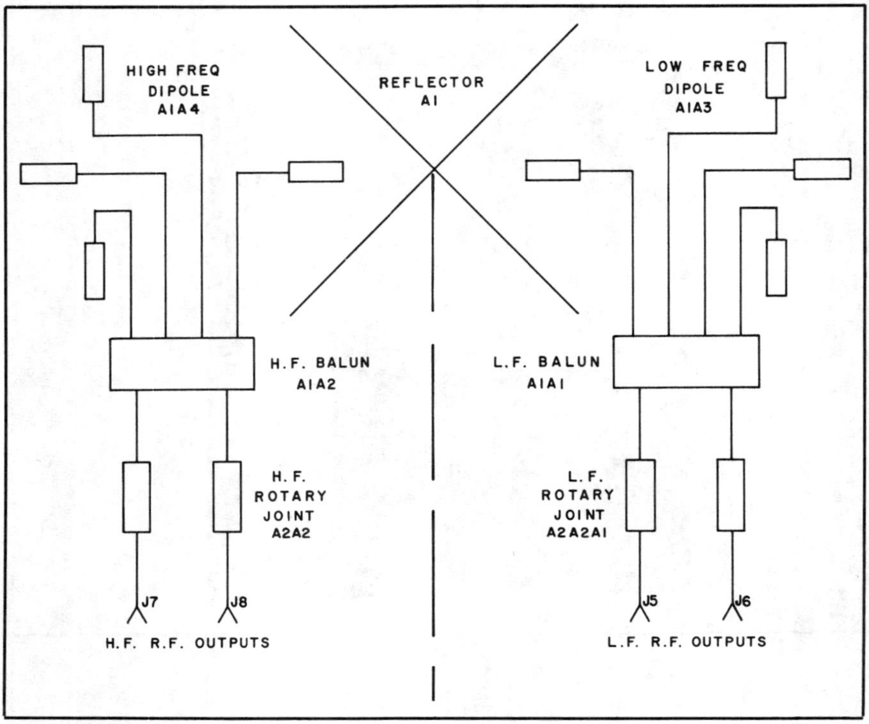

(a)

(b)

FIG. 5-12 AS-1023/SLR UHF ESM Receive Antenna

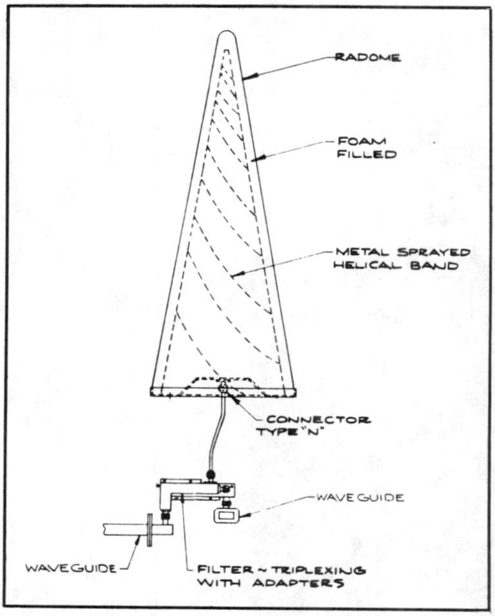

FIG. 5-13 AS-1174/SLR ESM Receive Antenna

FIG. 5-14 AS-2918 A/S SHF ESM Receive Antenna

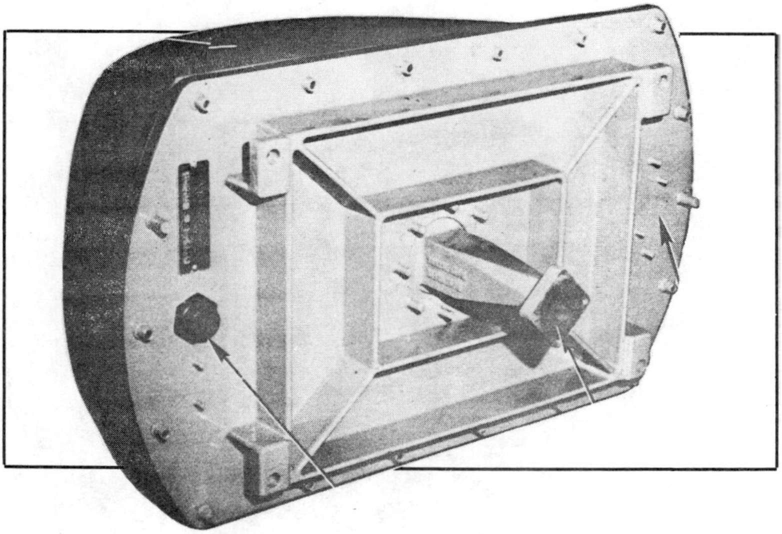

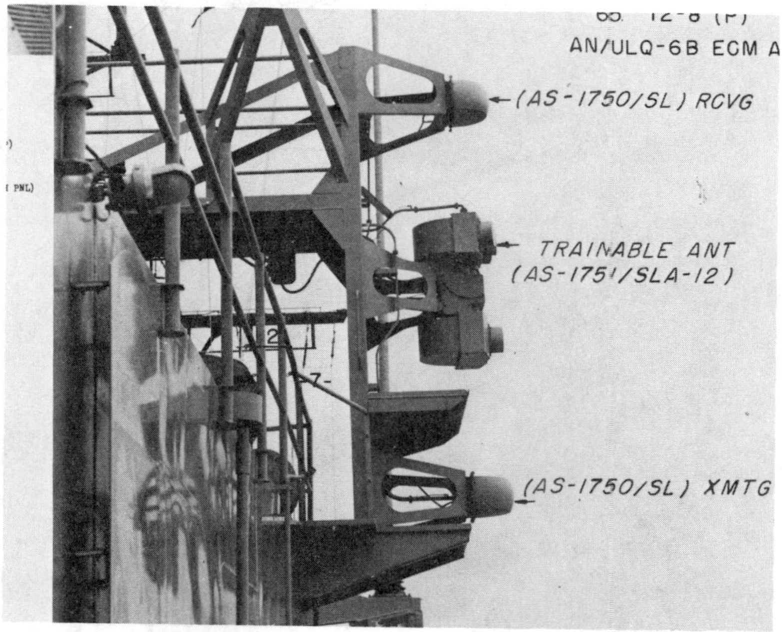

FIG. 5-15 AS-1750 A/SL ECM Antennas

FIG. 5-16 AS-1751A/SLA-12 ECM Antennas

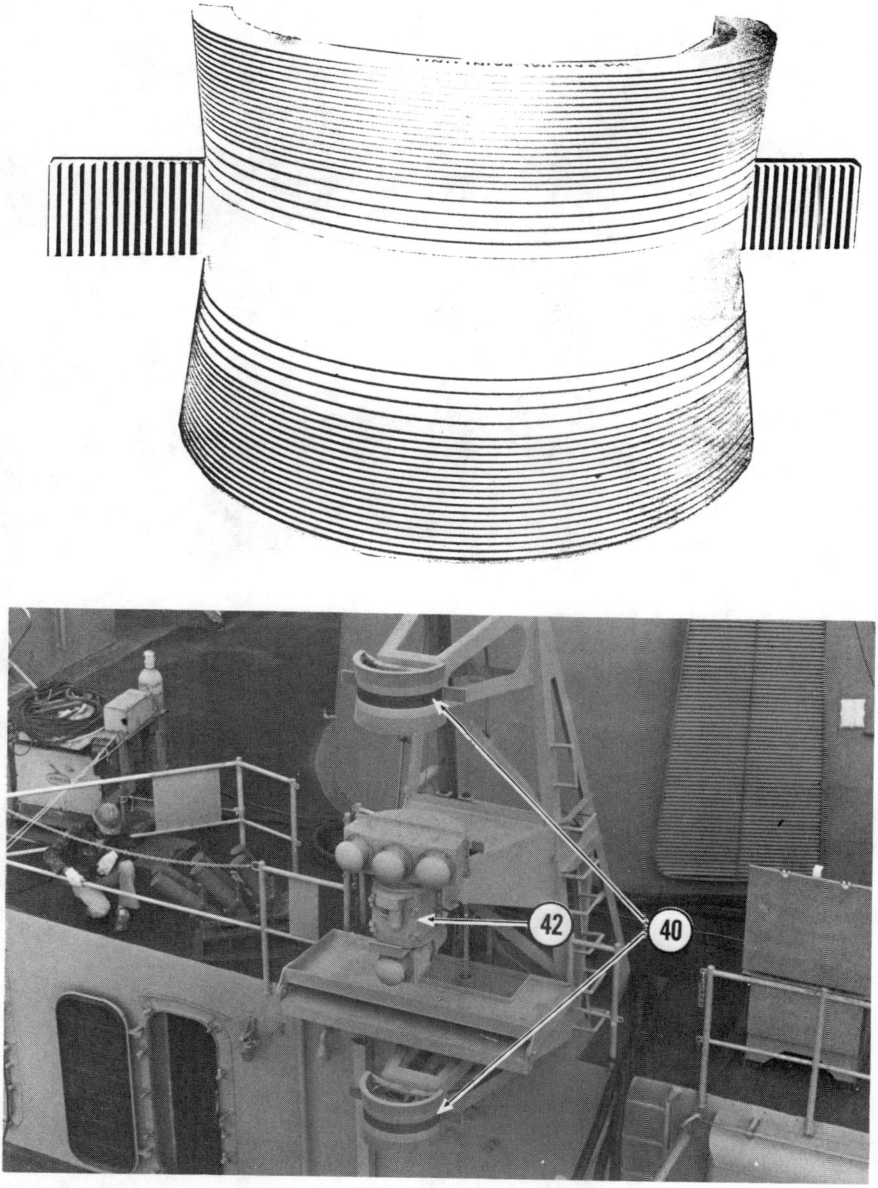

FIG. 5-17 AS-2430/SL ECM Antennas (#40 is AS-2430 system; #42 is AN 2431/SLA-15 trainable antenna)

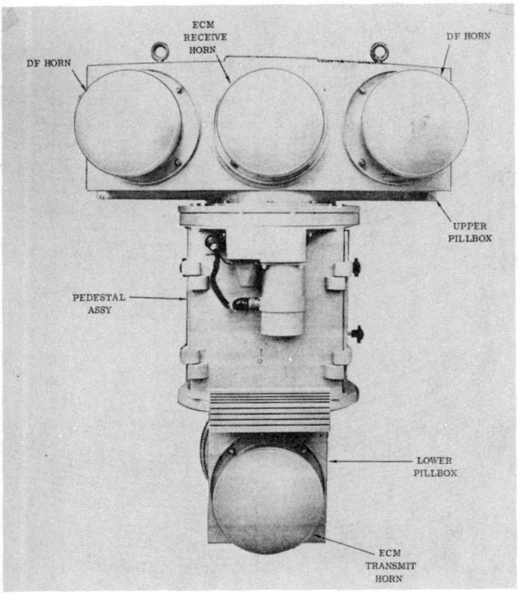

FIG. 5-18 AS-2431/SLA-15 ECM Antennas

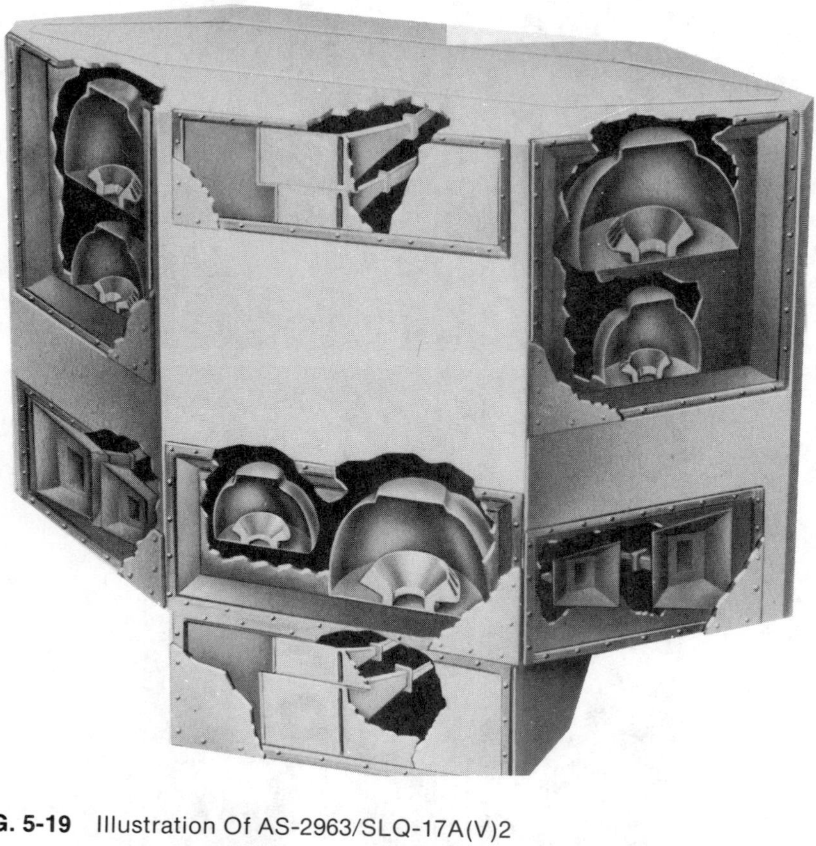

FIG. 5-19 Illustration Of AS-2963/SLQ-17A(V)2

FIG. 5-20 AS-2963/SLQ-17A(V)2 ECM Antenna (installed)

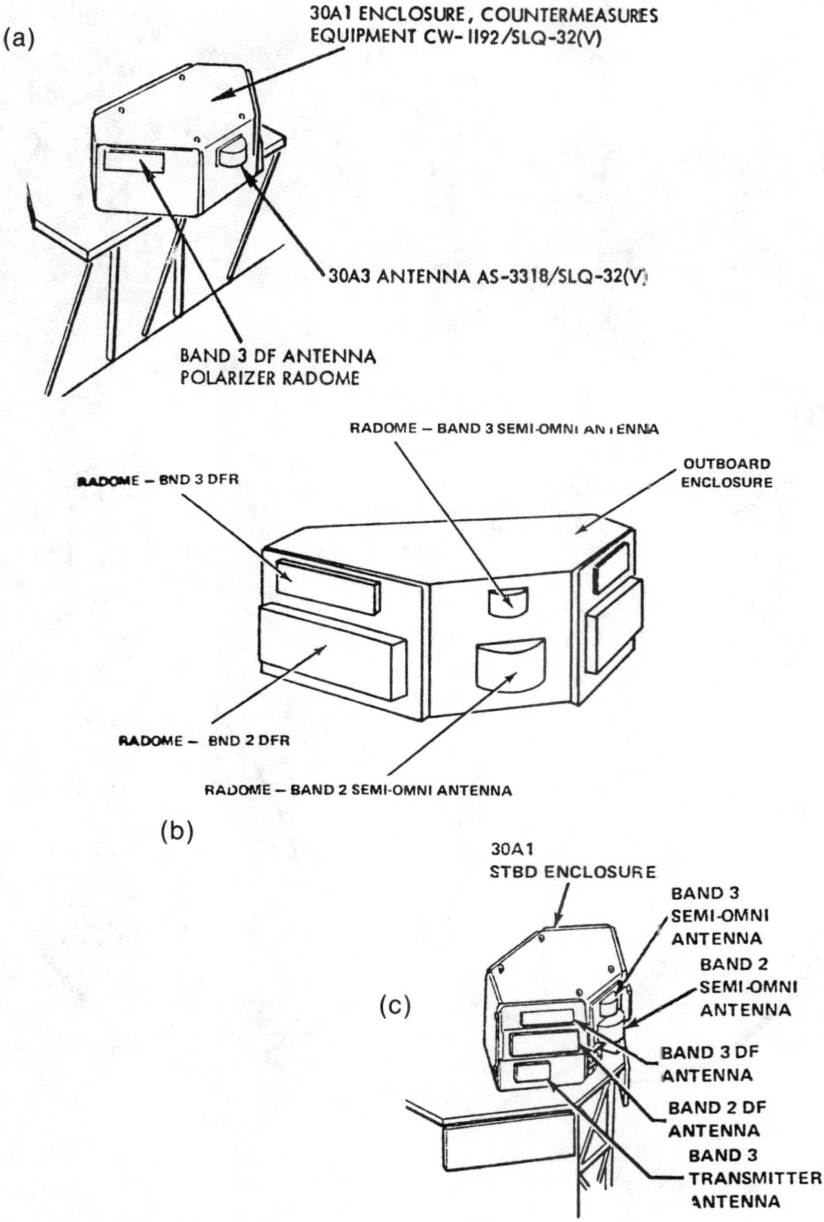

FIG. 5-21 Pictorial Variants Of AN/SLQ-32 Antenna Assembly

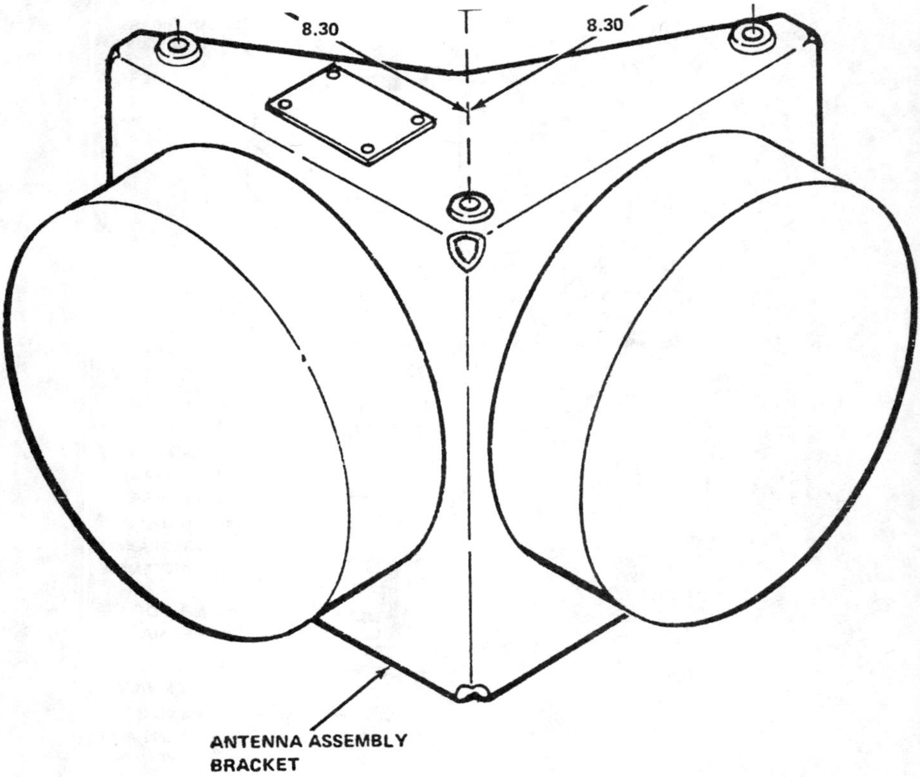

FIG. 5-22 AS-3316/SLQ-32 Band 1 ESM Antenna

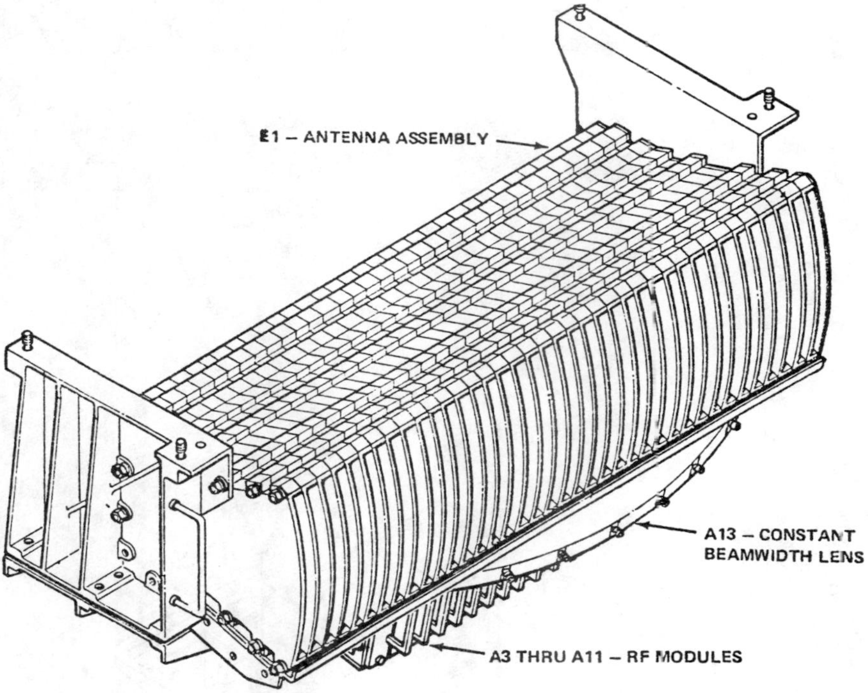

FIG. 5-23 SLQ-32 Band 2 DF Antenna Array

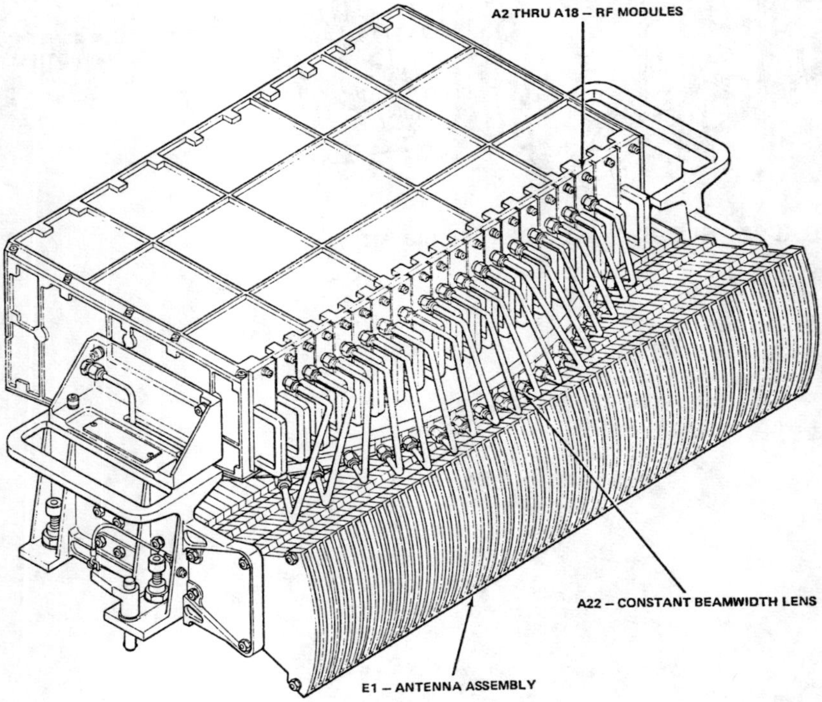

FIG. 5-24 SLQ-32 Band 3 DF Antenna Array

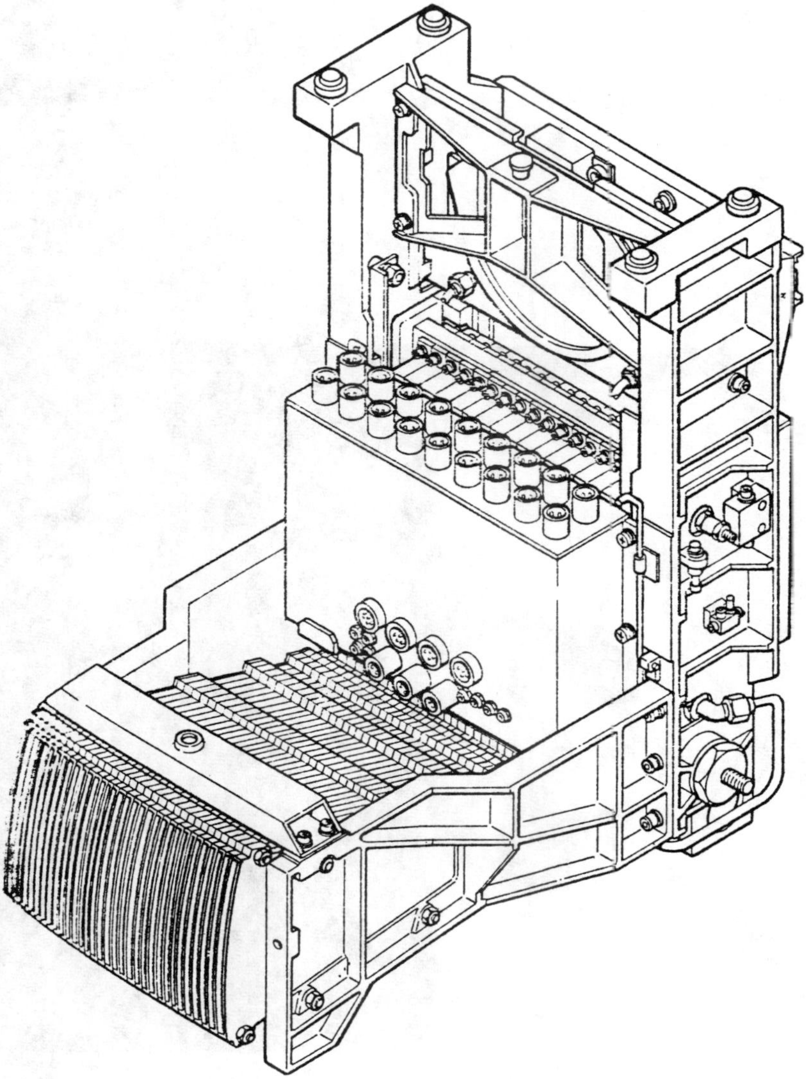

FIG. 5-25 SLQ-32(V)3 Band 3 ECM Transmit Array

FIG. 5-26 AN/SLQ-32(V)3 EW Antenna Assembly

Chapter 6

IFF Antennas

6-0 INTRODUCTION

From ancient times to well into the Twentieth century the need for adequate identification of possible enemy, as well as probably friendly, ships sighted on the open seas was satisfied by such visual means as flags, pennants, markings, and signaling. However, with the introduction of radar to make sea operations practical and safe in fog or at night and to facilitate the firing of naval guns with pin-point accuracy at targets unseen by the naked eye, there arose an imperative requirement for positive identification through the aid of electronics. The mere appearance of a reflecting object on a radar scope bore no distinguishing characteristics to merit identification as friend or foe. Moreover, with the rapid rise in the use of aircraft as a military device, the problem of quick and sure identification became acute. This was dramatized in the late 1930s when planes returning to home-base aircraft carriers, relying solely on coded visual techniques, were frequently hampered by poor visibility.

Realizing the urgency for providing immediate identification in a dawning age of electronic warfare, the Navy, through its Naval Research Laboratory (NRL), pioneered the proposing and developing of a series of radio identification systems for the U.S. armed forces.[1] By 1937 NRL had produced America's first "radio recognition" set, the Model XAE. This equipment transmitted coded radio signals from aircraft to ship, and, in reply, received verifying transmissions from the ship. In this manner the pilot of a plane, when in the presence of allied forces, activiated his set to radiate omnidirectional signals so as to avoid being fired upon by friendly ships.

Continued efforts by NRL led to improved *identification friend or foe* (IFF) techniques; for example, by employing pulsed systems to work integrally with shipboard radar equipment. Thus, radar pulses received by an airplane were automatically repeated back to the sending source in a prearranged sequence, so long as the plane carried the proper transponder. By World War II, IFF equipment operating in the VHF and UHF bands had been widely adopted. The chief deficiency of the early IFF systems was that they were too easily simulated and jammed by the enemy. Therefore, elaborate means of quick-change coding were resorted to.

Early in the war, a concentrated joint effort between the United States and Great Britain resulted in developing the renowned Mark III IFF system. Installed on virtually every ship and plane in the Allied forces, it was used extensively throughout the war years. Work continued in the post-war era to produce further equipment improvements, incorporated in such systems as the Mark V which operated at higher frequencies (950-1150 MHz) with sharper antenna characteristics, better target resolution, and enhanced agility (12 rapid-change channels) to avoid jamming. Following the Mark V, a combined military-civil effort resulted in the Mark X system. First available in 1959, the Mark X is still in use by commercial aviation.

Over the years a succession of naval shipboard IFF systems has been employed, principally the AN/UPX series such as the UPX-5, 9, 11, 12, 17, and the currently-used Mark XII (UPX-23, 25, 27, and 28). In general the antennas used with shipboard IFF systems have been of either two types: fixed omnidirectional or rotating slotted-waveguide arrays.

6-1 FIXED OMNIDIRECTIONAL IFF ANTENNAS

ANTENNA AS-177B/UPX

a. TYPE/USE — The AS-177 is a small, fixed, omnidirectional UHF antenna used with naval shipboard IFF systems such as the AN/UPX-25 and AN/UPX-28. It is often employed as the main transmit and receive antenna for transponder equipment, and to transmit sidelobe suppression reference pulses for interrogator sets.
b. PHYSICAL CONFIGURATION — [See Figure 6-1].
c. DIMENSIONS — (1) Height: 20⅛ inches; (2) Width: 6½ inches; (3) Depth: 6½ inches; (4) Weight: 7 pounds.
d. FREQUENCY RANGE — 1010 to 1110 MHz (optimized 1030-1090 MHz).
e. INPUT IMPEDANCE/VSWR — 50 ohms, VSWR not greater than 2:1 over frequence range.
f. RF POWER RATING — 5 kW peak at 1 percent duty cycle.
g. POLARIZATION — Vertical.
h. TYPE OF FEED — RG-219/U coaxial cable.

i. PRIMARY POWER REQUIRED — None.
j. INPUT CONNECTOR — Type LC, UG-154A/U.
k. INSTALLATION REQUIREMENTS — The AS-177 should be mounted vertically on a yardarm or other suitable mast platform using a 1½-inch extra heavy pipe for support. The antenna is seated firmly on the pipe then clamped tightly by two bolts and nuts.
l. SPECIAL CONSIDERATIONS — The antenna should be installed in a clear area as high as possible within the physical constraints of the ship. The length of the coaxial transmission line between the antenna and associated equipment should not exceed 150 feet. Greater lengths will introduce excessive attenuation.
m. REFERENCE — Technical Manual, Antenna AS-177B/UPX, NAVELEX-0967-434-9010.

6-2 STEERABLE DIRECTIONAL IFF ANTENNAS

ANTENNA AS-1065/UPX

a. TYPE/USE — The AS-1065 is a rotatable, high gain, directional UHF (L-band) transceive antenna used with naval shipboard IFF Interrogator systems such as the AN/UPX-25. The antenna may be mounted on a pedestal and operated independently as shown in Figure 6-2(a), or it may be employed integrally with radar antennas such as the AN/SPS-48 depicted in Figure 4-23. Based on an unconventional pillbox design, the AS-1065 antenna features very low side and back lobes, and wideband impedance matching. This is achieved by folding the pillbox along its axis of propagation, causing two layers to be formed, with the radiating element contained in one layer and the aperture in the other, so that no energy is reflected back into the feed. Radiation is a simulated cosecant-squared pattern in the elevation plane. A 1/16-inch thick fiberglass radome covers the antenna to protect it from the sea environment.

Referring to Figure 6-3, detailed theory of the antenna is as follows: Assuming that the RF signal is polarized as shown, the signal is directed against the parabolic curve where a 180-degree phase reversal occurs at the bend. Any energy reflected back into the lower layer is absorbed by RF absorbing material inside the antenna. Energy propagated through the upper layer is shaped by the parabolic curve and matched to free space by the dimensions of the aperature and the bend in the top surface. An additional impedance matching device to free space (not shown) is a 9/16-inch thick aluminum bar which is mounted along the length of the aperture one-quarter wavelength from the point where the top bend begins.

AZIMUTH PATTERNS SHAPING — In the azimuth pattern of the folded pillboxes, less than 30 dB sidelobes are present in the front

180 degrees. Back radiation is kept less than 30 dB by quarter-wave chokes along the back of the upper and lower lips of the aperture. Back radiation is further reduced by two baffles along the horizontal dimensions of the aperature. These baffles are held in place by metal webs between the sides of the flare and the baffles. The spaces between the baffles and the sides of the flare are waveguide regions. The webs form the "narrow" sides of the waveguide and the spacing between the webs determines the phase velocity. Since it is known that horns with a wide flare angle and short sides have considerable back radiation, the addition of baffles so changes the amplitude and phase taper that good radiation characteristics are maintained while back radiation is reduced.

ELEVATION PATTERN SHAPING — Since the antenna is to be used for the identification of aircraft, an approximate cosecant-squared power pattern is developed in the vertical plane by adjusting the phase and amplitude distribution over the vertical aperature. It is necessary to have an asymmetrical phase distribution for a cosecant-squared pattern and, for this reason, a non-symmetry is included by bending the top surface of the pillbox. By means of this bend and the baffles and webs, a simulated cosecant-squared pattern is propagated.

b. PHYSICAL CHARACTERISTICS — [See Figure 6-2].
c. DIMENSIONS — (1) Height: 18½ inches; (2) Width: 110½ inches; (3) Depth: 40 5/64 inches; (4) Weight: 83 pounds.
d. FREQUENCY RANGE — (1) 1030 MHz (transmit); (2) 1090 MHz (receive).
e. GAIN — 22.5 dB.
f. INPUT IMPEDANCE/VSWR — 50 ohms with VSWR 1.3:1.
g. RF POWER RATING — 10 kW peak.
h. POLARIZATION — Vertical.
i. TYPE OF FEED — Coaxial cable.
j. INPUT CONNECTOR — Type C, UG-568/U.
k. INSTALLATION REQUIREMENTS — The antenna is mounted by use of eight ⅜-16 NC-2A bolts through the mounting frame into the pillbox underside.
l. REFERENCE — Technical Manual, for Antenna As-1065/UPX, NAVELEX 0967-LP-958-8010.

ANTENNA AS-1688/SPS-48(V)

a. TYPE/USE — The AS-1688 is a rotatable, high gain, directional UHF (L-band) transceive antenna used with naval shipboard IFF systems. The antenna is designed as an integral part of the AN/SPS-48 radar antenna group OA-7028B and is mounted atop the AS-1686 phased array as shown in Figures 4-23, 4-49, and 6-4. A dual-aperture slotted array, the AS-1688 both radiates IFF interrogation signals and receives replies from transponders for identification by the ship. The fan-shaped beam is 45° wide in the vertical plane and only 9° in the horizontal. A bonded epoxy-fiberglass radome shields the slots from the sea environment.

b. PHYSICAL CONFIGURATION — [See Figure 6-4].
c. DIMENSIONS — (1) Height: 21¼ inches; (2) Width: 118⅜ inches; Depth: 21⅛ inches; (4) Weight: 100 pounds.
d. FREQUENCY RANGE — (1) 1010-1030 MHz (transmit); (2) 1070-1110 MHz (receive).
e. GAIN — 13 dB.
f. INPUT IMPEDANCE/VSWR - 50 ohms with VSWR 1.8:1.
g. RF POWER RATING — 10 kW peak.
h. POLARIZATION — Vertical.
i. TYPE OF FEED — Coaxial cable.
j. INSTALLATION REQUIREMENTS — Mounted atop the AN/SPS-48 radar by 18 screws.
k. REFERENCES — (1) Technical Manual, Overhaul and Repair for Radar Set AN/SPS-48, NAVSEA 0967-186-1170. (2) Technical Manual, Radar Set AN/SPS-48C(V), NAVSEA 0967-LP-604-6010 and 6030.

ANTENNA AS-2188/U

a. TYPE/USE — The AS-2188 is a rotatable, high gain UHF (L-band) antenna used with shipboard IFF systems. It is designed to be mounted integrally with naval radar antennas such as the AN/SPS-48 and 49, and, as shown in Figure 4-50, the AN/SPN-43A. The antenna is a broadside array comprised of an independent left and right hand section, each containing six full-wave slot radiators within a single reflector structure. The two sections are separately fed from a hybrid ring diplexer and directional coupler assembly. Twenty-four parasitic radiators arranged vertically in pairs inside box cavities act as the basic radiating elements. Each pair of elements is excited by a single vertical stub feedprobe and each box cavity performs as a half-wave slot. To generate the desired vertically polarized radiation pattern the probes toward the end of the antenna are made slightly shorter (less radiated power) than those toward the center.

The AS-2188 utilizes a dual radiation pattern to provide two distinct functions. A forward, narrowbeam directional pattern is used in transmitting pulsed interrogation signals. This pattern is synthesized in the sum excitation mode, in effect, when both sections of the antenna are fed in phase so that all the radiation fields from the individual elements are in phase with respect to each other. The result is that the fields add up in the front and cancel at the sides to produce a sharp 7° horizontal by 48° vertical beam in the forward direction.

The second, non-directional pattern is used for wideband interrogation sidelobe suppression (ISLS) to preclude IFF transponders from replying to interrogation signals in areas other than that covered by the directional main lobe. This is accomplished by switching in a supplementary difference mode of excitation whereby the left half of the array is fed 180° out of phase with the right section to produce a broad frontal pattern with a sharp center null. The broadness of the pattern is sufficient to cover the sidelobes of the sum pattern down to -30 dB below the main beam. However, the difference pattern has virtually no backward radiation to cover the small back lobe leakage of the sum pattern. Therefore, a separate, back-fill radiator consisting of a short, half-wave vertical dipole mounted in a 90° corner reflector is used to supply a broad back beam to cover the sum pattern back lobe.

To protect the antenna from the sea environment, the entire assembly is enclosed within a fiberglass radome.

b. PHYSICAL CONFIGURATION — [See Figure 6-5].
c. DIMENSIONS — (1) Height: 19 inches; (2) Width: 111⅜ inches; (3) Depth: 20½ inches; (4) Weight: 74 pounds.
d. FREQUENCY RANGE — (1) 1030 MHz (transmit); (2) 1090 MHz (receive).
e. GAIN — 18 dB.
f. INPUT IMPEDANCE/VSWR — 50 ohms with 1:3.5 VSWR.
g. RF POWER RATING — 10 kW.
h. POLARIZATION — Vertical.
i. TYPE OF FEED — Coaxial cable.
j. INSTALLATION REQUIREMENTS — The antenna is designed to be installed atop radar antennas, or on a suitable flat horizontal surface, using eight ⅜-24 bolts through the mounting plate holes under the AS-2188 antenna.
k. REFERENCE — Technical Manual, Antenna AS-2188/U, NAVELEX 0967-422-0010.

ANTENNA AS-3134/UPX

a. TYPE/USE — The AS-3134 is a fixed, electronically steerable, UHF (L-band) circular array designed as part of antenna group OE - 120/UPX for modern shipboard IFF systems such as the AN/UPX-29 interrogator set. Consisting of 64 cavity-backed dipole radiators arranged around a 12.5-foot diameter ring, the antenna is used for both transmit and receive with both directional and omnidirectional patterns. The directional beam can be steered 360° in azimuth, slewed ± 10° about any nominal bearing, or switched from one bearing to another nearly instantaneously. A control unit and programmer subsystem develop the beam forming and electronic steering, and, by automatic beam tilting, compensate for ship pitch and roll. The directional pattern is generated by simultaneously exciting 16 RF radiators in any 90° sector, with the result that the antenna can radiate a fan beam of 7° horizontal by 44° vertical beamwidth in any one of 1024 directions around the ring. During interrogate sidelobe suppression (ISLS) the RF energy is distributed equally among the 64 radiators to produce an omnidirectional pattern.

Each of the 64 radiator elements is contained in a rectangular cavity seven inches wide by twelve inches high and is fed by one of 64 phase-identical coaxial cables. To protect the array from the sea environment, a flat, epoxy-fiberglass, water-tight radome covers the individual cavities.

b. PHYSICAL CONFIGURATION — [See Figure 6-6].
c. DIMENSIONS — (1) Height: 16¼ inches; (2) Diameter: 150 inches; (3) Weight: 400 pounds.
d. FREQUENCY RANGE — (1) 1030 MHz (transmit); (2) 1090 MHz (receive).
e. GAIN — 19 dB.
f. INPUT IMPEDANCE/VSWR — 50 ohms with 1.4:1 VSWR.
g. RF POWER RATING — 10 kW peak.
h. POLARIZATION — Vertical.
i. TYPE OF FEED — Coaxial cable.
j. INPUT CONNECTOR — Type N.
k. INSTALLATION REQUIREMENTS — The antenna is normally installed around a mast as high as possible to be free of obstructions in the radiated field. It may be mounted so that its circumference is equidistant surrounding the mast, or on a platform offset from the mast. In either case, the antenna ring must be parallel to the ship's deck plane. Moreover, the programmer unit must be located very near the antenna.
l. REFERENCE — Technical Manual, OE-120/UPX Electronically Steered AIMS Shipboard Antenna System, NAVELEX, EE230-AC-OMI-010/E120-OE120.

ANTENNA AS-3430

a. TYPE/USE — The AS-3430 is a broadband, rotatable, directional UHF transceive antenna designed to be used with naval shipboard IFF systems. Intended as a replacement for the AS-1065, the AS-3430 is a log-periodic type array having a 7.5° horizontal and 50° vertical beamwidth. The antenna is normally installed atop the ship's search radar, e.g., the AN/SPS-48 or AN/SPS-43. It is enclosed in a fiberglass cover for protection from the sea environment.
b. PHYSICAL CONFIGURATION — [See Figure 6-7].

c. DIMENSIONS — (1) Height: 20 inches; (2) Width: 120 inches; (3) Depth: 38 inches; (4) Weight: 150 pounds.
d. FREQUENCY RANGE — (1) 6-800 MHz (DF); (2) 1030-1090 MHz (IFF).
e. GAIN — 17 dB.
f. INPUT IMPEDANCE/VSWR — 50 ohms with 1.4:1 VSWR.
g. RF POWER RATING — 5 kW peak.
h. POLARIZATION — Vertical.
i. TYPE OF FEED — Coaxial cable.
j. INSTALLATION REQUIREMENTS — Normally installed integrally with ship radar antenna.
k. REFERENCE — Technical Manual, NAVELEX, EE230-WA-OMI-010/W120-AS3430.

REFERENCES

1. Louis A. Gebhard, *Evolution of Naval Radio-Electronics and Contributions of the Naval Research Laboratory* (Washington, D.C.: Naval Research Laboratory Report 8300, 1979), p. 251.

FIG. 6-1 AS-177B/UPX Omnidirectional IFF Antenna

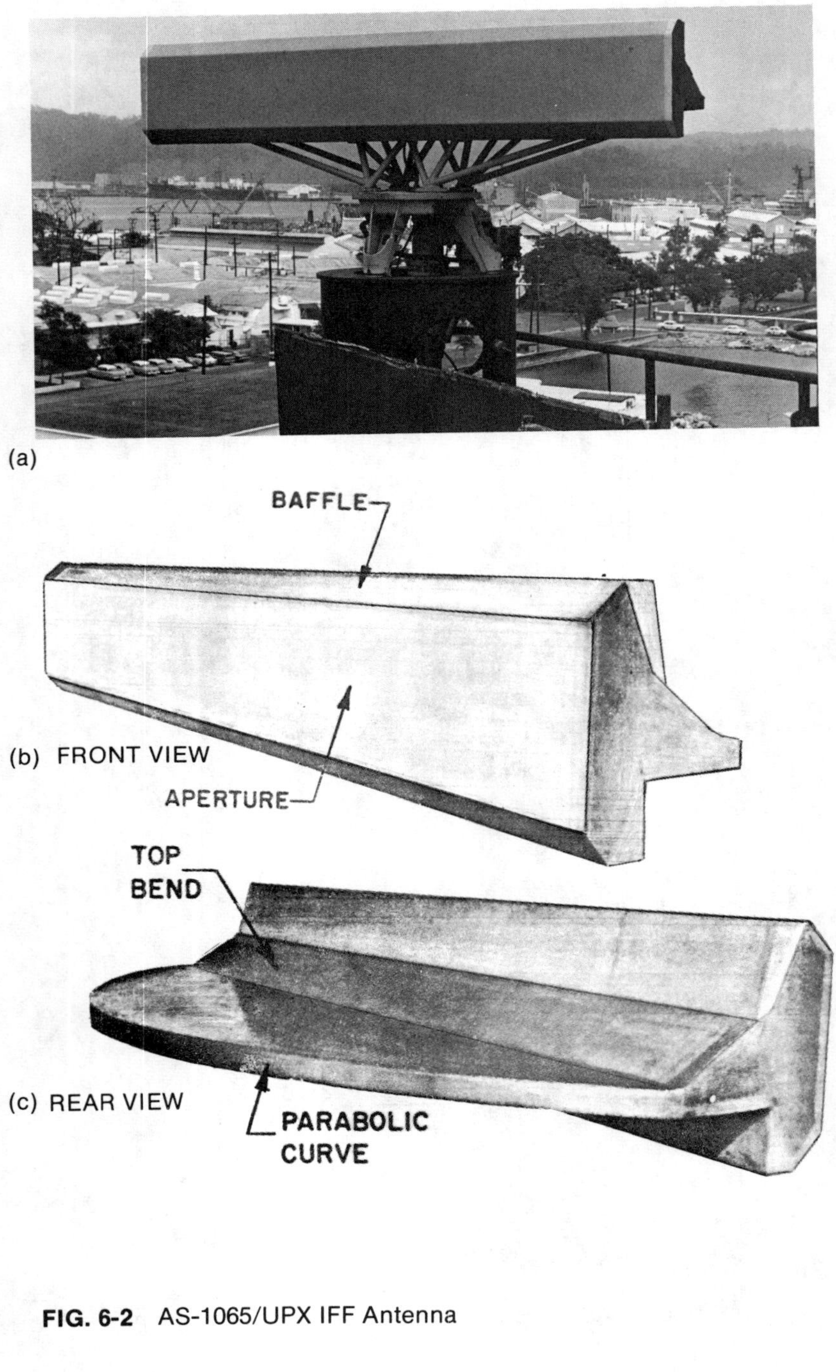

FIG. 6-2 AS-1065/UPX IFF Antenna

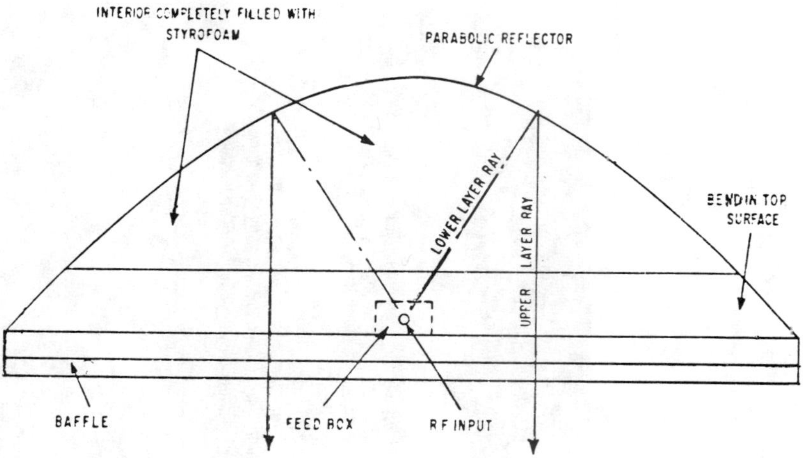

(a) Top View

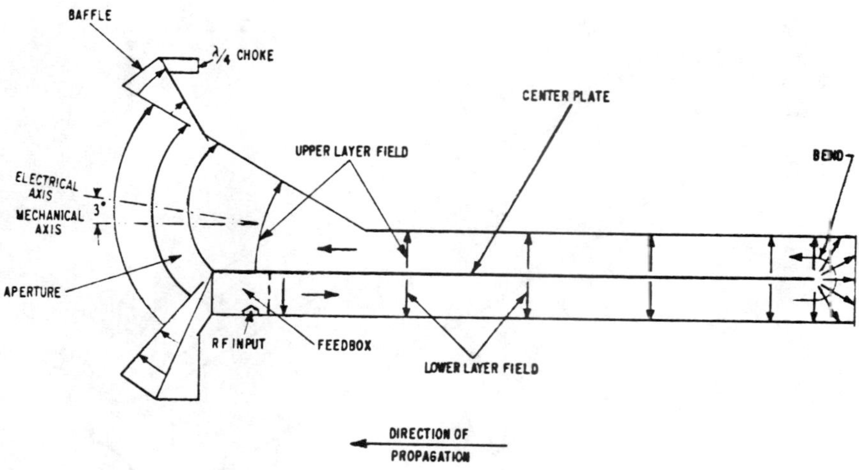

(b) Side View

FIG. 6-3 AS-1065/UPX Functional Representation

(a)

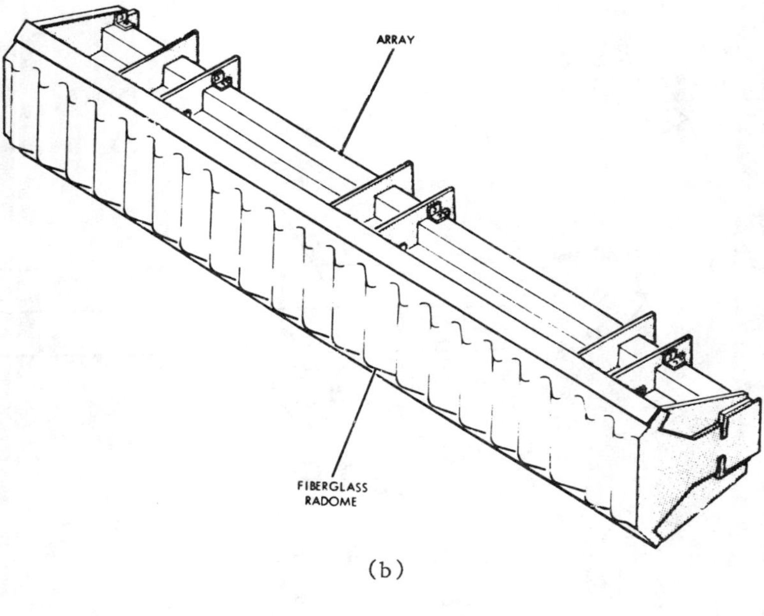

(b)

FIG. 6-4 AS-1688/SPS-48(V) IFF Antenna

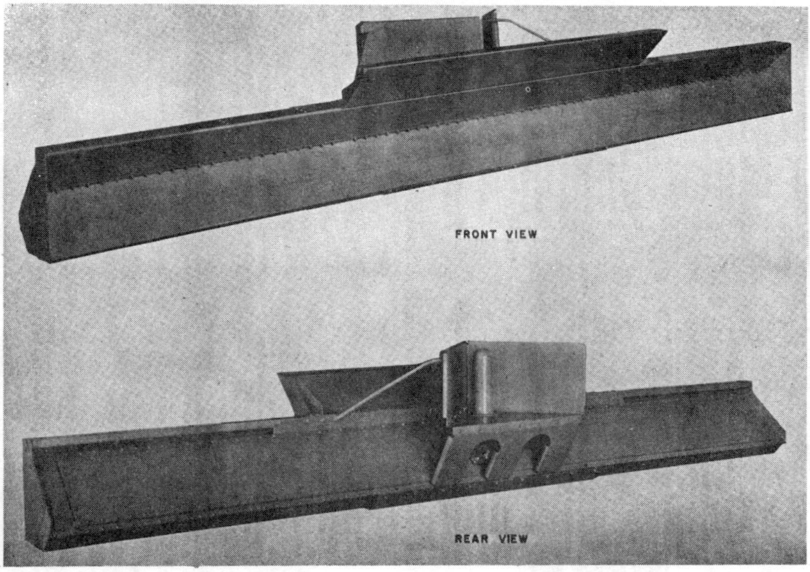

FIG. 6-5 AS-2188/U IFF Antenna

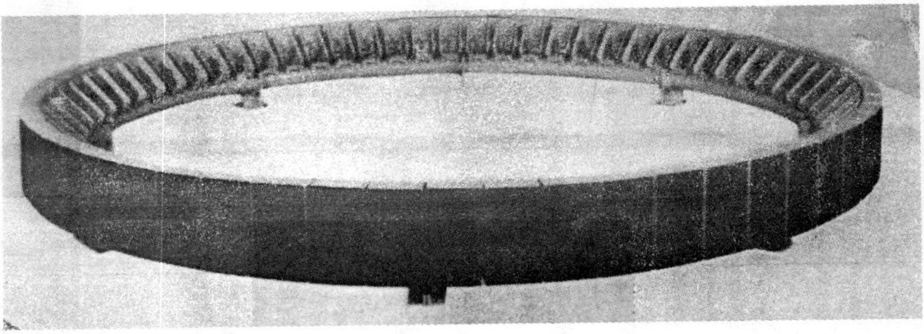

(a)

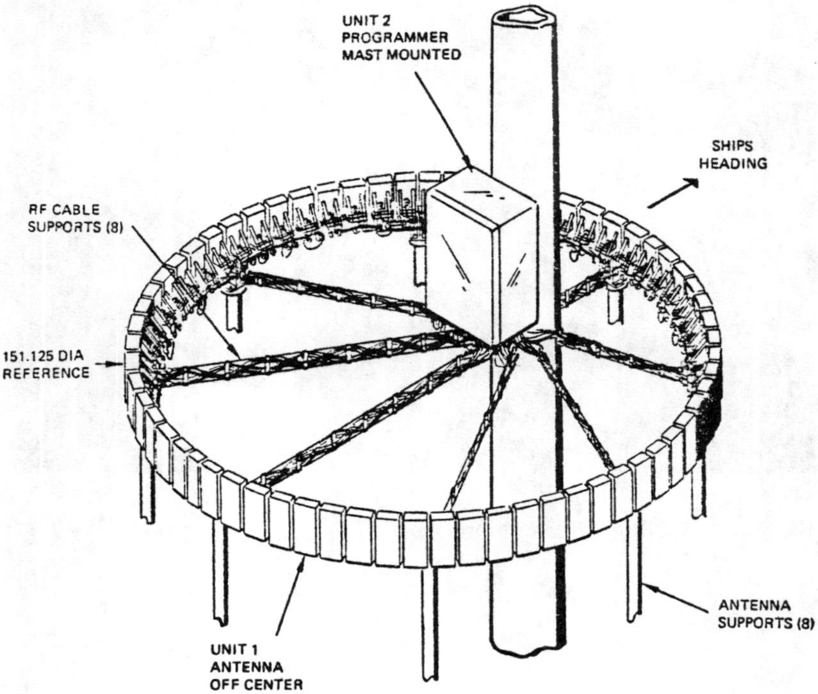

(b)

FIG. 6-6 AS-3134/UPX IFF Antenna

(a) FRONT

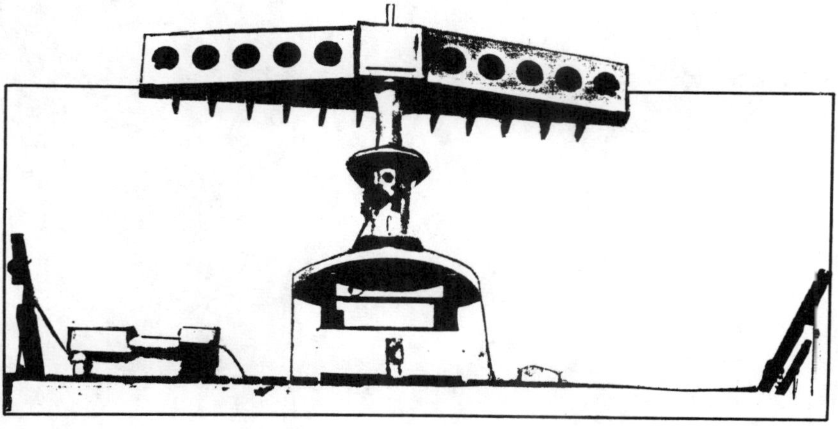

(b) REAR

FIG. 6-7 AS-3430 IFF Antenna

Chapter 7

Meteorology and Telemetry Antennas

7-0 INTRODUCTION

Weather — defined as the state of the earth's atmosphere with respect to wind, temperature, cloudiness, moisture, and barometric pressure — is, in terms of the sea, frequently thought of simply as either calm or stormy. Weather most certainly has provided a common thread of concern linking all mariners since man first ventured forth upon the waters. During times of peace it has been the crucial factor in countless episodes determining whether safe passage was gained from one point to another across the seas. In war, weather many times has dramatically influenced the course of history, as, for example, in 1588 A.D. when capricious coastal storms savagely wrecked the hopeless, limping Spanish Armada to ensure Elizabethan England's victory. Ship captains, war lords, and adventurers alike for ages have tirelessly studied collections of weather patterns recorded over varying seasons and courses in an attempt to anticipate weather conditions along their projected sea routes. Yet the very unpredictability of weather has perplexed shipping right up to the present time.

In our Twentieth century it was radio that first proved of great assistance in communicating to and from ships actual, as well as forecasted, weather conditions over various geographical locations. Then radar came to be heavily relied upon as an invaluable aid to ships traversing through fog, rain, and darkness. Still, because of rapidly changing weather conditions, forecasts continued to be too often unreliable and misleading. Furthermore, great expanses of the oceans were void of any means of providing weather information.

With the increasingly widespread use of naval air operations from carriers at sea, surer knowledge of long-range weather conditions began to be looked upon as essential.

During the 1950s, in response to the need for more accurate gathering and forecasting of weather information, the Navy and U.S. Weather Bureau collaborated to send aloft a series of trans-oceanic sounding (Transosonde) balloons. Carrying sensitive measurement instruments, these balloons travelled along high altitude paths transmitting meteorological data back to a network of earth stations for analysis and preparation of weather maps. The Transosonde project proved to be a valuable resource for expanding mankind's understanding of upper air influence on weather patterns.

Yet it was not until our most recent times, with the advent of weather satellites, that we were able finally to have before our eyes instantaneous, real-time pictures of actual environmental conditions, locally within our immediate sphere of interest and worldwide. Naval ships today, aircraft carriers in particular, have on-board equipment to track weather satellites to continuously receive, record, and display meteorological data. These systems help to ensure confident, safe operations at sea in all weather.

One prominent feature in the collecting of meteorological data by sensors aboard balloons, as discussed above, is that it involves a technique known as telemetering, or telemetry. That is, measurements made at a remote point are translated and converted to some convenient form for transmission back to earth for recording and analysis. Such is the nature of telemetry: the taking of measurements remotely, converting to a form facilitating transmission, and interpretation at a distant receiving post. The U.S. Navy makes use of a number of telemetry systems, notably as an aid in detecting submarine activity in the vicinity of ship operations and in the monitoring and evaluation of fired missile performance. Known as the AN/SKQ, SKR, SRQ, and SYR series of equipment, these telemetry systems and their associated antennas will be described below.

7-1 METEOROLOGY ANTENNAS

The principal meteorological data systems currently used aboard U.S. naval ships are those for tracking weather satellites, namely the AN/SMQ-6 and AN/SMQ-10 family of equipment. However, an older balloon-borne radiosonde set, the AN/SMQ-1 is still employed to a limited extent.

7-1.1 AN/SMQ-1

The AN/SMQ-1 radiosonde receiving set is a shipboard UHF meteorological system used to receive, display, and record signals relayed from balloon-borne radiosonde transmitters. Data received from the radiosonde is in pulse form modulated at a rate controlled by the temperature, pressure, and relative humidity of the atmosphere through which the balloon passes. A small omnidirectional dipole antenna is used aboard ship for receiving the radiosonde information.

ANTENNA FGP-151B-N-NP AN/SMQ-1

a. TYPE/USE — The FGP-151B-N-NP is a fixed, omnidirectional shipboard UHF antenna used with the AN/SMQ-1 series of radiosonde receiving sets. The antenna is a folded-dipole mounted vertically above a groundplane comprised of six radial spokes. This antenna replaces the AT-775/SMQ-1 vertical half-wave concentric dipole formerly used with the AN/SMQ-1.
b. PHYSICAL CONFIGURATION — [See Figure 7-1].
c. DIMENSIONS — (1) Height: 7¾ inches; (2) Diameter: 17½ inches (overall); (3) Weight: 20 pounds (estimate).
d. FREQUENCY RANGE — 310 to 410 MHz.
e. INPUT IMPEDANCE — 50 ohms.
f. RF POWER RATING — Not applicable, receive only.
g. POLARIZATION — Vertical.
h. TYPE OF FEED — Coaxial cable.
i. INPUT CONNECTOR — UG-29 B/U.
j. INSTALLATION REQUIREMENTS — The antenna base is threaded to seat upon a one-inch standard pipe. Mounting should be high and clear to afford omnidirectional coverage.
k. REFERENCE — Technical Manual, Receiving Set, Radiosonde AN/SMQ-1, AN/SMQ-1A, 3, NAVAIR 50-30SMQ1-5 [except antenna type FGP-151B-N-NP information].

7-1.2 AN/SMQ-6(V)

The AN/SMQ-6(V) is a meteorological data system used to receive, reproduce, and store weather information transmitted by satellite. There are four major units to the shipboard terminal: a receiver, a tape recorder, a facsimile recorder, and a steerable antenna for tracking the satellite. Three modes of operation are available from the receiver: (1) an automatic picture transmission (APT) mode to reproduce photos of cloud and land formations televised from the weather salellites; (2) a direct read-out infrared (DRIR) mode to record infrared pictures generated by scanning cloud and land masses (particularly at night) with infrared detection equipment mounted on the satellite; and (3) a mapping mode to recreate wire transmissions of weather maps and charts. The steerable VHF antenna installed aboard most U.S. aircraft carriers and a new VHF/UHF antenna employed by the later model AN/SMQ-6D system are described below.

ANTENNA AS-2192/SMQ-6(V)

a. TYPE/USE — The AS-2192 is a steerable, dual-helix directional VHF antenna used with the AN/SMQ-6(V) shipboard meteorological data system. Designed especially for the high RF electromagnetic radiation environment of naval ships, the antenna consists of two helices of five and one-half circular loops fabricated from ¾-inch aluminum tubing. Each helix is attached to a centerpole (running along the helix axis) by sixteeen fiberglass rods. A groundplane is fastened to one end of each helix. The antenna assembly is supported by a large pedestal containing servo system and drive motors that rotate the antenna up to 700 degrees in bearing and 90 degrees in elevation.

b. PHYSICAL CONFIGURATION — [See Figure 7-2].

c. DIMENSIONS — *Helix* (each) (1) Length: 100 inches; (2) Diameter: 28 inches; (3) Weight 165 pounds.
Antenna Pedestal (1) Height: 48 inches; (2) Diameter: 42 inches (base); (3) Weight: 460 pounds.

d. FREQUENCY RANGE — 136 to 138 MHz.

e. INPUT IMPEDANCE/VSWR — 50 ohms.

f. RF POWER RATING — Not applicable, receive only.

g. POLARIZATION — Circular.

h. TYPE OF FEED — Unbalanced, coaxial cable.

i. PRIMARY POWER REQUIRED — All servo and control power supplied from control console.

j. INPUT CONNECTOR — Type N.

k. INSTALLATION REQUIREMENTS — The base of the antenna pedestal has eight equally spaced bolt holes on its mounting flange for securing the assembly to the ship's structure. The antenna location should be selected to provide maximum unobstructed satellite viewing area. Therefore, the location must be free of obstructions within the antenna azimuth and elevation swing circles.

l. SPECIAL CONSIDERATIONS — The operating frequencies of this system require line-of-sight propagation paths. If the antenna is masked by ship structures or rigging, limited coverage at certain elevations and/or azimuths may be expected.

m. REFERENCES — (1) Technical Manual, Receiver-Recorder, Meteorological Data, AN/SMQ-6(V), NAVAIR 50-30SMQ6-1. (2) NAVAIR Drawing 67A32F25.

ANTENNA AS-3185/SMQ-6D(V)

a. TYPE/USE — The AS-3185 is a lightweight, steerable, dual-frequency, directional antenna used with the AN/SMQ-6D shipboard meteorological system to receive weather data from polar orbiting and geostationary satellites. The antenna is part of antenna group OE-181A/SMQ-6C(V) which, in addition to the antenna, consists of an antenna control unit and stabilizing pedestal. The antenna, designed to operate both at VHF and UHF, is capable of continuous rotation in azimuth and through 110° in elevation to maintain line-of-sight tracking of the satellite. A four-element dipole array spaced equally around the antenna perimeter comprises the VHF portion. The UHF sensor consists of an aluminum-mesh parabolic dish fed by a center element. A stabilized base pedestal supports the antenna assembly to compensate for ship's pitch and roll.
b. PHYSICAL CONFIGURATION — [See Figure 7-3].
c. DIMENSIONS — *Antenna* (1) Height: 48 inches; (2) Width: 48 inches; (3) Depth: 26 inches; (4) Weight: 30 pounds. Pedestal (1) Height: 47 inches: (2) Width: 21 inches; (3) Depth: 14 inches: (4) Weight: 175 pounds.
d. FREQUENCY RANGE — (1) VHF antenna: 136.5 MHz; (2) UHF (L-band) antenna: 1691 MHz.
e. GAIN — (1) VHF antenna: 8.5 dB; (2) UHF antenna: 22.5 db.
f. INPUT IMPEDANCE—50 ohms with 1.5:1 VSWR maximum.
g. POLARIZATION — (1) Right-hand Circular (VHF antenna); (2) Linear Field Variable Orientation (UHF antenna).
h. RF POWER RATING — Not applicable, receive only.
i. TYPE OF FEED — RF coaxial cable.
j. PRIMARY POWER REQUIRED — 120 vac, 60 Hz, one-phase.
k. INSTALLATION REQUIREMENTS — The antenna must be so located as to allow a clear line-of-sight tracking of the satellite; i.e., 360° in azimuth and from horizon to zenith. Mounting is accomplished by sixteen ⅜ x 2-inch bolts through the base of the pedestal.
l. REFERENCE — Technical Manual, Antenna Group OE-181A/SMQ-6C(V), NAVAIR 50-30SMQ6-6.

7-1.3 AN/SMQ-10

The AN/SMQ-10 is a shipboard meteorological terminal that receives, processes, and displays real-time S-band data from DMSP (Defense Meteorological Satellites Program). These satellites supply high resolution infrared and visual imagery of the earth's surface and lower atmosphere to ships for weather information. The SMQ-10 is designed to operate in the high RF interference environment encountered aboard naval ships, processing received weather data into a hard-copy output. Intended particularly for aircraft carriers, the system makes use of two identical high-gain, steerable antennas, one placed on the port side and one on the starboard side of the ship's flight deck canted downward at 20°. An associated unit of the SMQ-10, the antenna positioning subsystem, simultaneously points and controls the two antennas so that they will track the orbiting satellite.

ANTENNA AS-3063/SMQ-10

a. TYPE/USE — The AS-3063 is a high-gain, scannable, S-band shipboard antenna used with the AN/SMQ-10 system to receive meteorological data from weather satellites. Consisting of a six-foot diameter parabolic reflector with an offset conical feed, the AS-3063 is designed to have a sharp 5° beamwidth with very low sidelobes so as to be immune to the characteristically high radio frequency interference associated with aircraft carrier topside environments. A Y-yoke pedestal assembly supports the antenna and provides the means for continuous, stabilized positioning to track the satellite. Full hemispherical coverage is accomplished by utilizing two antennas, one mounted on each side of the flight deck. A radome housing protects the antenna from the sea environment.
b. PHYSICAL CONFIGURATION — [See Figure 7-4].
c. DIMENSIONS — (1) Height: 90 inches (including pedestal); (2) Width: 72 inches (antenna diameter); (3) Depth: 154 inches (with pedestal); (4) Weight: 1847 pounds each (total assembly including pedestal).
d. FREQUENCY RANGE — 2.2 to 2.3 GHz (S-band).
e. GAIN — 30.5 dB minimum.
f. RF POWER RATING — Not applicable, receive only.
g. POLARIZATION — Right-hand circular.
h. INPUT IMPEDANCE — 50 ohms.
i. TYPE OF FEED — RG-214/U coaxial cable.
j. PRIMARY POWER REQUIRED — (1) 208 vac, 400 Hz, three-phase; (2) 440 vac, 60 Hz, three-phase.
k. INSTALLATION REQUIREMENTS — Cable length from the antenna pedestal to antenna junction boxes (RG-214) shall not exceed ten feet. Cable length from the antenna junction box to the below-deck equipment (RG-233/U) shall not exceed 500 feet. Mounting of the AS-3063 antennas is very involved; detailed criteria are given in Reference (2) below.
l. REFERENCES — (1) Technical Manual, Receiver-Recorder Set, Meteorological Data, AN/SMQ-10, NAVAIR 50-30SMQ10-1. (2) Shipboard Installation Data for AN/SMQ-10, Navy Space Systems Activity publication NSSA-R-60-75-2, Revision B.

7-2 TELEMETRY ANTENNAS

7-2.1 AN/SKQ-2

The AN/SKQ-2 receiving-recording-scoring set is a shipboard UHF telemetry receiving and processing system used by the Navy to record data from missiles and associated target drones for evaluation of missile and fire control system performance. A combined antenna and frequency converter unit receives the telemetry signals simultaneously in either of two UHF channels (one for missile monitoring and one for drone scoring data) for processing by the telemetry set.

ANTENNA AS-2535/SKQ-2

a. TYPE /USE — Part of antenna group OE-80/SKQ-2, the AS-2535 is a steerable, high-gain, directional, dual-band UHF shipboard antenna designed for use with the AN/SKQ-2 telemetry receiving system. The antenna consists of a parabolic dish reflector with a radome-covered feed held by six strut-rods. Supported on a pedestal, the antenna assembly can be trained manually by handwheels at the below-deck antenna control unit of the console group, or steered automatically by means of synchro data from a fire control tracking antenna. The resulting 7° sharp beam may be slewed 270° in azimuth and from -22° to +95° in elevation.
b. PHYSICAL CONFIGURATION — [See Figure 7-5].
c. DIMENSIONS — (1) Height: 70 inches (with pedestal); (2) Diameter: 60 inches; (3) Weight: 495 pounds (with pedestal).
d. FREQUENCY RANGE — (1) 2200 to 2290 MHz (Channel 1 missile telemetry; (2) 1760 to 1850 MHz (Channel 2 target scoring).
e. GAIN — (1) 21 dB (Channel 1); (2) 18 dB (Channel 2).
f. INPUT IMPEDANCE/VSWR — 50 ohms with 1.5:1 maximum VSWR.
g. RF POWER RATING — Not applicable, receive only.
h. POLARIZATION — Right-hand circular.
i. TYPE OF FEED — Coaxial cable.
j. PRIMARY POWER REQUIRED — (1) 115 vac, 60 Hz, one-phase; (2) 115 vac, 400 Hz, one-phase; (3) 20 vdc, 300 watts.
k. INSTALLATION REQUIREMENTS — Varies depending upon the type of ship, location, placement and number of missile launchers, but usually is mounted on a mast aft of midship for adequate field of view.
l. REFERENCE — Technical Manual, Telemetric Data Receiving-Recording-Scoring Set AN/SKQ-2, NAVSEA OP-3770.

7-2.2 AN/SKQ-3

The AN/SKQ-3 receiving-recording-scoring set is a portable shipboard UHF telemetry receiving and processing system used in temporary installations aboard destroyers and frigates to record data from missiles and associated target drones for evaluation of overall surface missile system effectiveness. Two versions are available: an AN/SKQ-3(V)1, utilizing a single tracking antenna; and the AN/SKQ-3(V)2, having two tracking antennas to simultaneously receive data from two missiles at different frequencies. Additionally, there are two other variations available with the further capability to receive point defense signals by employing AS-2757/SKQ-3(V) type directional horn antennas: the SKQ-3(V)3 uses one horn and the SKQ-3(V)4 has a two-horn configuration.

In each of the SKQ-3 systems a combined antenna and frequency converter unit receives telemetry signals simultaneously in either of two UHF channels, one for missile monitoring and one for drone target scoring data, for processing by the telemetry set.

ANTENNA AS-2727/SKQ

a. TYPE/USE — Part of tracking antenna group OE-133/SKQ, the AS-2727 is a steerable, high-gain, dual-band directional UHF shipboard antenna designed for use with the AN/SKQ-3 telemetry receiving system. The antenna consists of a parabolic dish mesh reflector with a feed system of five helices. Supported by a pedestal, the antenna assembly can be trained manually by handwheels on the antenna control unit, or "remote-automatic" from signals generated in the ship fire control system. Azimuth steering is through 270° and elevation coverage is from -14° to +100°.
b. PHYSICAL CONFIGURATION — [See Figure 7-6].
c. DIMENSIONS — (1) Height: 60 inches (with pedestal); (2) Diameter: 60 inches; (3) Weight: 240 pounds (total).
d. FREQUENCY RANGE — (1) 2200 to 2290 MHz (missile telemetry); (2) 1760 to 1850 MHz (target scoring).
e. GAIN — (1) 23 dB minimum (2200 - 2290 MHz); (2) 18 dB minimum (1760 - 1850 MHz).
f. INPUT IMPEDANCE/VSWR — 50 ohms with 2:1 maximum VSWR.
g. RF POWER RATING — Not applicable, receive only.
h. POLARIZATION — Right-hand circular.
i. TYPE OF FEED — Coaxial cable.
j. PRIMARY POWER REQUIRED — (1) 115 vac, 60 Hz, one-phase; (2) 115 vac, 400 Hz, one-phase; (3) 50 vdc, 0.5 and 2.5 amps.

k. INSTALLATION REQUIREMENTS — Varies depending upon the type of ship, location, placement and number of missile launchers; but usually is mounted near ship's centerline, or whichever end of the ship the launcher is located, high enough for clear field of view.

l. REFERENCE — Technical Manual, Telemetric Data Receiving-Recording-Scoring Set AN/SKQ-3(V), NAVSEA OP-3996.

7-2.3 AN/SKQ-3B

At the time of this writing the Navy is employing an improved version of the SKQ-3 system called the SKQ-3B. This model has the ability to frequency scan, sector scan, acquire, track, receive, record, process, and display telemetry data from a single transmitter. Little is known of the principal SKQ-3B antenna currently except that it is a rotatable, directional, radome-covered planar array of twelve crossed dipoles. It has a wide acquisition beam of 40° in azimuth and 90° in elevation, but, upon lock-on, switches to a narrow beam mode covering 12° in bearing and 42° in elevation for auto-track. The RF group, of which the antenna is the chief element, is 41.2 inches high and 25.25 inches in diameter and weighs 140 pounds. To provide an even wider area of simultaneous coverage, a separate horn antenna having a 30° x 30° beam is mounted adjacent to the main antenna when this added capability is desired.

7-2.4 AN/SKR-4 (LAMPS MK I)

The AN/SKR-4 is an FM telemetry-receiving set used as the shipboard portion of the MK I Light Airborne Multipurpose System (LAMPS). Consisting of two RF links, one a VHF sonobuoy-to-aircraft data link, the other an S-band aircraft-to-ship multiplex link, the system supplies anti-submarine warfare (ASW) information to the ship operators. In the sonobuoy uplink, as many as four sonobuoys can be employed simultaneously to feed data to a tactical aircraft. The aircraft then transmits a composite signal of the multiplexed sonobuoy data and one voice circuit to the ship. Selection of three RF channels is available from the SKR-4 and, for increased capacity, twenty channels from the model SKR-4A.

The system is comprised of three major shipboard units: a receiver-demultiplexer; the receive antenna; and a small, lightweight preamplifier (AM-6493) mounted topside next to the antenna through a short length of cable.

ANTENNA AS-2743/SKR-4

a. TYPE/USE — The AS-2743 is a fixed, omnidirectional, UHF (S-band) cylindrical receive antenna used with the AN/SKR-4 (and 4A) shipboard telemetry set. It is installed at or near the top of the mainmast to provide 360° azimuthal coverage with a 15° vertical beamwidth. The associated preamplifier must be mounted very near to the antenna.
b. PHYSICAL CONFIGURATION — [See Figure 7-7].
c. DIMENSIONS — (1) Height: 19 inches; (2) Diameter: 3 inches (5½ in. base); (3) Weight: 3 pounds.
NOTE — Preamplifier AM-6493, mounted near the antenna, is 3¼ inches high x 10 inches wide x 6 inches deep and weighs 14 pounds.
d. FREQUENCY RANGE — 2200 to 2290 MHz.
e. GAIN — 7 dB.
f. INPUT IMPEDANCE — 50 ohms with VSWR 1.6:1.
g. RF POWER RATING — Not applicable, receive only.
h. POLARIZATION — Vertical.
i. TYPE OF FEED — RG-219/U coaxial.
j. INPUT CONNECTOR — Type N.
k. INSTALLATION REQUIREMENTS — The antenna should be mounted near the top of the mast for as clear a view as possible. The base mounting flange has eight equally spaced 0.205 inch diameter holes.
l. REFERENCE — Technical Manual, Telemetric Data Receiving Set AN/SKR - 4 & 4A, NAVELEX 0967-LP-595-6010.

7-2.5 AN/SRQ-4 (LAMPS MK III)

The AN/SRQ-4 radio terminal set (RTS) is a shipboard equipment used to provide a secure, two-way (full duplex), digital data microwave communications link between the ship and LAMPS MK III equipped helicopters. In this manner the surveillance and attack capabilities of the host ship are extended well beyond the ranges attainable by other on-board systems. The SRQ-4 consists of a receiver-transmitter processor unit, a high-gain directional antenna, a small, short-range omnidirectional antenna, and an antenna control-monitor unit. The RTS uplink transmits command, control, and voice information from the ship to the LAMPS helo over a dual antenna system, and receives downlink digital sensor data, and command, control, and voice communications from the helo via the same shipboard LAMPS antennas. A crypto device within the RTS encodes and decodes the data. Encryption, combined with use of the narrowbeam directional antenna, impedes hostile interception of the LAMPS tactical data.

ANTENNA AS-3274/SRQ-4

a. TYPE/USE — The AS-3274 is a high-gain, rotable, directional, microwave antenna designed for use with the shipboard AN/SRQ-4 LAMPS MK III system to transmit and receive tactical data information. Comprised of a parabolic section reflector and dual-feedhorn, the antenna scans continuously 360° in azimuth and from -30° to +90° in elevation. To compensate for ship's pitch and roll, the antenna is mounted on a stabilizing yoke assembly. A large radome encloses the entire assembly to protect it from the sea environment.
b. PHYSICAL CONFIGURATION — [See Figure 7-8].
c. DIMENSIONS — (1) Height: 56⅞ inches (with radome); (2) Diameter: 61⅝ inches (radome); (3) Weight: 183½ pounds (with radome).
d. FREQUENCY — G-band.
e. GAIN — [Classified].
f. RF POWER RATING — [Classified].
g. POLARIZATION — Circular.
h. TYPE OF FEED — Waveguide, RG-95/U.
i. PRIMARY POWER REQUIRED — 115 vac, 60 Hz, three-phase.
k. INSTALLATION REQUIREMENTS — The antenna should be mounted on a platform high on the ship mast to allow a clear, unobstructed field of view. Any severe blockage must be compensated for by installation of a second AS-3274 antenna to provide 360° total azimuth coverage. To allow maintenance access, the hatch opening must have a minimum of 64¼ inches from the geometric center of the radome for door clearance. The maximum waveguide length from antenna to receiver-transmitter unit is 200 feet.
l. REFERENCE — Technical Manual, Radio Terminal Set AN/SRQ-4, NAVELEX 0967-LP-640-0010.

ANTENNA AS-3275/SRQ-4

a. TYPE/USE — The AS-3275 is a small, fixed, omnidirectional microwave antenna designed for close range operation with the shipboard AN/SRQ-4 LAMPS MK III system to transmit and receive tactical data information. The antenna consists of a rectangular waveguide input, a transition to a circular waveguide, a circular polarizing section, a radiating aperture, and a divergent lens. A fiberglass covering protects the antenna from the sea environment.
b. PHYSICAL CONFIGURATION — [See Figure 7-8].
c. DIMENSIONS — (1) Height: 7⅛ inches; (2) Diameter: 4 inches; (3) Weight: 3 pounds.
d. FREQUENCY RANGE — G-band.
e. GAIN — [Classified].
f. RF POWER RATING — [Classified].
g. POLARIZATION — Circular.

h. TYPE OF FEED — Waveguide, RG-95/U.
i. PRIMARY POWER REQUIRED — None.
j. INSTALLATION REQUIREMENTS — The antenna should be installed as high as possible to be clear of obstructions for full 360° coverage. It is mounted in a vertical position using eight 8-32 UNC screws through the base flange plate. The maximum waveguide length from the AS-3275 to the AS-3274 antenna is 50 feet.
k. REFERENCE — Technical Manual, Radio Terminal Set AN/SRQ-4, NAVELEX 0967-LP-640-0010.

7-2.6 AN/SYR-1

The AN/SYR-1 is a shipboard communications tracking set used to receive downlink information from missiles relative to position, velocity, and performance. Data may be received on either of two RF channels, each operating independently of the other with separate antenna beams. That is, the SYR-1 is capable of handling two missiles simultaneously, one on each channel, in a continuous mode, or up to twenty missiles on a time-shared basis. After the data has been processed by the AN/SYR-1, it is routed to other telemetry equipment on board for recording, display, and evaluation.

ANTENNA AS-3393/SYR

a. TYPE/USE — The AS-3393 is an electronically steered, phased array receiving antenna designed for use with the AN/SYR-1 shipboard telemetry system. For signal acquisition, a wide search beam pattern is used, and upon acquisition, a narrow scan pattern is generated to track the missile. Four antennas are employed per system, each covering a 90° sector, so that the total azimuthal coverage is 360°. Moreover, each antenna provides two simultaneous and independently steered beams.
b. PHYSICAL CONFIGURATION — [See Figure 7-9].
c. DIMENSIONS — (1) Height: 24½ inches; (2) Width: 49 inches; (3) Depth: 25⅓ inches; (4) Weight: 256 pounds.
d. FREQUENCY RANGE — [Classified].
e. GAIN — 13 dB.
f. INPUT IMPEDANCE/VSWR — 50 ohms with VSWR 1.5:1 maximum.
g. RF POWER RATING — Not applicable, receive only.
h. POLARIZATION — Vertical.
i. TYPE OF FEED — RF coaxial cable.
j. PRIMARY POWER REQUIRED — (1) 115 vac, 400 Hz, one-phase; (2) 115 vac, 400 Hz, three-phase.
k. INSTALLATION REQUIREMENTS — The four antennas are mounted port and starboard in sectors about the superstructure to provide full 360° azimuthal coverage. A typical arrangement is illustrated in Figure 7-9(a).

l. REFERENCE — Technical Manual, Communication Tracking Set AN/SYR-1, NAVSEA OP 4548.

7-2.7 CV-ASWM

CV-ASWM is a specialized shipboard command support system placed aboard naval aircraft carriers as an anti-submarine warfare (ASW) nerve center. The system functions as a compact, integrated, automatic control point for rapid target detection, classification, evaluation, and decision. Real-time sonobuoy data is relayed to the ship via patrolling ASW aircraft, where it is received by the CV-ASWM's multichannel UHF antenna.

ANTENNA AS-3503/SKR

a. TYPE/USE — the AS-3503 is a fixed, omnidirectional, multichannel S-Band UHF antenna used with the shipboard CV-ASWM system for receiving relayed ASW data. A conformal array of 6 cavity-backed slot antennas, the array is circular and consists of two hinged halves so as to wrap around and clamp onto an 18-inch (or smaller) ship mast. Each half combines to form an overall circular pattern (in azimuth), having a 12° vertical beamwidth. An epoxy-fiberglass radome protects the array from the sea environment.
b. PHYSICAL CONFIGURATION — [See Figure 7-10].
c. DIMENSIONS — (1) Height: 24 inches: (2) Diameter: 15.81 inches (O.D.): (3) Weight: 60 pounds.
d. FREQUENCY RANGE — 2200 to 2290 MHz.
e. GAIN — 4 dB minimum.
f. INPUT IMPEDANCE/VSWR — 50 ohms with VSWR of 2.0:1 or less.
g. RF POWER RATING — 100 watts CW.
h. POLARIZATION — Vertical.
i. TYPE OF FEED — Coaxial cable.
j. INPUT CONNECTOR — Type N.
k. INSTALLATION REQUIREMENTS — To afford omnidirectional line-of-sight reception, the antenna is mounted around the mast, near the top, to be free of surrounding obstructions.
l. SPECIAL CONSIDERATIONS — Masts smaller than 15 inches in diameter require standoff spacers welded to the mast to accommodate the 15-inch (I.D.) AS-3503 antenna. The antenna does not have to be installed concentrically; it may be offset to allow cabling and mast hardware to pass up through the inner ring spacing.
m. REFERENCES — (1) Drawing, Outline and Mounting Dimensions, Antenna Assembly AS-3503/SKR, Naval Air Development Center Drawing SA-TS-27263. (2) Ball Aerospace Systems Division Specification Summary AS-3503/SKR.

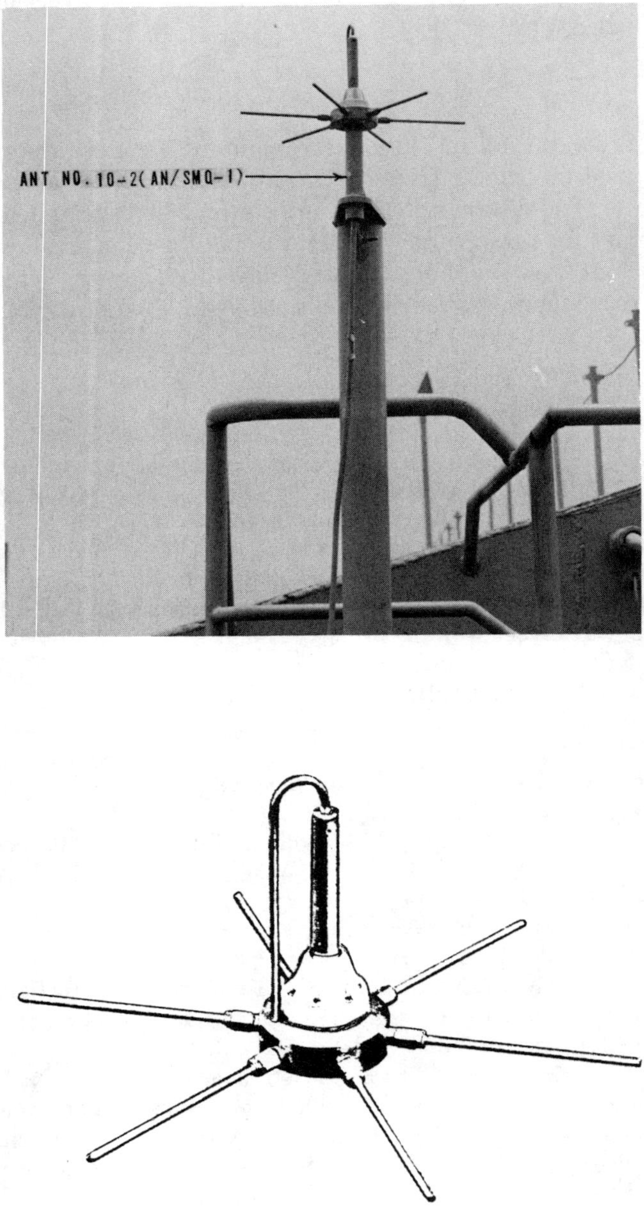

FIG. 7-1 FGPJ-151B-N-NP Meteorology Antenna

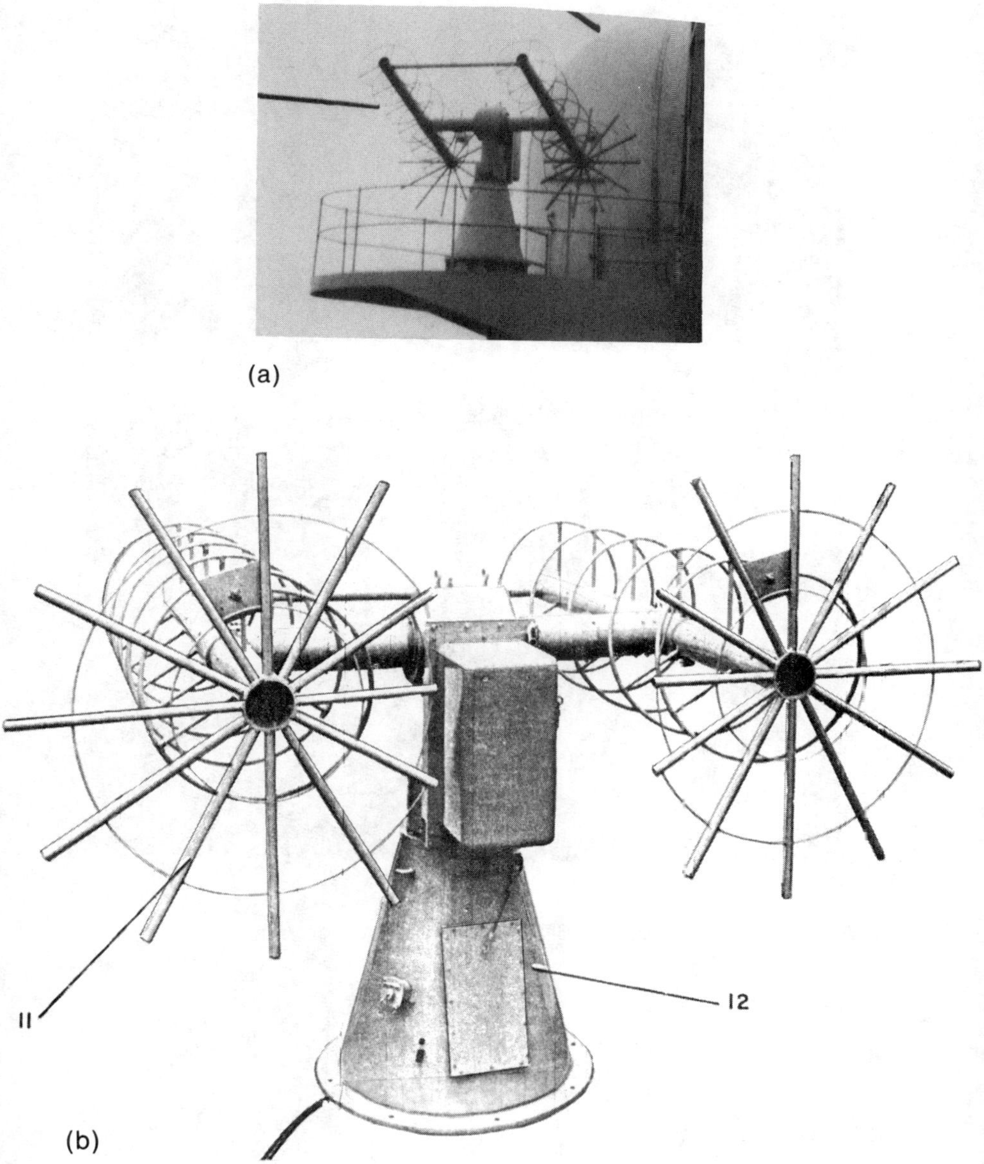

FIG. 7-2 AS-2192/SMQ-6(V) Meteorology Antenna

FIG. 7-3 AS-3185/SMQ-6D(V) Meteorology Antenna

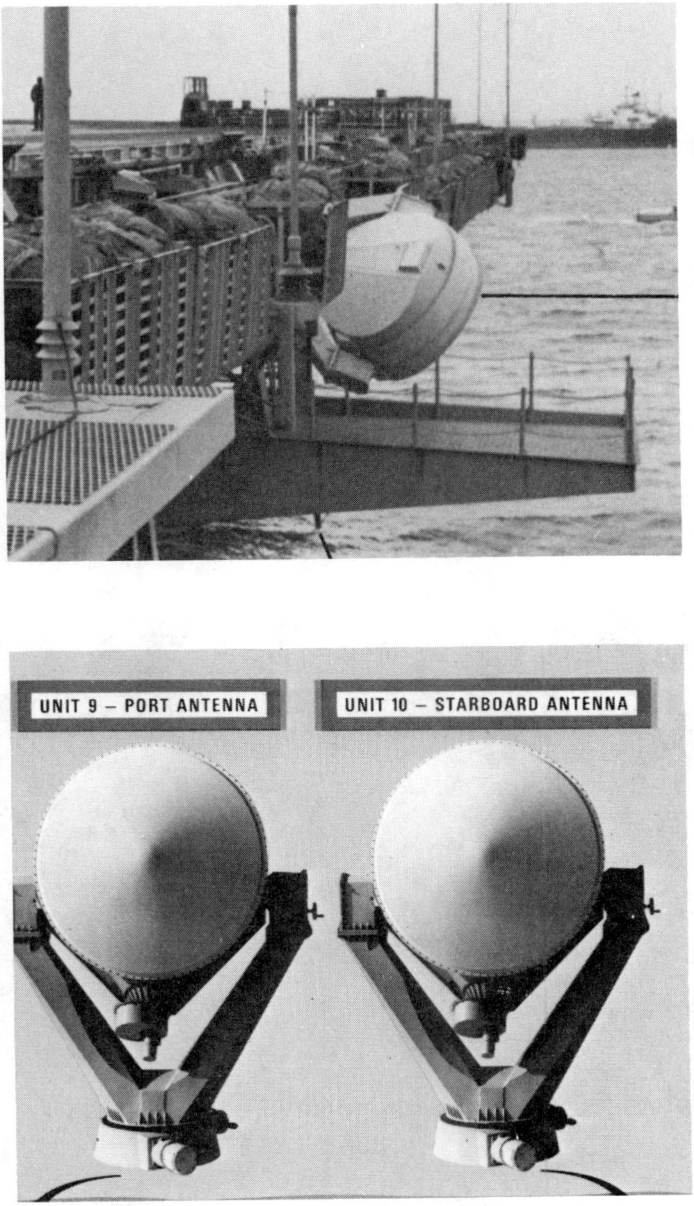

FIG. 7-4 AS-3063/SMQ-10 Meteorology Antenna

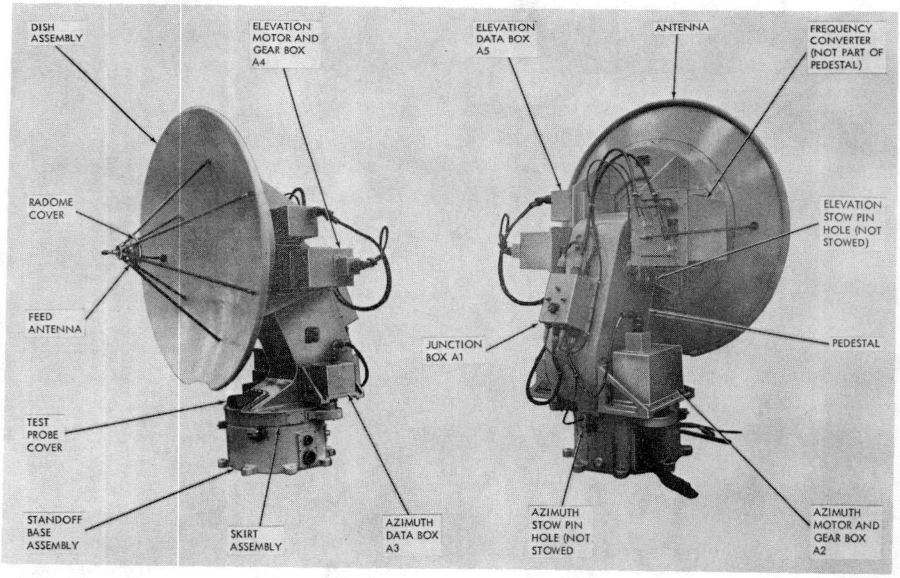

FIG. 7-5 AS-2535/SKQ-2 Telemetry Antenna

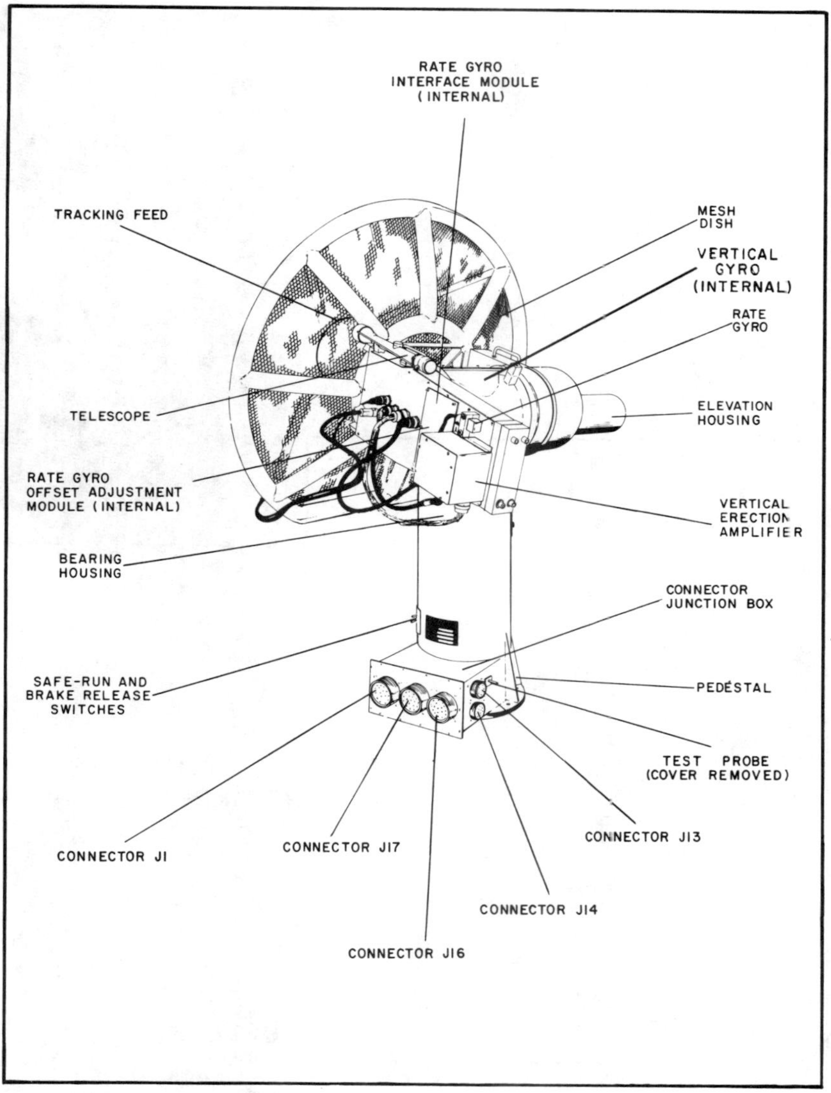

FIG. 7-6 AS-2727/SKQ Telemetry Antenna

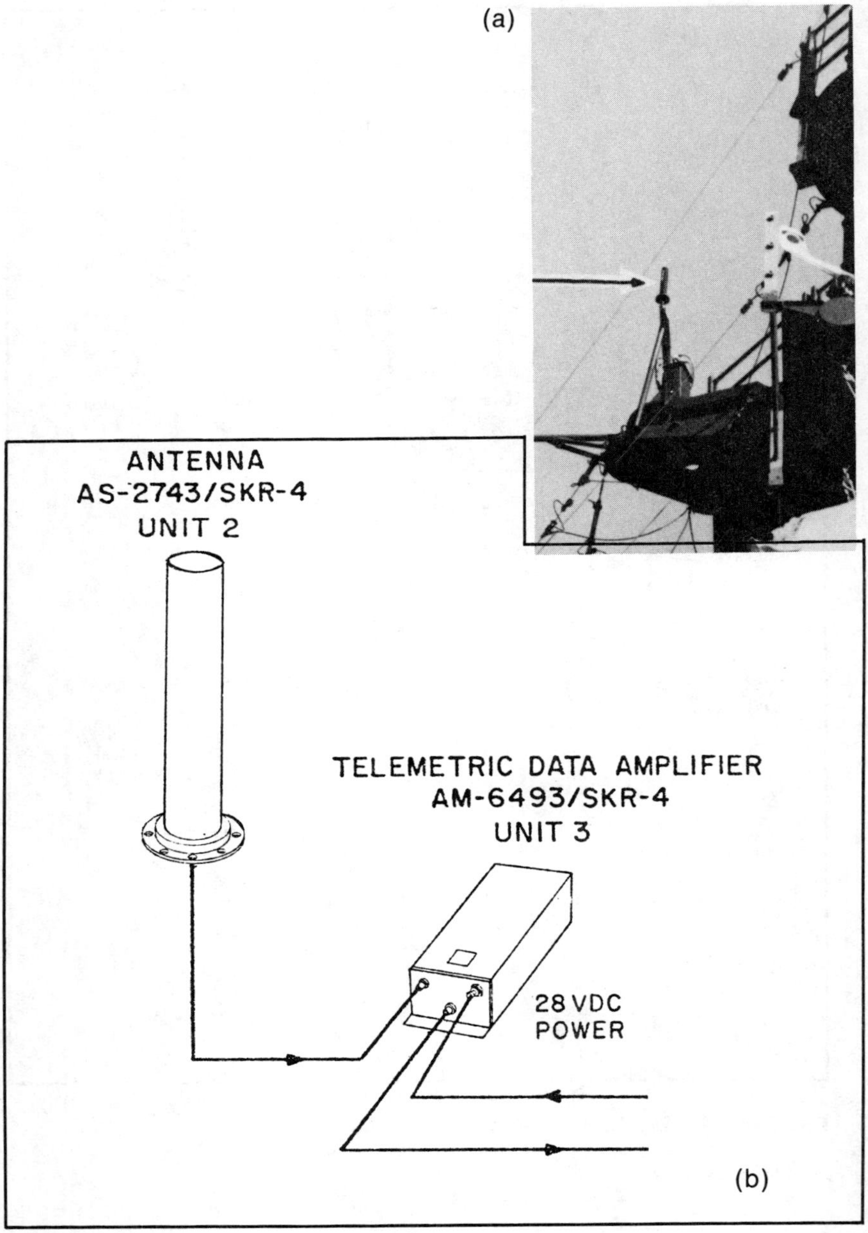

FIG. 7-7 AS-2743/SKR-4 LAMPS MK I Telemetry Antenna

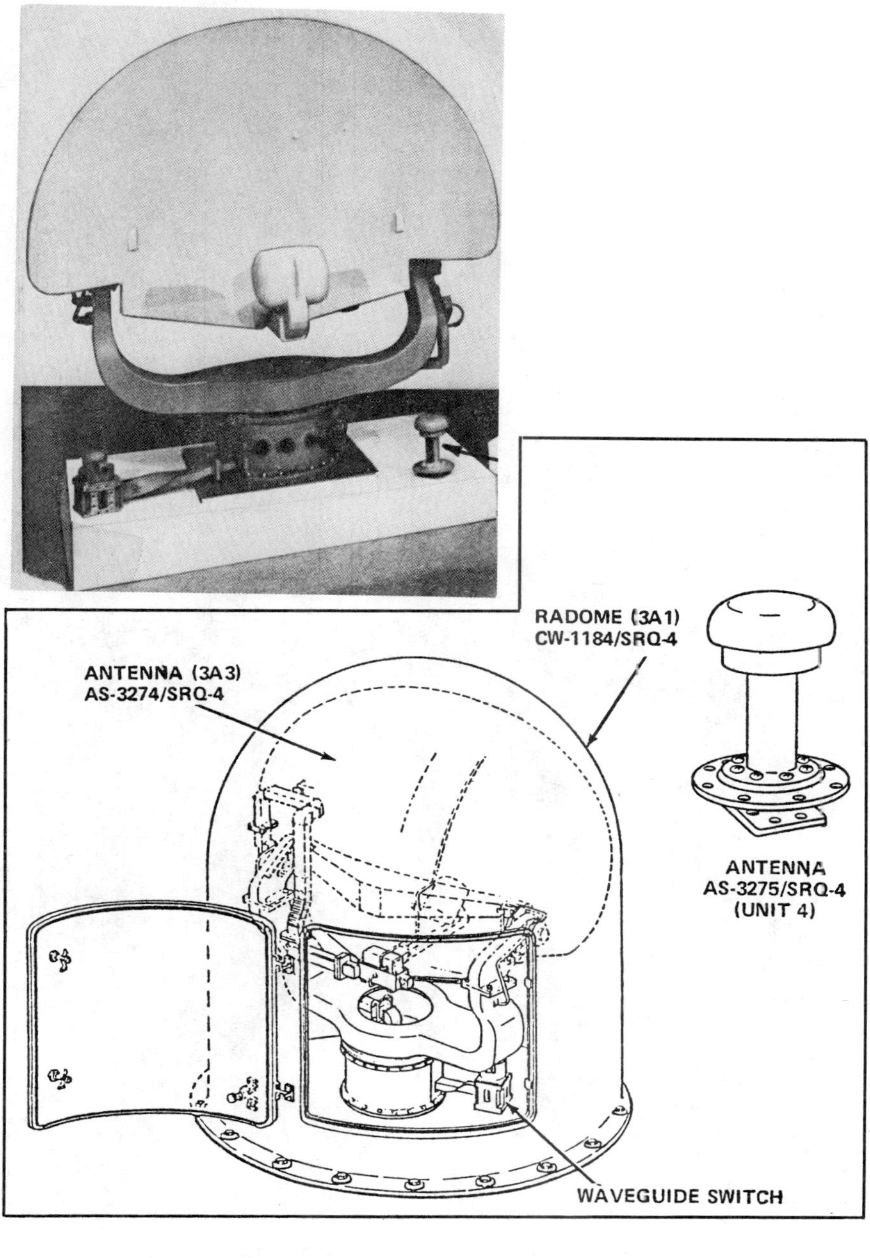

FIG. 7-8 LAMPS MK III Telemetry Antenna

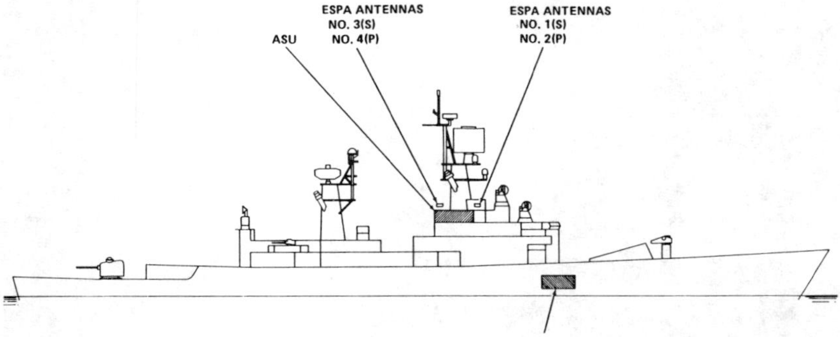

(a) Typical Placement

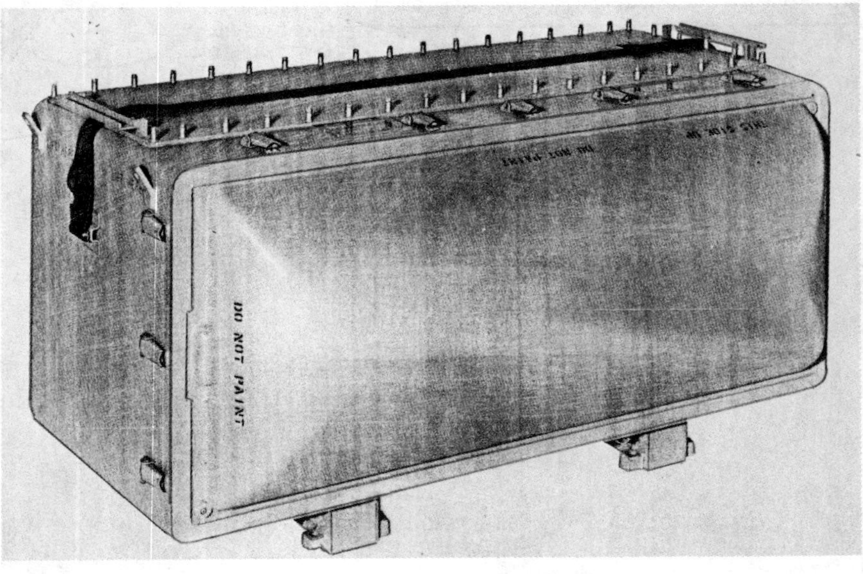

(b) Single Antenna Unit

FIG. 7-9 AS-3393/SYR Missile Tracking Telemetry Antenna

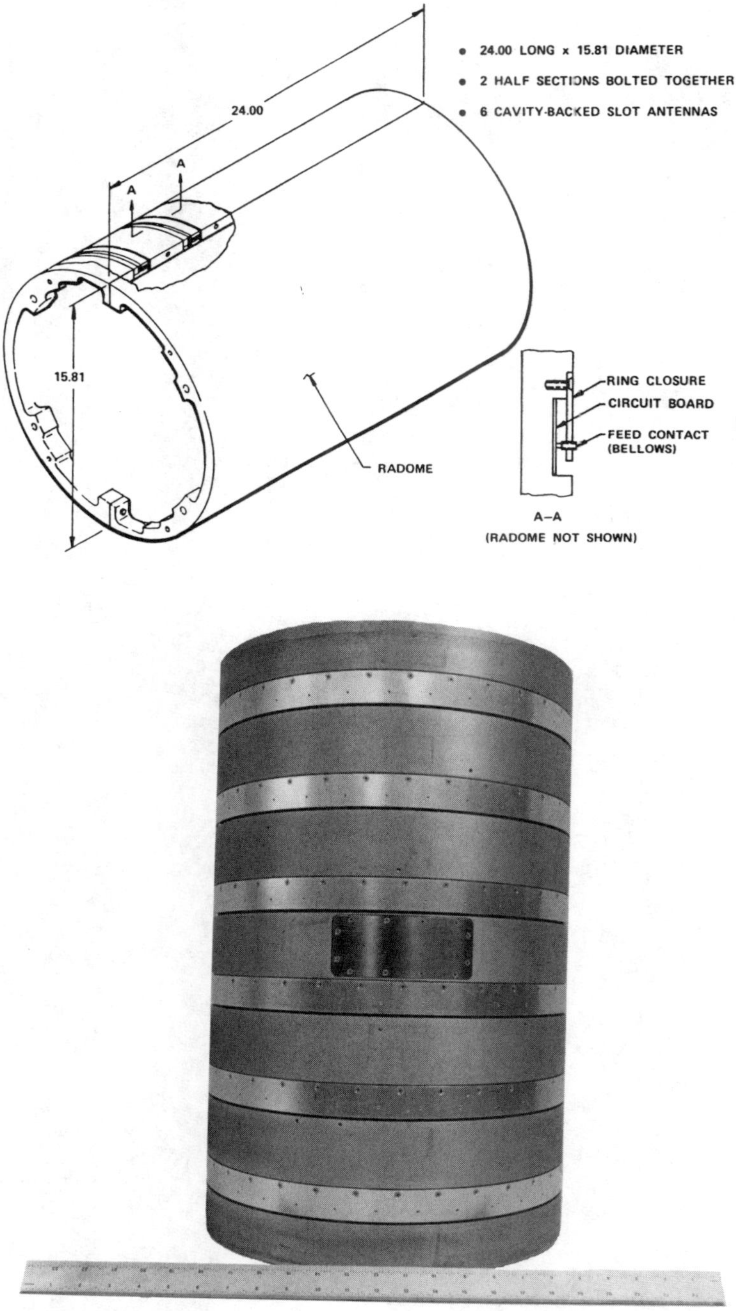

FIG. 7-10 AS-3503/SKR Telemetry Antenna (Courtesy of Ball Aerospace)

Chapter 8

Shipboard Antenna Systems Integration

8-0 THE PROBLEM IN GENERAL

In the introduction to this book, note was made of the difficulties involved in finding suitable locations aboard ship for the numerous, varied types of antennas needed to fulfill mission requirements. The chief cause of these difficulties, as was pointed out, is the severely limited space available topside. However, the intrinsic electromagnetic nature of antennas is itself a major cause of the problem. Characteristically, an antenna functions at its best when alone, well isolated from any of its own kind, from any other electromagnetic devices, and from any objects nearby which may interfere with good performance. Unfortunately, isolation in the shipboard environment is virtually impossible, interference is everywhere evident, and antennas are peculiarly sensitive to interference in many forms.

Shipboard antenna interference may be categorized primarily as that due to blockage, coupling, RF emission, and high-level radiation.

Blockage — When an antenna must be placed near to a large object such as a mast, portions of the superstructure, or other antennas, a corresponding sector of its intended coverage will be shadowed out. If the antenna is to be used for receiving, that blocked sector is unuseable and is either surrendered as such or a second antenna is added to fill in the gap (complementary coverage). If instead the antenna is employed for transmitting then its radiated energy, unless prevented from doing so, will impinge upon the offending obstacle, causing reflections and scattering (and quite likely coupling and reradiation), thereby distorting, perhaps significantly, the radiation pattern.

Coupling — If the nearby obstacle possesses certain electromagnetic characteristics, for example if it is made of metal, or is another antenna, then mutual coupling with this parasitic element will result, altering the impedance as well as the pattern. The effect on the system VSWR (transfer of energy) may be so great as to drastically reduce the antenna's utility.

RF Emissions — Reception of undesired emissions is one of the most frequently encountered forms of interference aboard naval vessels. It is the natural consequence of a limited platform crowded with so many radiators. Unwanted signals are generated on ship as harmonics, intermodulation products, noise spikes, and broadband noise, to be picked up by onboard sensors used for receiving distant, and often much weaker, signals. Suppression of electromagnetic interference (EMI control) and protecting the ship's own receiving equipment is a challenging, unending process in topside design and integration.

High-level Radiation — In many instances, particularly on warships, high power radiators are required for carrying out ship missions. Often the emitters are microwave, posing biological hazards to personnel, but even at lower frequencies the high energy levels are a threat to onboard fuel and explosives. A common problem is that of HF transmitting antennas inducing currents in nearby metallic structures of sufficient electrical height (fractional multiples of wavelengths at various frequencies) to cause RF burns to personnel coming in contact. Restricted locations must be allocated aboard ship in which to place these high power emitters so as to minimize the dangers of RF radiation.

Clearly, in the naval shipboard case, all the above conditions inherently must exist. There can be little or no true isolation of antennas. There will always be sources of interference close by — several in fact — which will indeed adversely affect antenna (and, therefore, mission) performance. Consequently, it is this high potential for interference that the shipboard antenna systems engineer must anticipate, and with which he must cope, during the design and integration process.

8-1 THE INITIAL PROCESS

Three words highly descriptive of the design and integration process are *iterative*, *competition* and *compromise*. The first, iterative, is most apparent in new ship design programs, but the third, compromise, is a key factor, whether in terms of new designs or alterations to active fleet ships. In either event, any antenna authorized for installation aboard a Navy surface ship must compete for platform space. Normally the result is some degree of compromise and degraded performance ranging from mild to severe.

In the new ship design process, which is the most detailed and therefore the most illustrative case, the effort begins with the top level specifications (TLS) derived from the operational requirements for the ship as issued by the Chief of Naval Operations (CNO). The TLS describes and develops candidate systems, and options, proposed to satisfy the operational (mission) requirements. Space allocations and a general hull arrangement are defined, along with initial trade-off studies involving various configurations. Additionally, specific equipment and interface needs are identified in support of the missions.

8-2 PRELIMINARY STUDIES

The engineering problem facing the antenna systems designer is, then, to somehow place each antenna in the topside to (a) provide good coverage — that is, to avoid blockage of the radiation pattern, (b) realize maximum intended range for each antenna's purpose of communications, or navigation, or radar target search, detection, acquiring, tracking, illumination, and weapon control, (c) avoid being susceptible to EMI, and (d) avoid being a radiation hazard to ship personnel, ordnance, and fuel. Only the most careful thought in placing the antennas can produce a topside integration which effectively achieves all of these objectives together.

Let us begin with a look at the ordinary functional classifications of antennas to be placed in the topside. For system design and integration purposes, shipboard antennas fall generally into one of three categories:

I. Omnidirectional antennas used mainly for communications, navigation, and passive reception to satisfy the need of ships and aircraft to maneuver independently of each other and fixed radio stations.

II. Directional antennas used for transmitting and receiving spatially concentrated energy in one direction at a time; e.g., radar, weapons control, ECM, and satellite communication, to radiate to or obtain information from remote sources.

III. Directional antennas used to determine bearing of incident radiation; e.g., direction finding, navigation, and electronic warfare (ESM).

To accommodate these three classes of antennas, four specific approaches are taken:

1. Broadband excitation of the masts and superstructure as in the case of high-frequency wire-rope fan type antennas.
2. Probe excitation of ship structures as with very low frequency tuner whips (e.g., Loran and Omega).
3. Tuned independent antennas such as 35-foot whips with base couplers.
4. Directional, independent, antennas and arrays such as those used for radar and weapons control.

Years ago the routine procedure for arranging shipboard antenna systems started with engineering judgment of the best possible layout of antennas in the topside and ended with experimental verification. However, as ships became increasingly sophisticated, so too did their antenna requirements. Ship commanders began to rely very heavily on electronic systems such as communications, radar, navigation, gunfire control, friend or foe identification, electronic countermeasures, and aircraft operations. Complex, intricate below-decks electronic equipments were found to be useless unless matched with satisfactory antenna performance. Former methods of antenna design and topside arrangements were no longer adequate. It was realized that shipboard antennas, rather than being independent systems, are, in fact, topside subsystems which must be tailored to each ship type in order to operate effectively within the constraints of very limited space and weight, high ambient RF fields, a highly corrosive atmosphere, and in competition with the many other users of a ship's topside. Careful, scientific, planning had to be adopted.

8-2.1 The Topside Design Team and EMC Advisory Board

During the early 1970s, the Navy inaugurated a new method of topside design by committee. Recognizing that there are many different contenders for topside space and mission requirements, with each competitor careful to protect a special interest, a panel of engineering and architectural experts (for each ship design) was formed to produce systematically a topside design in the aggregate. The product of course represents a compromise, the best compromise to meet jointly the needs of all competitors in support of the overall ship mission.

Used successfully now for well over a decade, this design committee is termed the Topside Design Information Exchange Team, or more commonly, the *TDIET*. Because of the crucial nature of electromagnetic compatibility and control of electromagnetic interference in meeting the ship missions, the antenna systems integration engineer serves as chairman of the TDIET. Other members include principal engineers, architects, and consultants from the multi-discipline design areas of naval combat systems, weapons arrangements, hull structures, communications, electronic warfare, and navigation lighting.

Concurrently, to ensure and enforce electromagnetic compatibility of the combat systems during the design process, the antenna systems engineer also chairs an adjunct committee known as the Electromagnetic Compatibility Advisory Board, or *EMCAB*. Members of the EMCAB are EM specialists from naval headquarters commands, naval laboratories, naval weapons centers, industrial activities, and universities.

Particularly noteworthy in this scheme is that everyday debilitating or annoying EMI problems being experienced in the fleet, and the particular resolution applied, are fed back in real time to the TDIET and EMCAB sessions. In this manner the designers are able to avoid or correct deficiencies so that these EMI problems will not be perpetuated in the topside design.

8-2.2 HF Antenna Systems Integration

Given the proposed outlines of a new hull, the antenna designer is concerned immediately with the interrelationships of major topside items: the height and shape of the superstructure, placement of the deck weapon systems, location and form of the stacks, quantity and physical structure of the masts, and available installation space for antennas. At the early stages of topside design, none of the above are fixed. Placement of large, high power HF antennas on deck will affect performance of the weapons, and *vice versa*. The quantity and weight of antennas proposed for mast mounting may determine the number of masts and will certainly influence the shape and height of any mast. Height and geometry of the superstructure above the main deck may influence greatly the radiation characteristics and impedance of certain antennas. Thus, each item impacts the location and performance of the others. In the beginning, only gross arrangements can be suggested, with alternatives proposed as options.

An obvious first step in the preliminary systems integration is a serious attempt to reduce the total number of antennas required. Because of simple practicalities the most likely candidate is HF communications. While the optimum theoretical shipboard communications antenna is a half-wave vertical element (quarter-wave monopole plus image over perfect ground) located in a clear site to provide low-angle omnidirectional coverage at each frequency used in the 2-30 MHz range, three problems immediately arise to thwart this idealistic concept: number, size, and siting. To employ a separate, single-frequency (quarter-wave) vertical antenna for each of the HF frequencies used would require a ridiculous number of antennas. Similarly inordinate would be the height of antennas at the lower end of the HF range; e.g., 123 feet high at 3 MHz. Finally, it is of course impossible to install more than one or two HF antennas in a truly clear topside site. It therefore becomes imperative to minimize the number of HF antennas required to fulfill the high frequency communications needs.

As a consequence of these purely practical considerations, the Navy has had to compromise by selecting the 35-foot whip antenna as its standard vertical monopole. A quarter-wave at 7 MHz, the 35-foot whip's efficiency decreases very rapidly below about 4 MHz, but a height much above 35 feet becomes difficult for shipboard application. By employing a base coupler, the whip may be variably tuned to any frequency within the 2-30 MHz range, thus satisfying the objective of having one antenna able to handle the entire HF band. However, this method permits but one transmitter unit to be used with each whip, so that several whips are needed with a multiple transmitter system.

A second method now used routinely to reduce the overall number of shipboard HF antennas is to make use of broadbanding and multicoupling. By dividing the HF band into three overlapping segments, namely 2 to 6, 4 to 12, and 10 to 30 MHz, and using broadbanded antennas (wire-rope fans, trussed or twin whips — see Chapter 2, Sections 2-1.1 and 2-1.3), several transmitters or receivers may be operated into a single antenna. Not only is the number of antennas reduced but, since multicouplers are also filtering devices, electromagnetic interaction is lessened through high frequency isolation. The design goal is to have minimum channel spacing so as to allow a maximum number of communications circuits. At the present time, through the use of modern shipboard HF multicouplers, minimum transmit-to-transmit separation is 5% and transmit-to-receive separation is 8%.

Because other antenna types must satisfy specific one-to-one missions (radar, EW, weapons control, navigation) it is less likely that their numbers can be reduced on any particular ship platform. (Consequently the need arises for multifunction antennas such as the AN/SPY-1 phased array system.) Therefore the effort to reduce antennas is normally at HF. It should be made clear at this point that a major input to the topside engineering design effort is based in large part on HF model range studies.

8-3 ANTENNA MODEL RANGE STUDIES

Either at a naval laboratory or specialized industrial facility, scaled models of the ship with variable antenna configurations are subjected to carefully controlled measurements to determine feasibility of the preliminary arrangement, or to recommend the best alternative. These models are usually 1/48th scale and made of sheet brass (Figures 8-1 and 8-2), and include all topside structural elements influencing the HF antenna characteristics. Based on tests made on the range (Figure 8-3), changes to the model's topside may be made quickly and easily to expedite the HF antenna systems design.

8-3.1 HF Antenna Scale Modeling

The emphasis during modeling is on design of the broadband HF transmitting antennas, normally one each to cover the 2-6 MHz, 4-12 MHz, and 10-30 MHz frequency bands for each ship. The main objective is to provide, and integrate into the ship hull, HF communications antennas with efficient, omnidirectional, radiation characteristics by attaining a 3:1 VSWR throughout the entire frequency band. Such efficiency is achieved by the physical form and resultant topside placement of the antenna itself, and by exacting calculations to derive the inductive-capacitive L, Tee, or Pi type matching network inserted in the transmission line at the antenna feedpoint for maximum transfer of energy. Through modeling experimentation, test, and analysis, the component values of the variable inductors and capacitors, and their configuration in the matching network, are of such a degree of accuracy that only final tuning adjustments need to be done at the time of installation in the actual shipboard environment (Figure 8-4).

The antenna range, with its rotating lead sheet turntable simulating the sea "ground," is used also during modeling for radiation pattern measurements. Polar plots are made at specified band frequencies, and varying elevation angle cuts, over 360 degrees of azimuth. The resulting plots graphically illustrate (see Figure 8-5) the degree of coverage along with pattern nulls and perturbations of the proposed ideally omnidirectional antenna for a particular frequency and elevation radiation angle. Full scale measurements at sea have over the years conclusively validated the scale modeling results.

The brass modeling design and testing of broadband HF receive antennas is far less demanding than for the transmit antennas. As noted in Chapter 2, Section 2-1.2, naval HF receive antennas are not required to be highly efficient. On the contrary, it is desirable that they be inefficient to the extent of just matching the threshold of the external minimum atmospheric noise level to the receiver system internal noise. The receive antenna is thus made as insensitive as possible to locally generated highlevel interference of the shipboard environment, while still being an effective receptor for its intended purpose. The primary aim of the modeling engineer then is to choose locations with good all around reception while placing the receive antenna on the ship as far as possible from the HF transmit antennas, to provide maximum isolation and electromagnetic decoupling. The experienced designer is aware that additional RF protection to the receive system is afforded by employing HF receive multicouplers having highly selective filter networks. Therefore, the modeling work overall, being more simply resolved for the HF receive case, is mostly directed toward finding suitable designs for the HF transmit antennas.

Model studies have proved to be accurate, cost effective, and an invaluable aid throughout the life of a ship when future alterations to the topside may require a new run of model measurements.

8-3.2 Computer Modeling

By the mid-1980s, antenna systems design and integration engineers began making good use of in-house computer-aided graphics to rapidly provide visual assessments of ship antenna and weapons placement. Employing VAX 11/780 Topside Synthesis Modeling (TSM) routines, the designer is able to quickly determine the validity of varying options selected in topside siting. The ship hull is displayed in modular, three-dimensional form (Figure 8-6) and, when a particular location for an antenna or weapon is chosen, polar and rectangular plots show graphically the radiation coverage or firing zone coverage available. Likewise, blockage due to superstructure and deck obstacles is immediately evident from these computerized plots (Figures 8-7 and 8-8). Using TSM in this manner the systems designer is able to readily illustrate those arrangements which are viable for further consideration. From there, additional computer-aided techniques are used to predict systems performance in terms of range, frequency, power, gain, and expected electromagnetic interference to and from each emitter and sensor.

At the laboratory antenna range, to complement brass modeling, the U.S. Navy is now heavily engaged in computer math modeling. Using numerical simulation techniques such as the Method-of-Moments (frequencies below microwave) and Geometric Theory of Diffraction (microwave frequencies), shipboard systems are being math-modeled with good success. Results have correlated well with brass model measurements. The speed and flexibility of computer programs to provide impedance, coupling, radiation patterns, and EMI reduction will eventually cause math-modeling to become extensively employed on the antenna range.

8-4 EMC AND PERFORMANCE POTENTIAL ANALYSES

Once a fairly firm complement is obtained of the number and types of antennas required, the next step in the topside design and integration process is to begin tentative placement of antennas and to anticipate the impact that arrangement will have in terms of overall ship's predicted performance potential, electromagnetic interference (EMI), radiation hazards to personnel (RADHAZ), to ordnance (HERO) and fuel (HERF). In fact, it is this competition with other systems (and structures) that is most difficult to resolve. A first approach that might come to mind is to locate all antennas as high as possible, in the clear, for all-around (omnidirectional) transmission and reception. The masts and yardarms would seem the best choice. Unfortunately, as seen by Figure 8-9, there are problems with this choice: (a) communications engineers, weapons control engineers, radar engineers, navigation-aid engineers, and EW engineers all have the same hopes; (b) it is undesirable to have transmit and receive antennas in the same frequency band collocated — one mode, either transmit or receive — has to be placed elsewhere (normally, the transmit antennas are installed in the vicinity of the transmitter equipment room in order to minimize cable length attenuation losses); and (c) some antennas, particularly transmit antennas in the low portion of the HF band, do not function well when high above the water, as their radiation patterns begin to split up, forming multiple lobes in the elevation plane. To compound the problem even further, the yardarm and masts are used also to support flag halyards (which frequently become entangled in the antennas), commissioning pennants, navigation lights, and wind speed indicators. Moreover, some communications antennas, especially in the HF range, are much too large and heavy for mast mounting.

As a result only antennas that absolutely require such locations can be mast-mounted. For example, air-to-ground UHF communications antennas, TACAN, and DF antennas are installed high above the sea to get the maximum possible line-of sight range and have an azimuthal radiation pattern which is as nearly circular as possible. For large, heavy antennas, other locations must be sought, and competition for real estate begins to get quite difficult. On any ship there are areas which are immediately eliminated; e.g., helicopter operation areas, vertical replenishment zones, gun arc-of-fire zones, missile launching zones, cargo and boat handling zones, and visual navigation zones. Additionally, antennas should not be installed on stacks or next to fuel handling areas or ordnance stowage. For the antenna system designer, installation space seems to evaporate.

Isolation between antennas is maximized to the greatest extent possible. Separation of communications receiving antennas from high power transmitting antennas is necessary to prevent overloading the receivers and the generation of intermodulation products within the receivers. Isolation not adequately afforded by physical separation is compensated by frequency separation and filtering. It is also advisable, and in some cases essential, that isolation be provided between antennas of different functions; e.g., communications and radar, or search radar-to-navigation radar. A typical case is the requirement for certain satellite communications antennas to be located well away from ship-to-ship UHF transmitting antennas. Of course, the very heart of the problem lies in the lack of flexibility in isolating antennas, due to the physical limitations of the ship platform. Consider the important evolution of RADHAZ concerns as a critical case in point:

8-4.1 Radiation Hazards to Personnel and New Exposure Levels

Recognition and Early Developments

By the end of World War II, significant advancements had been realized in the ability to generate high power at the higher, microwave, frequencies. With the continued development and installation of high power shipboard radars for enemy detection and missile guidance, the need for control of RF emissions became evident. There was a growing awareness of RF radiation hazards in general — hazards to personnel, interference between various electronic equipments, and a number of unexplained accidents involving electro-explosive devices, all of which were suspected to have been caused by stray RF radiation.

The Navy's first programs for the control of electromagnetic radiation hazards to personnel began in the late 1950s. Early studies, and laboratory tests using small animals, demonstrated that at various levels of irradiation there were numerous observable effects, both thermal and athermal, with the major concern being the resulting rise in overall body temperature. The capacity of the human body to absorb heat was found to depend upon many factors, including physical size, frequency of radiation, and time of exposure. The body's ability to dissipate heat depends on such additional features as environmental air circulation rate, clothing, and blood circulation around the heated areas.

Aware of the potential danger of thermal effects, the U.S. Navy adopted standard exposure limits based on the power density of the radiated beam and the exposure time in the radiation field. The continuous exposure level selected was 10 milliwatts per centimeter squared (mW/cm^2) within the frequency range of 100 through 100,000 MHz. The limit for intermittent exposure was that the incident energy level not exceed 300 millijoules per square centimeter per 30-second interval.

In 1977, the standard was revised by lowering the frequency range to 10 MHz and extending the averaging time to 0.1 hour, while retaining the 10 mW/cm^2 continuous exposure limit.

Continuing biomedical research over the intervening years has shown that the permissible exposure levels (PELs) for personnel should account for frequency dependence and body resonance, both of which are functions of physical size. Additionally, the biological effects have been determined to be a function of what is termed the specific absorption rate (SAR) expressed in watts per kilogram (W/kg). As a consequence, the threshold for adverse biological effects is now recognized to be above 4 W/kg, and, with a safety factor of 10 allowed, the accepted SAR is 0.4 W/kg for the whole body, averaged over any 6-minute period.

During this time the American National Standards Institute (ANSI) C95.1-82 and the American Congress of Government Industrial Hygienists (ACGIH) developed recommendations for radio frequency protection guides based on the 0.4 W/kg SAR. However, there is a difference in the upper and lower frequency coverage and also in the permissible exposure to radiation of frequencies above 100 MHz. The ANSI standard is applicable for the general population, which includes small children, whereas the ACGIH standard is meant for the normal workplace where only full-size persons are permitted, which accounts for the difference in PELs.

New Navy RADHAZ Standards

In July 1985, the Navy, in an effort to provide an additional margin of safety, and at the same time minimize operational impact, set more stringent hazard limits for personnel exposure. These new levels are based upon a combination of the ANSI and ACGIH recommendations and have provisions for restricted and unrestricted access areas. Restricted access areas are those where persons less than 55 inches in height are excluded, which encompasses ships at sea. Unrestricted areas are those accessible to the general population, which would include ships in port.

Direct measurement of the specific absorption rate is presently possible only by using laboratory instruments and careful measurement techniques; however, for the purpose of compliance with the 0.4 W/kg SAR, derived equivalents in units of power density (mW/cm^2), volts squared per meter squared (E^2/m^2), or amperes squared per meter squared (A^2/m^2), may be used.

Impact on Topside Antenna Systems Siting

Because the new exposure limits are based upon radiation frequency, the limits for the 2–30 MHz frequency range now vary from a high of 100 mW/cm^2 at 3 MHz, to a low of one mW/cm^2 at 30 MHz, as illustrated in Figure 8-10. Considering that most of the Navy's HF transmitters are capable of operating at 1000 watts RF output, and that the space for shipboard HF antenna placement is limited, compliance with the new standards will be difficult to achieve. For example, when transmitters are operated at full power, the exposure level can be exceeded out to a distance of approximately 12 feet from the antenna. Adherence to the more rigid exposure limits must be realized by locating the antenna in an inaccessible area, or restricting personnel from the deck area surrounding the antenna, or reducing the transmitter power, or a combination of the three. It is hoped, moreover, that the operational impact can be reduced by the development and use of programmable power reduction devices to automatically limit power when required, and automatic alarms to warn personnel to remain clear of specified areas while the antennas are radiating under conditions where the limits would be exceeded. In any event, the tougher new exposure standards will cause topside design and integration to be an even more difficult challenge.

8-4.2 System Performance Trade-offs

Taking all the electromagnetic interrelationships into account, consultation with all other topside subsystems designers must be maintained through the TDIET on a routine basis. Weight and moment estimates of the deck-mounted, and particularly the mast-mounted, antennas and ancillary equipment must be provided to mast designers. Optical blockages to microwave antennas are determined. Blast and thermal environment parameters must be derived and the impact upon antenna systems analyzed. Where necessary, zone cut-outs must be incorporated into weapons systems. The total topside EM system performance potential is predicted, degradation allocated where required, and recommendations to improve EMC provided. In short, performance compromises must be reached among all major topside systems — not only antennas, but weapons and deck operations as well.

8-5 CANDIDATE ANTENNA ARRANGEMENTS

As a result of working closely with all the various engineering participants — hull, machinery, deck arrangements, weapons, electrical — candidate topside configurations are proposed. The options fulfill each of the requirements to the greatest extent possible; however, it is recognized that no single solution is capable of meeting all requirements. The trade-off studies determine those options most nearly meeting requirements, with the risks inherent in selection of each. Recommendations are made, with documented rationale for the selection, including the identification of any risks and deficiencies of the resultant system. The final selection of a particular topside configuration then becomes the baseline arrangment drawing, which itself is refined over and over, with changes necessitated by policy decisions involving such diverse factors as ship cost, size, and manning.

8-6 POST-DESIGN PHASE

The shipboard antenna system design and integration process has not reached completion. During both the shipbuilding and active fleet life, revisions are made, ranging from simple addition of platforms and structural reinforcements to major changes in ship equipment complement. Such modifications more often than not affect antenna characteristics, usually adversely. Therefore, the process goes on; in assessing the impact on topside systems performance, in minimizing the effects of EMI, and in preventing radiation hazards — virtually for the entire lifetime of each ship.

Examples of topside changes which may seriously degrade antenna performance include addition of deck houses; extensions to bridge wings; modifications of mast and yardarm configurations; additions, deletions, or relocations of antennas; and changes in weapons systems. Since each antenna has been tailored to its specific environment, such alterations may have a dramatic effect upon antenna performance. Ultimately, then, it is the ship's performance and its mission capability that suffer. There is, therefore, a need for rigid configuration control of the topside. This is accomplished by designating a single control point to maintain and issue the topside antenna systems drawings. Periodic onboard surveys are made to determine actual "as is" arrangements, and then to feed back the noted deviations and additions to the configuration control point so that the ship drawings may be updated and kept current.

How well the total, integrated shipboard antenna system will function in support of the ship missions is only determined when put to the test of actual operations. At that time the quality of the overall topside design, incuding the initial planning, the model range studies, the EM compatibility analyses, the coordinated iterative efforts to reach compromises, and the EM performance assessments, in derivation of the topside antenna arrangement, may be finally and fairly judged.

8-7 A LOOK AT FUTURE NEEDS

Throughout its long and distinguished electronics history the U.S. Navy has used, for the most part, conventional (one might even say traditional) type antennas for its surface ships. That is, shipboard antennas are ordinary in the sense that they are the usual type long wires, monopole whips, discone cages, vertical dipoles, planar arrays, and parabolic dishes seen on land sites and vehicles. For shipboard use they of course have been specially ruggedized and weathersealed to withstand the harsh sea environment. Thus adapted, these mainly conventional antennas have served the Navy quite well for many decades.

There are, however, some noteworthy disadvantages in employing conventional antennas in surface ships. For one thing they are easily recognized and identified by any specialist trained in the science of electromagnetics. Experts need hardly more than to view the form of the antenna, and scale its size, to quite accurately surmise the antenna's frequency, gain, and probable use as radar, weapons control, communications, navigation, or countermeasures. One then might hazard a fair guess as to how critical the antenna might be to the ship mission, and whether it acts alone or has auxiliary antennas for system redundancy.

For another thing, conventional shipboard antennas, prominently exposed and abundant in the ship topside, are too vulnerable to battle damage.[1] They are, moreover, vulnerable to the deleterious effects of ownship missile blast, gun overpressure, and metal corrosion accelerated by constant exposure to mixes of sea spray and fuel exhaust contaminates. Frequent, difficult maintenance is required to keep the antennas in good order.

Finally, in an effort to gain clear views, the antennas, many of which are quite large, must be placed upon pedestals, barbettes, atop deck houses, and on mast platforms, resulting in significant undesirable contribution to ship weight and moment.

How then might we ameliorate such drawbacks? Can these noteworthy disadvantages be overcome in future naval ship design? Serious consideration to do just that is warranted.

To begin with, thought should be given to moving away from having so many antennas placed about the open deck and superstructure as independent, isolated topside items. Instead, there should be a move toward having antennas integrated *within* the hull and superstructure. For example, antennas could conform to the shape of the ship, flush with the hull, by being recessed within. Such a notion does not necessarily represent new technology. To satisfy the natural laws of aerodynamics, most aircraft antennas are recessed neatly within the airframe. Submarines, when preparing to submerge, retract their antenna masts, pulling them safely inside. Space exploration vehicles make use of exotic antennas that are carefully stowed within the satellite and launching rocket, thus able to withstand enormous thrust and vibration forces.

Upon reaching desired orbit, they cleverly unfurl and blossom forth in space.

Cannot naval ships, older in history than airplanes, submarines, and spacecraft, make use of similar technology? Mission-critical shipboard antennas, at least, could be better protected from combat damage, and identification by the curious, through such innovative techniques as:

1. *Flush-Mounted Antennas* — This concept is well-used by the aircraft industry with obvious success. With a hull as large as a warship, there would be many opportunities to employ a wide variety of sensors recessed within the contour of the main hull, superstructure, and mast. Examples readily adaptable might be: cavity-backed planar spirals, corner reflector arrays, horn arrays, planar linear arrays, imbedded wires, and microwave lenses.

The AN/SPY-1 is an excellent example of this concept, but many EW antennas could likewise be so integrated. A case in point is the AN/SLQ-32. If the SLQ-32 faces are currently flush with the front of a very large "box" housing, why not do away with the box itself and place the SLQ-32 within the hull so that only the antenna faces are exposed? It would be feasible to do this also with SATCOM and some HF communications antennas. Serious research would very likely uncover many additional adaptations for flush mounting.

2. *Mast Shroud* — The Navy has sponsored some investigation into covering the entire mainmast with a fiberglass cowling. This method would provide protection from ballistic fragmentation to principal radar antennas used for surveillance and navigation which are not housed at present within individual radomes. Of course a host of other mast-mounted antennas would also benefit from the shroud. An added everyday benefit is complete protection from the sea environment, providing longer life and ease of maintenance. The current drawbacks to this program are cost and difficulty of providing a composite material with varying "radar windows" in the shroud for the different radar wavelength transparencies. Nevertheless, the technology is feasible.

3. *Radome Covers* — Mission-essential sensors can be placed within radomes as a means for protection. However, as evident from a look at ships with an AN/SPQ-9, or AN/SPN-35, to house antennas such as the AN/SPS-48, SPS-49, SPS-43, SPG-60, or SPN-43 in this way would require massive volumes of topside space where space and weight are normally at a premium.

4. *Retractable Multifunction Mast Array* — This technique is borrowed from submarine technology and is functionally well-established, already

encompassing VLF receive, MF/HF transceive, UHF line-of sight transceive, UHF SATCOM, and UHF NAVSAT integrally stacked on a single pole. A shipboard version of such a mast array could feasibly add further a circular phased array search and navigation radar, TACAN, and IFF. One or two of these masts could be stowed below deck, strategically located in a small well or "silo" and then cranked upwards to project above deck when so required for essential operations.

5. *Clam Shells* — Similar in concept to the retractable mast, certain critical antennas could be located about the main deck, or atop deck houses, in a recessed mode until needed, whereupon a hatch-like cover would spring up and become a shaped reflector, or a corner reflector, set at a predetermined angle for a primary radiator elevated into place when the hatch opens. Such antennas could be designed for air traffic control, EW, communications, SATCOM, NAVSAT, and even weapons control, to be used in crisis situations.

6. *Unfurlable Antennas* — This technology is heavily relied upon in space satellite engineering. Similar means could be adapted to shipboard, with emergency reflector-type sensors furled and stowed about the superstructure and mast, ready to pop out when called upon during topside casualties.

7. *Portable Redundant Antennas* — Perhaps the simplest immediate solution is to provide spare antennas not meant to withstand the rigors of shock, vibration, thermal, blast overpressure, and corrosive sea environment of permanently installed topside antennas. Such emergency antennas could range from simple coiled wire and thin fiberglass whips to portable lightweight 5-foot diameter parabolic dishes similar to those employed by TV crews for hauling about on vans and small trailers from location to location. Any of these type antennas could be run up on deck and clamped down in position, to take over during casualties and serve until replaced again by permanent fixtures.

The above ideas are but a few of perhaps many potential measures that might be employed for better protection of ship antenna systems. The Navy has been slow to adopt such techniques because generations of development have gone into perfecting conventional designs that are rugged, practical, and work well. It is always difficult to promote a shifting away from the tried and proved. Yet, such a shift in thinking must indeed be urged. The benefits are too important to the future health of the fleet to ignore.

REFERENCE

1. RADM Julian S. Lake, USN (ret.), "On Bringing Back The Battleship," *Defense Electronics*, March 1982, p. 22.

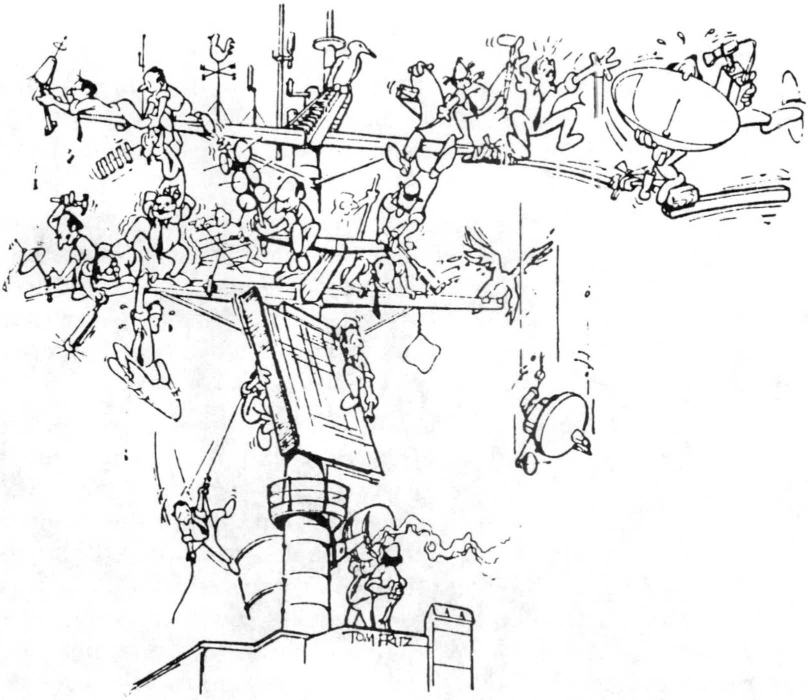

Tom Fritz

FIG. 8-1 Scaled Brass Models (1/48)

FIG. 8-2 Scaled Brass Model — Mast Areas

FIG. 8-3 Scaled Model On Test Range

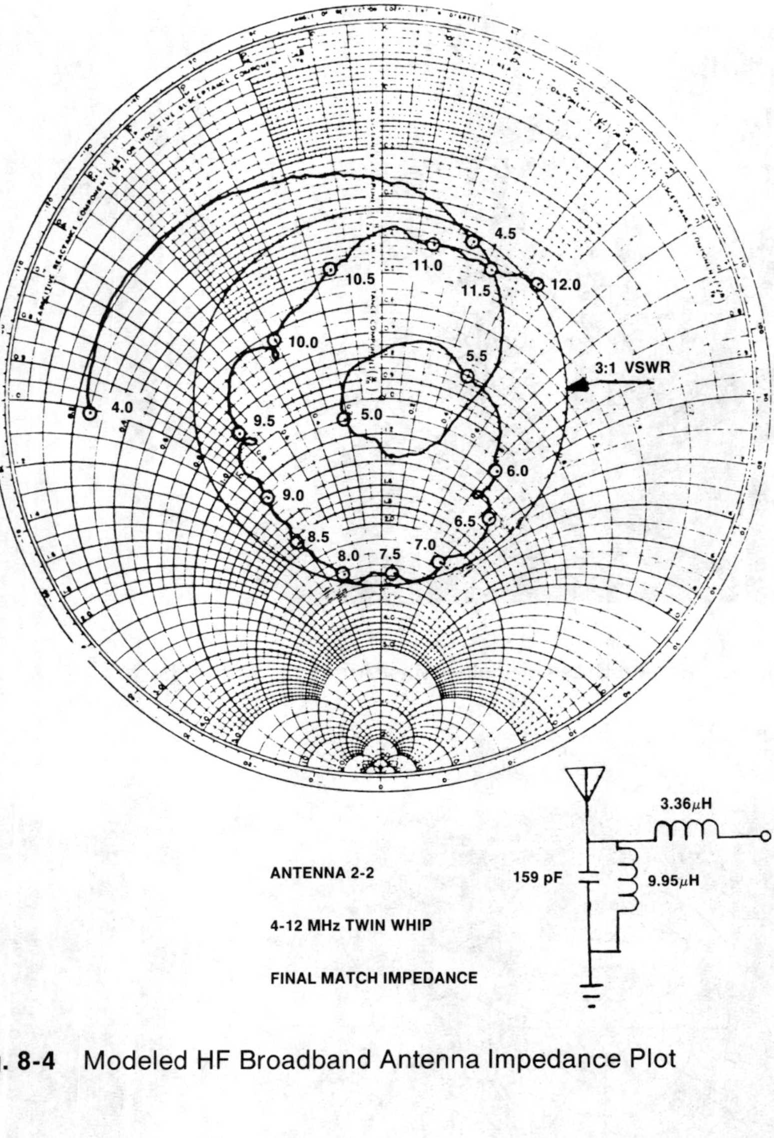

Fig. 8-4 Modeled HF Broadband Antenna Impedance Plot

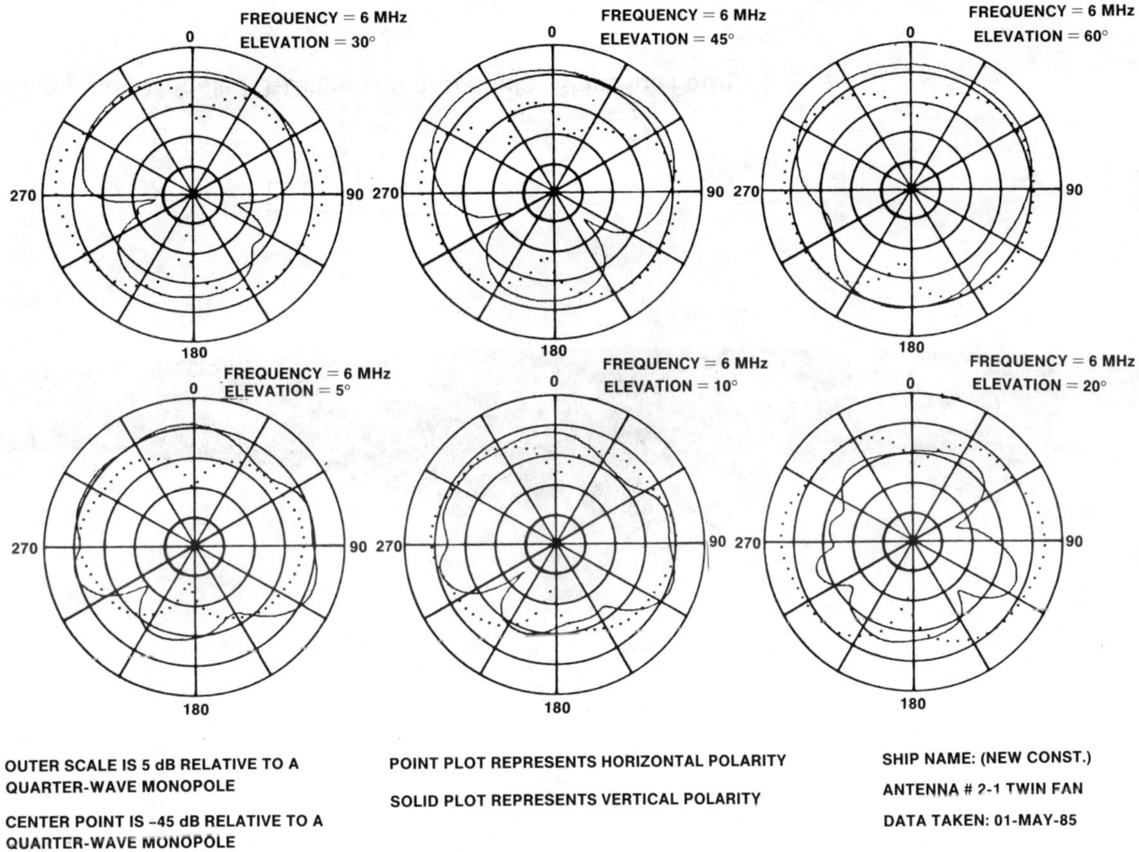

Fig. 8-5 Modeled HF Broadband Antenna Radiation Pattern Plots

Fig. 8-6 Computer-Modeled 3-Dimensional Modular Ship Hull Form

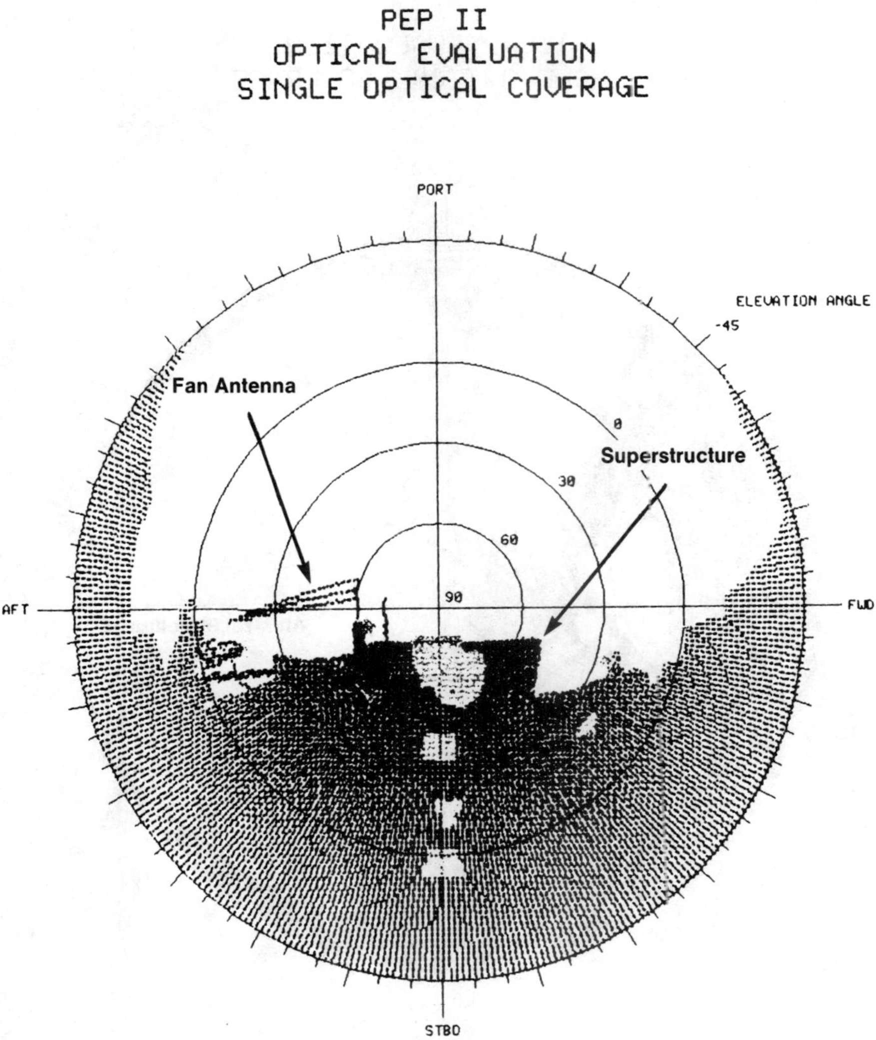

Fig. 8-7(a) Computer-Modeled Sensor Coverage (Blockage) Polar Plot

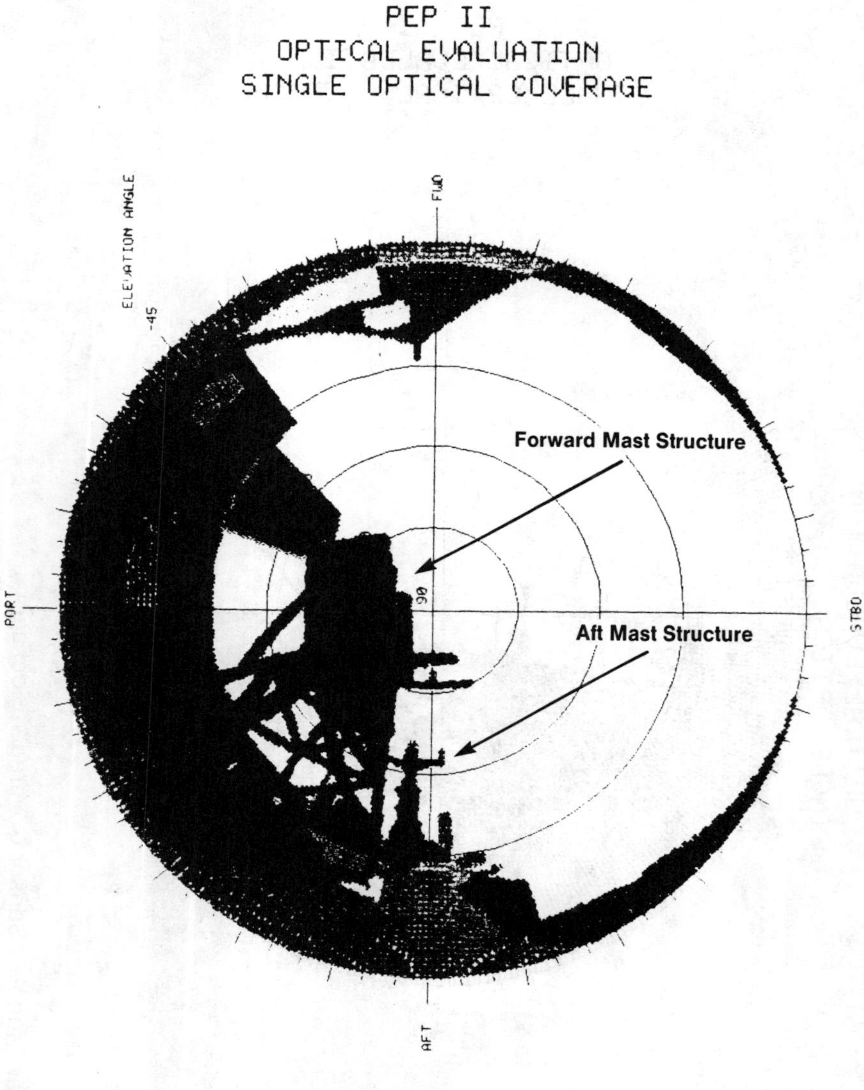

Fig. 8-7(b) Computer-Modeled Sensor Coverage (Blockage) Polar Plot

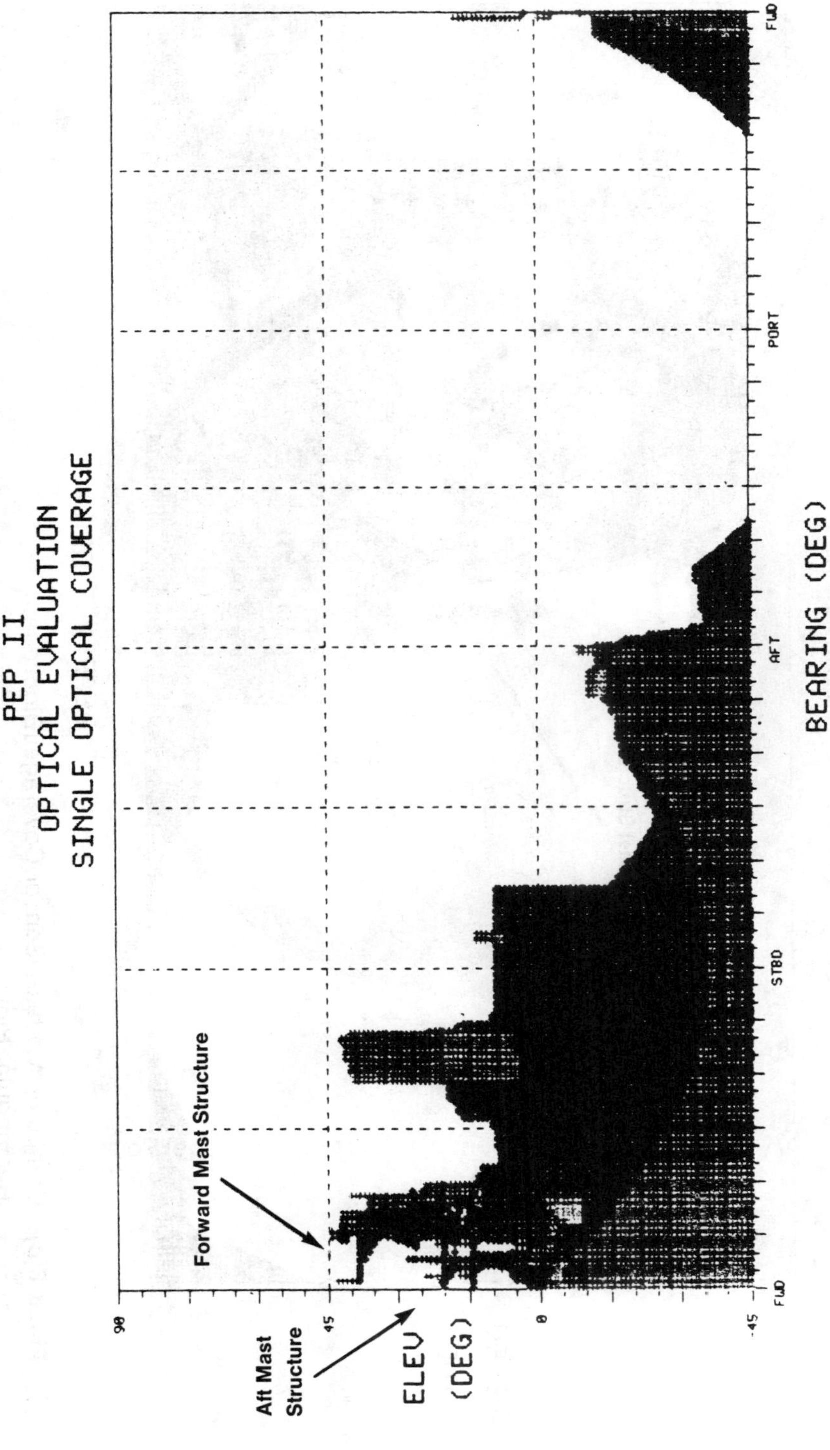

Fig. 8-8(a) Computer-Modeled Sensor Coverage (Blockage) Rectangular Plot

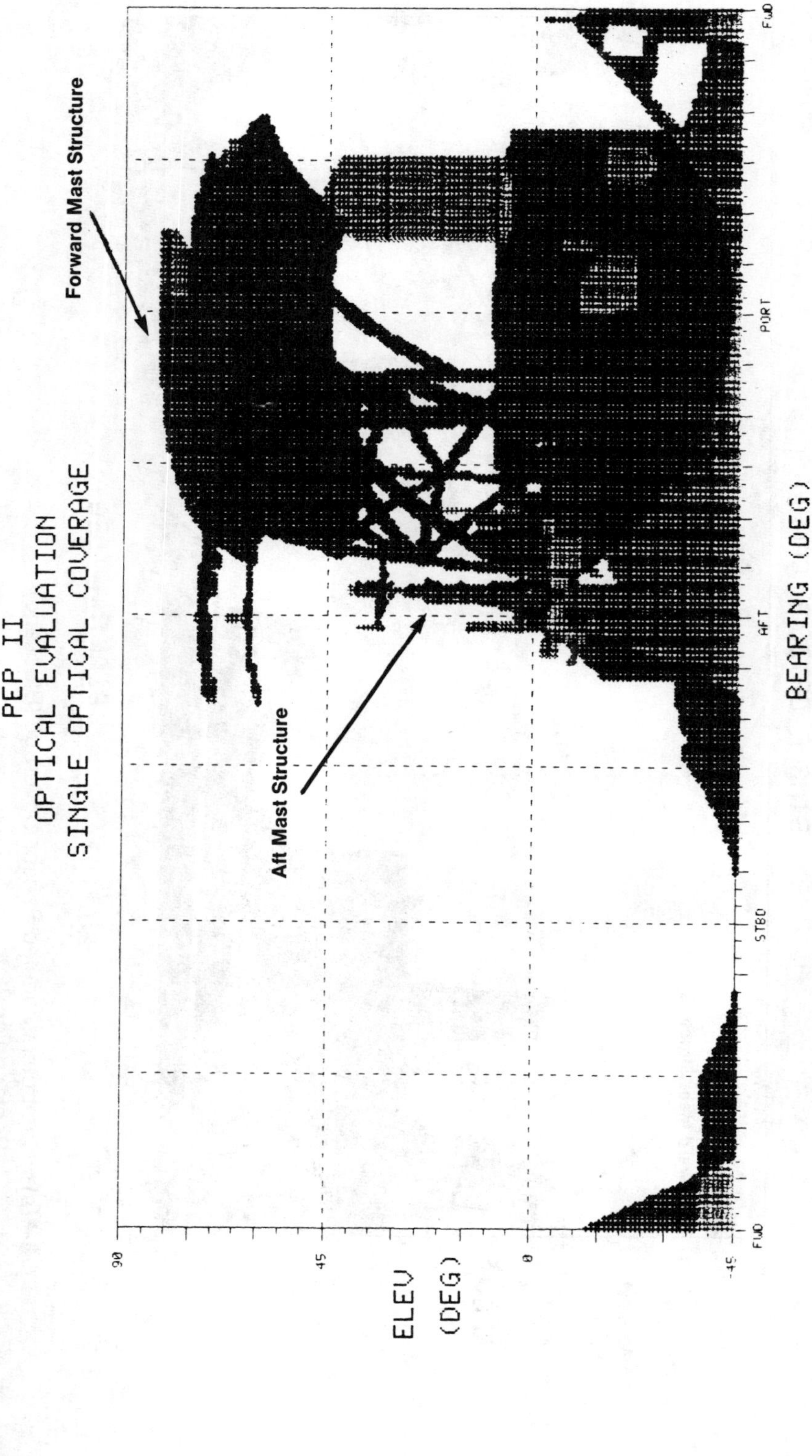

Fig. 8-8(b) Computer-Modeled Sensor Coverage (Blockage) Rectangular Plot

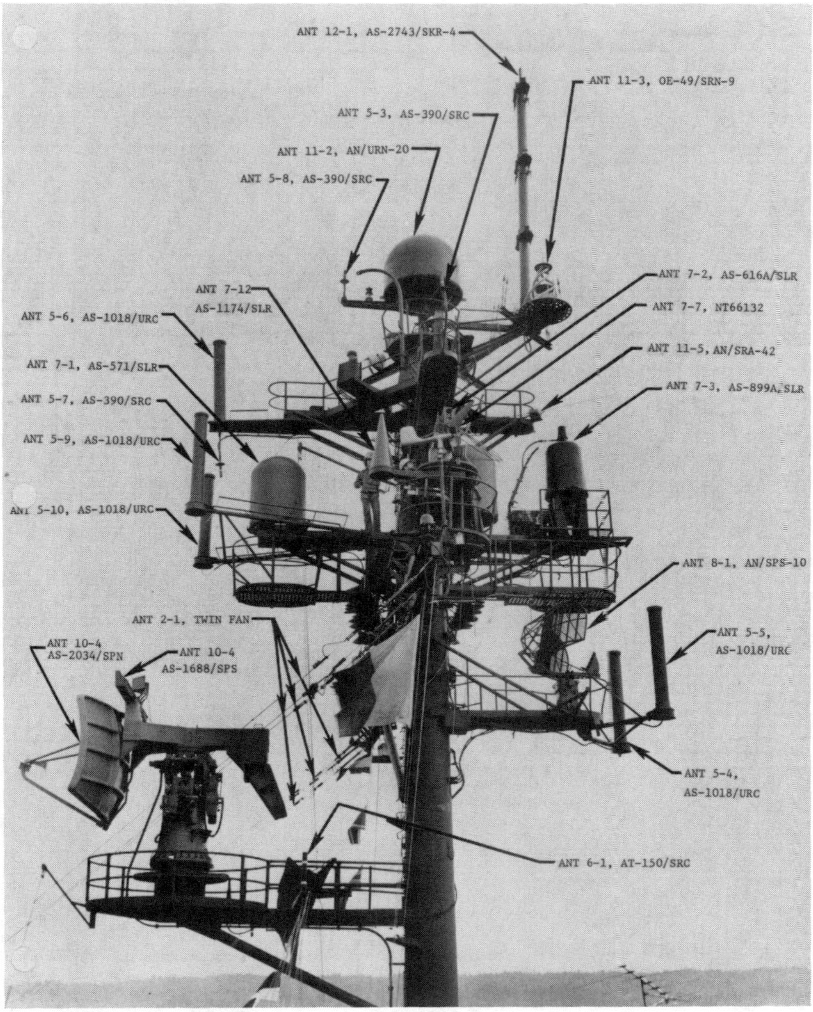

Fig. 8-9 Mast Area of Typical Warship

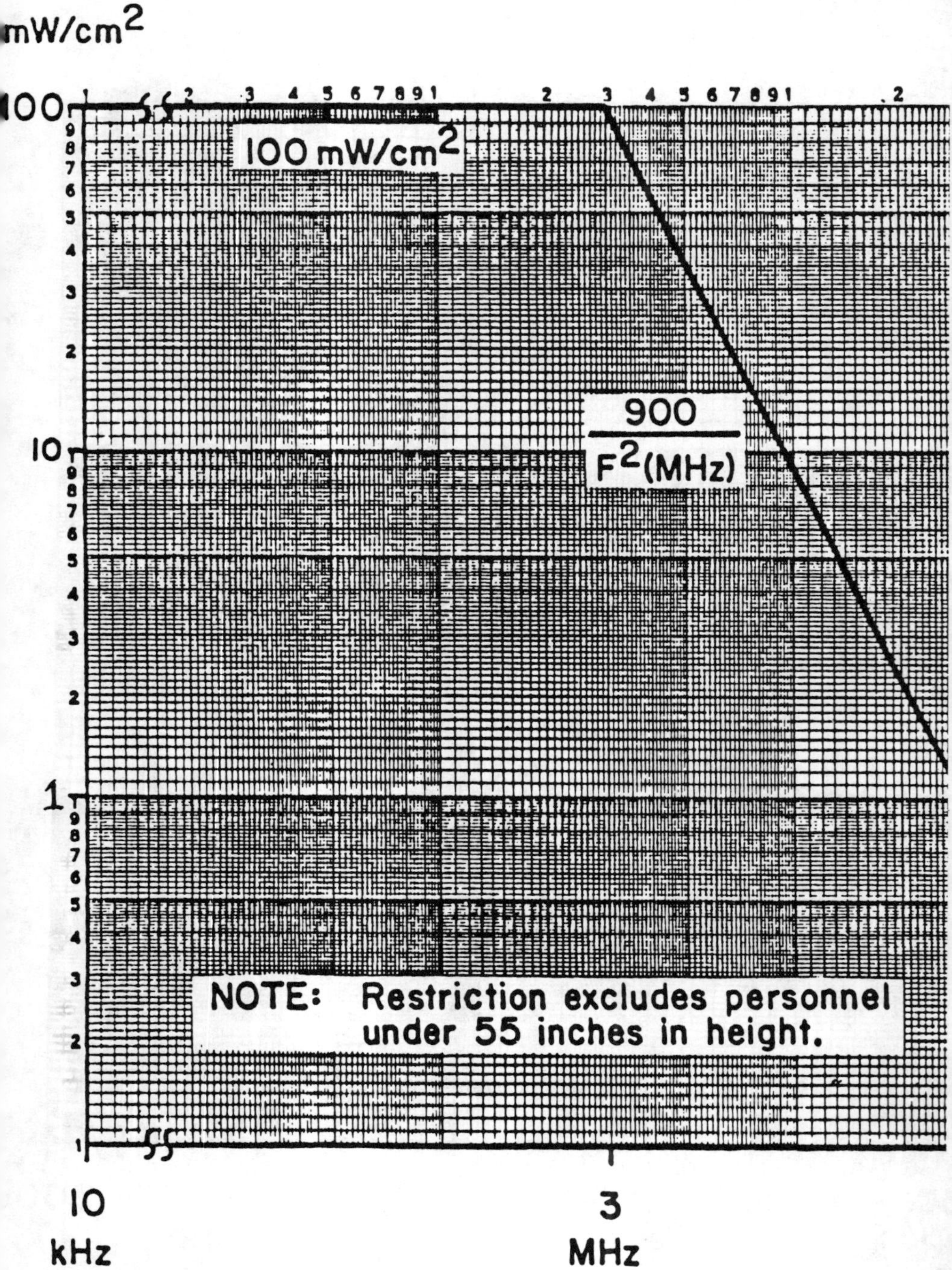

Fig. 8-10 New RF Radiation Exposure Limits

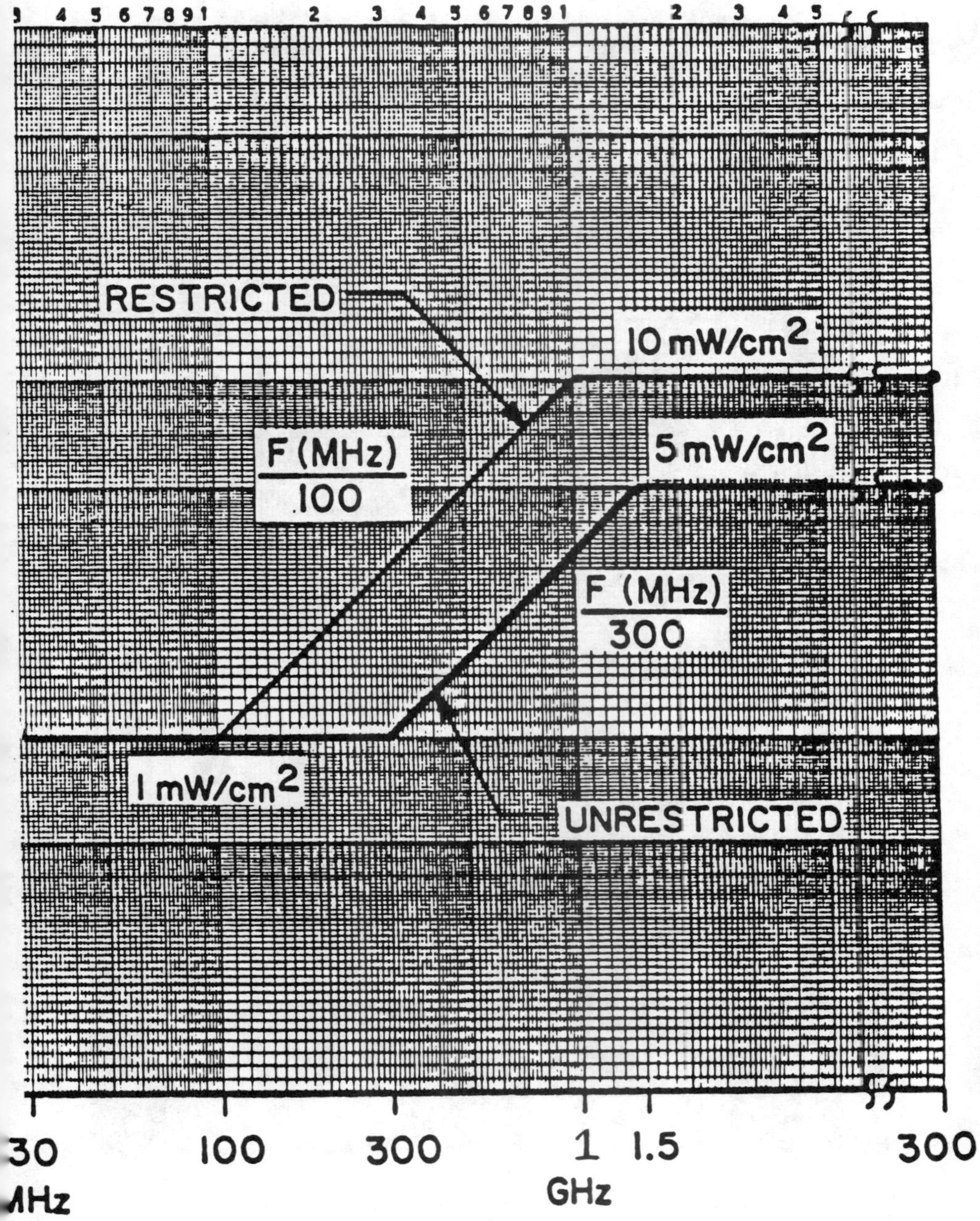

RESTRICTED
10 mW/cm2
F (MHz) / 100
5 mW/cm2
F (MHz) / 300
1 mW/cm2
UNRESTRICTED
30 MHz
100
300
1
1.5
GHz
300

Brass model of Iowa Class Battleship, *USS New Jersey* (circa 1982).

BIBLIOGRAPHY

The following texts are recommended as references for those readers desiring mathematical treatments and specialized study in the general theory of antennas.

1. J.D. Kraus, *Antennas*, New York: McGraw-Hill, 1950.
2. H. Jasik, *Antenna Engineering Handbook*, New York: McGraw-Hill, 1961.
3. L.V. Blake, *Antennas*, Dedham, MA: Artech House, 1984.
4. S.A. Schelkunoff and H.T. Friis, *Antennas: Theory and Practice*, New York: John Wiley & Sons, 1952.
5. E.A. Laport, *Radio Antenna Engineering*, New York: McGraw-Hill, 1952.
6. M.I. Skolnik, *Radar Handbook*, New York: McGraw-Hill, 1970.
7. E.A. Wolff, *Antenna Analysis*, New York: John Wiley & Sons, 1966.
8. S. Silver, *Microwave Antenna Theory and Design*, New York: McGraw-Hill, 1949.
9. C.H. Walter, *Traveling Wave Antennas*, New York: McGraw-Hill, 1965.
10. R.W.P. King, *The Theory of Linear Antennas*, Cambridge, MA: Harvard University Press, 1956.
11. R.S. Elliott, *Antenna Theory and Design*, Englewood Cliffs, NJ: Prentice-Hall, 1981.
12. W.L. Stutzman and G.A. Thiele, *Antenna Theory and Design*, New York: John Wiley & Sons, 1981.
13. C.A. Balanis, *Antenna Theory, Analysis, and Design*, New York: Harper & Row, 1982.
14. A.W. Rudge, *et al.*, *The Handbook of Antenna Design*, London: Peter Peregrinus, 1982.
15. R.E. Collin, *Antennas and Radiowave Propagation*, New York: McGraw-Hill, 1985.

INDEX